AF832318

Integrated Science

Volume 36

Editor-in-Chief

Nima Rezaei, Tehran University of Medical Sciences, Tehran, Iran

The **Integrated Science** Series aims to publish the most relevant and novel research in all areas of Formal Sciences, Physical and Chemical Sciences, Biological Sciences, Medical Sciences, and Social Sciences. We are especially focused on the research involving the integration of two of more academic fields offering an innovative view, which is one of the main focuses of Universal Scientific Education and Research Network (USERN), science without borders.

Integrated Science is committed to upholding the integrity of the scientific record and will follow the Committee on Publication Ethics (COPE) guidelines on how to deal with potential acts of misconduct and correcting the literature.

Todd McElroy
Editor

Decision-Making in Life and Work

Foundations, Strategies, and Current Neuroscience

Springer

Editor
Todd McElroy
Florida Gulf Coast University
Fort Myers, FL, USA

ISSN 2662-9461 ISSN 2662-947X (electronic)
Integrated Science
ISBN 978-3-031-88414-6 ISBN 978-3-031-88415-3 (eBook)
https://doi.org/10.1007/978-3-031-88415-3

© The Editor(s) (if applicable) and The Author(s), under exclusive license to Springer Nature Switzerland AG 2025

This work is subject to copyright. All rights are solely and exclusively licensed by the Publisher, whether the whole or part of the material is concerned, specifically the rights of translation, reprinting, reuse of illustrations, recitation, broadcasting, reproduction on microfilms or in any other physical way, and transmission or information storage and retrieval, electronic adaptation, computer software, or by similar or dissimilar methodology now known or hereafter developed.
The use of general descriptive names, registered names, trademarks, service marks, etc. in this publication does not imply, even in the absence of a specific statement, that such names are exempt from the relevant protective laws and regulations and therefore free for general use.
The publisher, the authors and the editors are safe to assume that the advice and information in this book are believed to be true and accurate at the date of publication. Neither the publisher nor the authors or the editors give a warranty, expressed or implied, with respect to the material contained herein or for any errors or omissions that may have been made. The publisher remains neutral with regard to jurisdictional claims in published maps and institutional affiliations.

This Springer imprint is published by the registered company Springer Nature Switzerland AG
The registered company address is: Gewerbestrasse 11, 6330 Cham, Switzerland

If disposing of this product, please recycle the paper.

Preface

Decision-making is a fundamental human process that permeates nearly every aspect of our lives. Understanding the complex interplay of cognitive, emotional, social, and biological factors that shape our decisions has been a central focus of psychological research for decades. This edited volume, *Decision-Making in Life and Work: Foundations, Strategies, and Current Neuroscience*, brings together cutting-edge research and diverse perspectives to provide a comprehensive exploration of the decision-making process.

The book spans a wide range of topics, from foundational theories and replications to the impacts of brain physiology, working memory, uncertainty, and group dynamics on choice. The introductory chapter lays the groundwork by reviewing seminal theories and contemporary frameworks, setting the stage for subsequent chapters that delve into specific aspects of decision-making in greater depth.

Several chapters focus on the neurological underpinnings of decision-making, while others explore cognitive factors, applied contexts such as medical and legal settings, and the social dimensions of choice. Methodological issues, including a critical review of replications in judgment and decision-making research, are also addressed.

By integrating theoretical foundations with the latest empirical findings and real-world applications, this volume aims to push the boundaries of our understanding of human decision-making. It is my hope that the insights compiled here will stimulate further research and inform evidence-based strategies for optimizing choices and judgments across diverse domains. I am grateful to the expert contributors who shared their knowledge and perspectives to make this book possible.

Fort Myers, USA Todd McElroy

Contents

About the Editor

Todd McElroy, Ph.D. is Associate Professor of Psychology at Florida Gulf Coast University. His primary research interests are in decision-making, the psychology of physical activity, and examining factors that influence complex choices. Dr. McElroy takes an integrative approach, investigating decision-making through the study of personality differences, situational factors, biological influences, and their interactions.

His work has uncovered fascinating insights on the roles of sleep, circadian rhythm, glucose levels, regret, cognitive reflection, and hemispheric processing on our decisions and judgments. Dr. McElroy's recent research also includes the relationship between physical activity and happiness, romantic regret, and a new model of the psychological determinants of physical activity called the Mind-Motion Matrix.

With over 50 peer-reviewed publications to date, Dr. McElroy continues to advance our understanding of the complex interplay of factors shaping human judgment and decision-making. In 2024, he published the book *The Science of Laziness: Achieving More by Doing Less* as part of *The Self-Enhancement Science Series*. His integrative approach and novel findings offer valuable insights for both research and real-world applications.

Laying the Groundwork: Theories and Applications in Decision Making

1

Todd McElroy

We think, each of us, that we're much more rational than we are. And we think that we make our decisions because we have good reasons to make them. Even when it's the other way around. We believe in the reasons, because we've already made the decision.

Daniel Kahneman

Abstract

This chapter provides a comprehensive overview of decision-making theories and their applications, emphasizing the dual-process framework that distinguishes between intuitive and deliberative cognitive systems. It explores the historical context and evolution of decision-making research, highlighting classical theories such as expected utility and prospect theory, while also discussing contemporary models like dual-process theories and fuzzy-trace theory. The chapter underscores the significance of understanding individual differences in decision-making, illustrating how personal experiences and cognitive styles influence choices. Additionally, it focuses on the neural mechanisms underlying decision-making processes while highlighting advancements in neuroscience that illuminate the brain regions involved in various types of decisions. By integrating insights from psychology, neuroscience, and behavioral economics, this

T. McElroy (✉)
Department of Psychology and The Water School, Florida Gulf Coast University, Fort Myers, FL, USA
e-mail: toddmcelroyfgcu@gmail.com

Integrated Science Association (ISA), Universal Scientific Education and Research Network (USERN), Fort Myers, USA

© The Author(s), under exclusive license to Springer Nature Switzerland AG 2025

T. McElroy (ed.), *Decision-Making in Life and Work*, Integrated Science 36,
https://doi.org/10.1007/978-3-031-88415-3_1

chapter lays the groundwork for further exploration into the complexities of human decision-making.

Graphical Abstract

Keywords

Decision-making • Dual-process theory • System 1 and System 2 • Cognitive biases • Fuzzy-trace theory • Neuroscience • Individual differences • Rationality • Behavioral economics

1 Purpose of the Chapter

The purpose of this lead chapter is to set the stage for the book *Decision-Making in Life and Work: Foundations, Strategies, and Current Neuroscience* by explaining the relevance of decision-making theories and their applications. This edited volume covers a wide range of topics from decision-making replications to the neuroscience of decision-making. The field is expansive, reflecting the multifaced role of decision-making in the human experience. By providing a comprehensive overview of the theories, strategies, and current neuroscience findings related to decision-making, this book should serve as a valuable resource for researchers, practitioners, and students alike. By bringing together diverse perspectives and cutting-edge research, this volume offers a multifaceted understanding of the complex processes underlying human decision-making.

1.1 Overview of the Field

Decision-making research has a rich history, with contributions from various disciplines such as psychology, economics, neuroscience, and computer science. The field has witnessed significant advancements in recent years, driven by the development of new theories, methodologies, and technologies. Understanding the mechanisms and factors that influence decision-making is crucial, as the choices we make have far-reaching consequences in both personal and professional contexts.

Classical decision-making theories, such as expected utility theory and prospect theory, have laid the foundation for our understanding of how individuals make choices under uncertainty. These theories assume that people are rational agents who seek to maximize their expected utility by weighing the probabilities and values of different outcomes. However, research has shown that human decision-making often deviates from these normative models, as people are subject to various cognitive biases and heuristics.

Contemporary theories of decision-making, such as dual-process theories and fuzzy-trace theory, have emerged to account for these deviations and provide a more refined understanding of the cognitive processes involved in decision-making. Dual-process theories propose that decision-making involves two distinct systems: an intuitive, fast, and automatic System 1, and a deliberate, slow, and controlled System 2 [1]. Fuzzy-trace theory, on the other hand, emphasizes the role of gist representations and intuitive reasoning in decision-making, particularly in the context of real-world decision-making [2].

In recent years, neuroscience has made significant contributions to our understanding of the neural basis of decision-making. Studies using neuroimaging techniques, such as functional magnetic resonance imaging (fMRI) and electroencephalography (EEG), have identified the brain regions and networks involved in various aspects of decision-making, such as value computation, risk assessment, and action selection [3]. This research has also shed light on the impact of factors such as handedness, sleep, and working memory on decision-making processes [4].

The application of decision-making research extends beyond the laboratory, with important implications for various domains such as healthcare, law, and public policy. For example, understanding the factors that influence medical decision-making can help improve patient outcomes and reduce healthcare costs [5]. Similarly, the use of neuroscience evidence in legal contexts has the potential to inform our understanding of criminal responsibility and sentencing decisions [6].

In the rapidly evolving field of decision-making research, integrating insights from various disciplines and applying them to real-world problems is key. This edited volume brings together cutting-edge research in neuroscience alongside pragmatic applications of decision-making principles from professionals in the field. By combining theoretical advancements with practical implementations, this interdisciplinary approach establishes a foundation for significant future developments in understanding and optimizing decision-making processes across various domains, from personal choices to organizational strategies.

2 Dual-Process Theories of Decision Making

As we navigate the complexities of daily life, our minds are constantly making decisions, both large and small. But how do we actually make these choices? According to one of the more prominent dual-process theories proposed by Daniel Kahneman [7], our thoughts flow through two distinct channels: one fast and intuitive, the other slow and deliberate.

2.1 Overview of Dual-Process Models

The basic premise of dual-process models is that we have two cognitive systems, each with its own strengths and weaknesses [8]. System 1 is automatic, emotional, and largely unconscious. It relies on mental shortcuts and gut feelings to quickly assess situations and guide our actions. System 2, in contrast, is controlled, logical, and conscious. It carefully analyzes information to reach rational conclusions but requires more time and mental effort [9].

This distinction between intuitive and reflective thinking has deep roots in psychology, with early inklings in the work of William James [10]. But it was in the 1970s and 80 s that the idea crystallized into the dual-process framework we know today, thanks to pioneering researchers like Amos Tversky, Daniel Kahneman, Steven Sloman, and Keith Stanovich [11, 12].

2.2 System 1 & 2: Automatic and Controlled Processes

Let's take a closer look at System 1. This is our evolutionary older mind, shared with many animals. It is always operating in the background, effortlessly generating impressions, impulses, and intuitions [3]. We rely on System 1 for innumerable everyday activities. In fact, most of our activities and routines are System 1 controlled. This includes a very wide variety of tasks from detecting anger in a voice to answering the question: "what is 2 + 2".

Recent neuroimaging studies have begun mapping the neural correlates of System 1 [13, 14]. This research finds that activation in brain regions associated with emotion, like the amygdala and ventromedial prefrontal cortex, are associated with System 1. There is also involvement of areas linked to memory and habit formation, such as the basal ganglia.

In contrast, System 2 is slow, effortful, and governed by a system of rules [1, 3]. It is the part of our mind that carefully weighs evidence, analyzes arguments, and makes deliberate, thoughtful choices. We engage System 2 when we need to make complex decisions and perform mental tasks, like solving a challenging math problem or filling out a tax form.

Brain imaging reveals that System 2 relies heavily on regions associated with executive control and working memory, especially the dorsolateral prefrontal and parietal cortices [13, 14]. These areas light up when we focus attention, inhibit impulses, plan ahead, and reason abstractly.

2.3 Interaction Between Systems

While we can study System 1 and 2 separately, in reality they constantly interact with one another to produce the human experience [11]. Most of the time, these two systems work in harmony, creating an efficient and highly successful team that gives as a real advantage when we interact with our environment. System 1 efficiently processes the world and guides our intuitions and feelings. System 2 monitors things happening in the world around us and steps in to include more thoughtful analysis when needed [1].

However, there are situations where the two systems can come into conflict with each other [15]. For example, imagine that you are on a diet and are confronted with a delicious piece of cake, you're favorite kind. Your System 1 screams "Eat it!" while your System 2 tries to resist the temptation, elucidating the positive and negative tradeoffs associated with this behavior. Often, in such cases, the fast emotional impulses overpower the more thoughtful, rational self-control.

This interplay between systems also helps explain many cognitive biases; systematic errors in judgment to which we all fall prey [1]. For example, the anchoring bias occurs when exposure to an arbitrary number influences our later estimates. For example, if asked whether Gandhi lived to be more or less than 144 years, people will give a much higher estimate of his true age than if they are asked the same question with an anchor of 35. Such biases arise when we overly rely on quick System 1 heuristics and fail to engage our deliberate System 2 analysis.

2.4 Empirical Evidence Supporting Dual-Process Theories

Over the past few decades, researchers have amassed a wealth of evidence for dual-process models. Brain imaging reliably shows distinct neural signatures for intuitive vs. reflective thinking [9, 13]. Behavioral studies demonstrate how factors like time pressure, cognitive load, and emotion all may influence the balance between systems [11]. There have also been illuminating case studies of patients with localized brain damage [13]. Individuals with injuries to prefrontal areas often show deficits in System 2 processes, like planning and impulse control. While at the same time System 1 functions remain largely intact. In contrast, amygdala damage can impair the automatic processing of emotion, highlighting the critical role of this brain region in the intuitive and emotional aspects of decision-making.

These case studies provide compelling evidence for the dissociation between the two systems, suggesting that they rely on distinct neural substrates and can be selectively impaired by brain lesions. Furthermore, the observations from patients with prefrontal damage underscore the importance of System 2 processes in regulating behavior and making deliberate, goal-directed decisions, while the findings from amygdala-damaged patients emphasize the central role of emotion and intuition in shaping our choices and actions.

2.5 Applications of Dual-Process Theories

The dual-process framework has proven remarkably generative across many aspects of human cognition and behavior [4]. In cognitive psychology, it helps explain phenomena like creativity, insight, and moral judgment [6]. For instance, the interplay between the intuitive and reflective systems can account for the sudden "Aha!" moments that characterize creative breakthroughs, as well as the deliberate, step-by-step reasoning that underlies analytical problem-solving. Similarly, dual-process theories suggest that moral judgments arise from both automatic, emotional responses (driven by System 1) and controlled, rational deliberation (guided by System 2), helping to explain the complex nature of moral decision-making.

Social psychologists have used the dual-process framework to understand persuasion, stereotyping, and interpersonal attraction [6]. They propose that persuasive messages can influence attitudes and behavior through either the peripheral route (by appealing to System 1 processes like emotion and heuristics) or the central route (by engaging System 2 processes like logical reasoning and careful evaluation of arguments). Likewise, stereotyping and prejudice are thought to involve automatic, implicit associations (System 1) that can be overridden by controlled conscious beliefs (System 2), highlighting the role of both systems in shaping social cognition and behavior.

Behavioral economists build on dual-process theories to explain irrational consumer choices and anomalies in financial markets [13]. They argue that many economic decisions are driven by fast, intuitive processes (System 1) that can lead to biases and heuristics, rather than the slow, rational deliberation assumed by traditional economic models (System 2). For example, the tendency to overvalue immediate rewards and undervalue future consequences (known as hyperbolic discounting) can be attributed to the dominance of System 1 in decision-making. Similarly, the influence of emotions and social cues on investment decisions can cause market inefficiencies and bubbles, challenging the assumption of perfect rationality in financial markets.

3 Fuzzy-Trace Theory

Fuzzy-Trace Theory (FTT) is a cognitive theory that describes the role of memory in decision-making. It has garnered significant attention in recent years [15–17]. Developed by Valerie Reyna and Charles Brainerd, FTT proposes that individuals process information using two distinct types of memory traces: verbatim and gist [18, 19]. Verbatim traces are precise, detailed representations of the information that we encounter in the world, while gist traces are more abstract representations. In gist processing, we capture the essential meaning or conceptual essence of the information [20].

The type of processing employed in a situation: verbatim or gist, can have profound implications for decision outcomes. This is particularly true if the decision situation involves risk [21]. When engaging in verbatim processing, decision-makers are more likely to be swayed by the specific details of the information they encounter, such as numerical magnitudes presented from the options [22]. In contrast, gist processing involves extracting the global contextualized meaning and then simplifying the decision information to its core essence [23]. According to FTT, verbatim and gist extraction operate in parallel. However, it is also the case that individuals predominantly rely upon gist for processing in daily life [24]. Gist representations are particularly effective in navigating situations characterized by ambiguity or uncertainty [25].

Once the informational gist has been mentally represented, value-related information stored in memory is activated. This set of ideals then guides judgments and decisions [26]. Ultimately, gist processing produces outcomes that are in alignment with the decision-maker's internalized beliefs and worldview [27]. In situations where the decision information is uncertain or ambiguous, people are especially prone to rely more heavily on gist processing; in this event, decisions are more driven by abstractions [16].

A classic illustration of how this processing dynamic unfolds in risky-choice decision-making comes from Reyna and Brainerd's seminal work [17]. They demonstrated that replacing exact numerical information in scenarios (e.g., "200 will die") with more vague, abstract language (e.g., "some people will die") yielded similar framing effect sizes [18]. Reyna and Brainerd explain this phenomenon by describing how individuals understand numbers vaguely in what they term "semantic terms," abstracting the most fundamental gist of categorical distinctions between quantities, such as "some," "all," or "none" [15].

3.1 Expanding Views of FTT

The application of FTT extends beyond the realm of risky decision-making, finding particular relevance in legal settings [24]. In the courtroom, jurors are often faced with complex, ambiguous, and emotionally charged information [25]. FTT suggests that jurors are likely to rely on gist processing when making decisions, basing their judgments on simplified categorical representations of the evidence presented [26]. This tendency to rely on gist can have significant implications for legal outcomes, as it may lead to decisions that are more influenced by preexisting beliefs and stereotypes than by the specific details of the case [27].

In conclusion, Fuzzy-Trace Theory offers a compelling framework for understanding the cognitive processes underlying decision-making, particularly in situations involving risk and uncertainty. By distinguishing between verbatim and gist processing, FTT sheds light on how individuals navigate complex decision scenarios, simplifying information and relying on abstract representations to guide their choices. As the chapters in this volume will demonstrate, FTT has far-reaching implications across various domains, from everyday decision-making to the high-stakes realm of legal judgment.

Recent research [28] has further explored the role of frame valence intensity in risky decision-making, comparing predictions from Prospect Theory (PT) and Fuzzy-Trace Theory (FTT) PT suggests that the frame of decision alternatives determines whether they are categorized as gains or losses, which is central to predicting risk preference. In contrast, FTT proposes that the level of frame valence should correspond to the valence intensity of the extracted gist representation, leading to variations in risky choice.

This study found that differences in frame valence led to predictable shifts in risky choice, with increasing positive valence associated with less risk-taking and increasing negative valence intensity associated with more risk-taking. These findings support FTT's prediction that the intensity of frame valence influences the gist representation and subsequent decision outcomes.

The study also highlighted the complex nature of valence intensity effects, particularly in negative framing conditions. While increasing negative valence intensity generally led to increased risk-seeking, this relationship was not always straightforward, suggesting that context and scenario-specific factors may interact with valence intensity to shape risk preferences.

This research extends our understanding of framing effects by demonstrating that both the direction and intensity of frame valence play a role in shaping risky decision-making. These findings challenge the traditional focus on the simple categorization of gains and losses in Prospect Theory, while providing support for the role of gist representations in decision-making as proposed by FTT.

This advanced understanding of frame valence intensity has important implications for decision-making research, highlighting the need for a more comprehensive model that incorporates both the directional and intensity aspects of framing. By integrating insights from both PT and FTT, researchers can develop a more complete framework for understanding the complex interplay between framing, context, valence intensity, and risky decision-making.

4 Decision-Making Processes and the Brain

In recent decades, the rapid development of neuroimaging techniques has revolutionized our understanding of the neural basis of decision-making. By providing a window into the living, active brain, tools like functional magnetic resonance imaging (fMRI), electroencephalography (EEG) and positron emission tomography (PET) have allowed researchers to map the complex neural circuitry underlying the choices we make [29].

Studies using these methods have identified a network of brain regions that play a critical role in decision-making, with the prefrontal cortex at the center [30]. The prefrontal cortex is a highly interconnected region that integrates information from various sensory, motor, and limbic areas to guide goal-directed behavior. Within the prefrontal cortex, the orbitofrontal cortex is particularly involved in assessing the reward value of options, by encoding the subjective value of potential outcomes and updating these valuations based on changing circumstances. This allows individuals to flexibly adapt their choices to maximize rewards and minimize punishments.

The dorsolateral prefrontal cortex, on the other hand, supports the executive control needed to carry out decisions. This region is critical for working memory, planning, and cognitive flexibility, enabling individuals to hold and manipulate information, consider future consequences, and switch between different strategies as needed. The anterior cingulate cortex also plays a key role in decision-making by monitoring for conflicts and errors. It detects situations where automatic, habitual responses are inappropriate or suboptimal, and signals the need for increased cognitive control to resolve these conflicts and optimize behavior.

In addition to these prefrontal regions, subcortical structures like the basal ganglia and amygdala are also important for decision-making. The basal ganglia, particularly the striatum, are involved in reinforcement learning and habit formation, allowing individuals to learn from the outcomes of their choices and develop automatic, efficient responses to familiar situations. The amygdala, on the other hand, mediates the emotional and motivational aspects of choice, by processing the affective significance of stimuli and guiding behavior towards appetitive (rewarding) and away from aversive (punishing) outcomes.

This research also highlights the distributed nature of the neural circuitry underlying decision-making. Findings show that multiple brain regions working in concert to enable adaptive, flexible decisions [31]. The prefrontal cortex serves as a central hub that integrates information from diverse sources, while subcortical structures provide important inputs related to learning, emotion, and motivation. By understanding how these regions interact and contribute to different aspects of decision-making, researchers can gain insights into the complex neural basis of this fundamental cognitive function.

4.1 Neural Pathways and the Dynamics of Decision-Making

At its core, decision-making involves assigning values to different options and selecting the one with the highest value. Neuroscientists have begun to uncover the neural currencies of value that guide this process. For example, dopaminergic pathways projecting from the midbrain to the striatum and prefrontal cortex play a key role in reward processing [32]. The nucleus accumbens and ventral striatum in particular show increased activity in response to rewarding stimuli, tracking the predicted value of options. Dopamine neurons themselves seem to encode reward prediction errors, firing when outcomes are better than expected [33].

But the brain must also contend with the uncertainty inherent in most real-world decisions. Take for example the insula, a region involved in processing bodily sensations, it is activated by risk and ambiguity [34]. Along a similar line of reasoning, the orbitofrontal cortex seems to integrate information about the probability and magnitude of potential outcomes to represent the expected value of choices [35].

Decisions also unfold over time, as the brain gradually accumulates evidence for deciding among different options. Computational models like drift diffusion have been used to describe how noisy sensory evidence is eventually integrated until a decision threshold is reached [36]. fMRI studies have identified neural correlates of this "evidence accumulation" process in parietal and prefrontal regions, with activity ramping up as more evidence is gathered [37]. At the same time, the basal ganglia region seems to play a key role in setting the decision threshold itself, determining when enough evidence has been collected to trigger a decision choice [38].

Neuroscience is also shedding light on the dual-process theory of decision-making, including the operation of its distinct systems for intuitive and deliberative choice. On one hand, fast, automatic responses seem to rely on subcortical structures and emotional circuitry. While on the other hand, the amygdala seems to play a key role in rapid evaluations of threat and reward [39]. An even greater contrast emerges for deliberative processing which appears to engage the prefrontal cortex, particularly the dorsolateral regions involved in working memory and cognitive control [40].

While research appears to show that there are common neural mechanisms underlying decision-making, there is also substantial variability across individuals. To address this, researchers have begun to identify some of the biological factors that contribute to these differences. One avenue focuses on how certain genes involved in neurotransmitter function seem to influence decision-making traits. For example, polymorphisms in dopamine receptor genes have been linked to differences in reward sensitivity and impulsivity [41], while variation in serotonin transporter genes may affect risk attitudes [42].

Even basic physiological traits may play a role in shaping decision processes. For instance, some studies suggest that left-handed individuals show more exploratory behavior, while strongly right-handed people tend to be more risk-averse [43], a trait that may have an underlying genetic component. Acute stress and hormones like cortisol and testosterone can also modulate decision-making, including "in the moment" decisions.

Building on these insights, researchers are exploring ways to alter decision-making by targeting specific neural systems. Current research is finding that specific drugs that affect key neurotransmitters can also powerfully influence choice behavior. Further, stimulants that increase dopamine levels, such as amphetamines, can also boost reward-seeking and risk-taking [43]. In addition, commonly prescribed serotonergic drugs appear to have contrasting effects, seemingly promoting caution and risk-aversion.

There is growing interest in using pharmacology to optimize decision-making in clinical and practical settings. For instance, stimulants like modafinil may improve decision speed and vigilance in high-pressure jobs like military aviation [44]. Anxiolytics that reduce physiological arousal, are a potential resource to help people make cooler, more deliberative choices under stress.

Disrupted decision-making is also a core feature of many neurological and psychiatric disorders, making it an important target for clinical neuroscience. Patients with Parkinson's disease, which is characterized by dopamine depletion, often demonstrate impairments in learning from rewards and making adaptive decisions [45]. In ADHD, reduced activity in prefrontal control systems may underlie impulsive and inconsistent choice behavior. Similarly, depression is associated with blunted neural responses to rewards, possibly leading to apathy and difficulty making positive choices. And in substance abuse and behavioral addictions, dysfunctional decision-making may arise from altered reward processing in the striatum and prefrontal cortex.

As the field of decision neuroscience advances, many new tools and frameworks are opening exciting avenues for future research. High-resolution fMRI is providing increasingly detailed maps of choice-related brain activity, and multimodal imaging. In the future, combining fMRI with EEG or PET, may yield even richer information about the spatiotemporal dynamics of decision processes [46]. Computational approaches are also bringing new rigor to the study of decision-making in the brain. Biologically realistic neural network models can simulate learning and choice behavior, bridging the gap between neural mechanisms and cognitive processes [47]. Machine learning tools are being applied to neural data to uncover hidden patterns and predict individual decisions from brain activity.

By integrating insights across levels of analysis, from molecules to mind, neuroscience is poised to unravel the complex neural codes and computations underlying the choices that shape our lives. While much work remains to be done, the rapid progress in this field promises to yield profound insights into the biological basis of human decision-making, with far-reaching implications for medicine, public policy, and our understanding of ourselves.

5 Sleep and Decision-Making

Sleep and circadian rhythms are intrinsic to all humans, occupying about 1/3 of our lives, while playing a pivotal role in various cognitive processes, notably decision-making. A growing body of research has demonstrated that sleep deprivation and suboptimal times of day can significantly impact the quality and nature of decisions that we make [48, 49]. This section will explore the relationship between sleep, circadian rhythms, and decision-making, highlighting the effects of sleep deprivation and circadian timing on specific decision-making processes.

5.1 Effects of Sleep Deprivation on Decision-Making

Sleep deprivation has been shown to impair information processing and memory consolidation, both of which can lead to suboptimal decision-making [50]. When individuals are sleep-deprived, they may experience difficulty in updating information and adapting to changing circumstances, which results in reduced cognitive flexibility [51]. Additionally, sleep loss has been associated with increased risk-taking behavior, as individuals may be more prone to making impulsive or risky choices when fatigued [52].

5.2 Impact on Specific Decision-Making Processes

Some studies have investigated the impact of sleep deprivation on various types of decision-making processes. For example, research has shown that sleep loss can influence moral judgments, leading to more utilitarian decisions in moral dilemmas [53]. Strategic decision-making, which involves considering multiple factors and anticipating future consequences, has also been found to be impaired by sleep deprivation [54]. Furthermore, economic and financial decisions may be affected by sleep loss, with individuals showing altered risk preferences and reduced ability to process complex information [55].

5.3 Circadian Rhythms and Decision-Making

In addition to sleep deprivation, circadian rhythms can also influence decision-making processes. Circadian rhythms refer to the internal 24-h cycles that regulate a wide array of physiological and cognitive functions, including alertness, attention, and cognitive performance [52]. Research has demonstrated that decision-making abilities fluctuate throughout the day, with performance often peaking in the morning and declining in the late afternoon and evening. Equally important, research also shows that substantial differences exist among people, with “Morning” and “Evening” chronotypes emerging [52].

Studies have shown that circadian timing can impact strategic reasoning, such that individuals tend to make more cooperative choices during their circadian peak compared to their circadian trough [53]. Similarly, research has found that people tend to engage in more deliberative and analytical thinking during the daytime, while relying more on heuristics and intuition during the night [54].

This interaction between sleep deprivation and circadian timing has also been explored in the context of decision-making. One study examined the combined effects of sleep restriction and suboptimal times of day on decision-making using a sample of college students. The researchers found that the impact of sleep deprivation on decision-making was more pronounced when individuals were tested at suboptimal times of day, highlighting the importance of considering both factors when assessing decision-making performance [55, 56].

6 Navigating Complexity, Uncertainty, and Regret in Everyday Life

Decision-making is not an abstract process relegated to boardrooms or laboratories; it pervades almost all aspects of everyday life. From the mundane to the monumental, our ability to make decisions defines the quality of our lives. Regret is one of the most powerful emotions tied to decision-making, it influences both future choices as well as emotional well-being [57, 58]. As humans, we often reflect on and "mentally recreate" past decisions. In a sense, we are haunted by the paths we did not take [59]. The psychological mechanisms behind regret shape our decision-making processes. By understanding these psychological processes, we will be better equipped to mitigate the paralyzing effects of bad outcomes [60].

Counterfactual thinking involves mentally simulating alternative outcomes or scenarios that differ from reality [61, 62]. It is a common cognitive process that allows individuals to explore "what if" situations and consider how things might have unfolded differently. While counterfactual thinking can serve as a useful tool for learning and growth, it can also lead to feelings of regret, particularly when people dwell on missed opportunities or suboptimal decisions. Regret, in turn, can have a significant impact on emotional well-being and future decision-making. However, it is possible to reframe these negative experiences through a process known as cognitive reappraisal. Through cognitive reappraisal, individuals can shift from ruminating on "If only I had done something differently" to learning from missed opportunities and improving future decisions with greater wisdom [61].

Regret also provides a bridge between System 1 (emotional, intuitive) and System 2 (deliberate, rational) thinking in dual-process theory [62]. While System 1 may lead to impulsive decisions that foster immediate regret, System 2 enables a reflective pause, allowing individuals to make choices that align more with long-term goals and values [63].

Uncertainty is inherent in decision-making. Whether we are investing in financial markets, choosing a career path, or making health-related choices, outcomes are rarely guaranteed [64]. How we perceive and react to uncertainty, whether with anxiety or curiosity, can deeply influence the quality of our decisions [65]. The dual-process model provides an excellent approach to understanding this duality: System 1 often prompts emotional responses like fear or overconfidence, while System 2 engages in more calculated risk assessment [66]. Real-world outcomes can both challenge and reinforce our initial attitudes toward uncertainty, often leading to shifts in our decision-making style over time [67].

Prospect theory's approach to behavioral economics allows for an understanding of how uncertainty is processed from a psychophysical perspective. It provides evidence-based reasoning for why humans are prone to certain cognitive biases, such as loss aversion and over-optimism [68]. This insight allows us to better navigate complex, unpredictable environments with resilience and wisdom [69]. Furthermore, it enables us to have foresight in predicting how people will respond in certain risky situations.

Few domains are as laden with weighty, often irreversible decisions as healthcare. The phrase, "If only I'd known then what I know now," underscores the profound impact of medical decision-making on both patients and practitioners alike [70]. Whether it is deciding on a treatment plan, or considering long-term prognoses, medical decisions often involve incomplete information and high stakes, life or death decision making [71].

The cognitive and emotional processes behind medical decisions involve both System 1 and System 2 thinking [72]. System 1 may drive decisions based on fear or hope that surround the medical procedure or prognosis. System 2 allows for more deliberative reasoning, such as understanding statistics that provide an analytic representation of the probabilities associated with the medical dilemma and advice [14]. The regret associated with these decisions can often reverberate for both patients and healthcare providers, raising ethical questions about informed consent and shared decision-making [73].

Decision aids in healthcare represent some of the latest technological tools designed to support patients in making informed, value-consistent decisions. These techniques are grounded in research on dual-process theory [74]. These tools seek to balance emotional responses associated with the personal aspect of medical decisions with rational analysis, helping patients avoid regret in these critical moments [75].

In the realm of law, decisions carry profound societal consequences, from individual freedom to the integrity of justice systems [76]. Recent advances in neuroscience have begun to permeate legal decision-making, offering insights into brain function that challenge traditional notions of guilt, intent, and responsibility [77].

With technologies like fMRI and EEG now capable of revealing brain activity tied to decision-making, judges and juries are increasingly presented with complex scientific data to inform their rulings [78]. However, the introduction of such evidence raises significant ethical and practical questions. Specifically, how do we interpret neurological data from a legal perspective? Can neurological data truly predict behavior? If so, to what extent should it influence sentencing and criminal responsibility [72]?

System 1 thinking might initially lead legal actors to view neuroscientific evidence as deterministic, while System 2 requires a more balanced interpretation, acknowledging both its potential contributions as well as its limitations [79]. The challenge is to balance the scientific allure of neuroscience with the deeply human element of judgment. By taking this approach, we help ensure that legal decisions remain grounded in both fairness and empirical rigor [80].

When considering the many factors that weigh on everyday decision-making, it becomes clear that the process is far from straightforward. From the weight of regret and uncertainty of outcomes to the gravity of medical choices and implications of neuroscience in the courtroom, decision-making is a deeply human endeavor [81]. By striving to incorporate both our emotional and rational systems, we can cultivate wisdom and humanitarian needs to make our decisions better align with our values and goals [82]. In doing so, we not only improve the quality of our own individual lives but also contribute to a society that is more compassionate, efficient and productive [18].

7 Being Both Different and the Same: Human Variability and the Future of Decision Research

The field of decision-making research long operated under the assumption that "people are people," and human variability is not a significant aspect of the decision-making process [83]. Early decision researchers, such as Edwards [74], assumed that decision-makers were completely informed, sensitive to infinitesimal changes in value, and could be counted on to be perfectly rational [84]. This approach to human decision-making helped create a general, mathematically-based model of decision-making. However, it did not account for the many ways in which humans vary in multiple attributes and, ultimately, how this may influence the processing and outcome of decisions [85].

The foremost studied theory in decision research, prospect theory, was designed to account for comparison of group effects, not individual differences [86]. Although the theory acknowledged that different individuals may have different psychophysical functions, specific predictions for how individual differences might interact with the theoretical assumptions were never fully developed [87].

As later research would show, investigating how factors that vary among individuals are associated with differences in decision-making is a critical component for more fully understanding how people think about decisions [88–90]. For example, individual limitations in cognitive ability are a likely suspect for much of the variability found within decision tasks [4]. Many personality differences play a role in decision-making, though this varies with the type of decision task [91].

One individual difference variable that has garnered research attention is numeracy, which describes differences in the understanding and ability to work with numbers [92]. This research has generally shown that highly numerate individuals possess more optimal decision-making skills in a variety of analytically-based situations [92].

Other individual difference variables, such as emotions and motivation, have also been examined in relation to decision-making [93, 94]. For example, higher levels of motivation and less negative emotions about financial decisions have been associated with better performance on financial decision-making tasks [95, 96].

Research has also shown that people can change their decision-making as they themselves change across their lifespan in cognitive development [19, 97]. Adolescents, for example, may engage in increased risk-taking due to differences in their mental representations of risky options [19]. Older adults, relative to younger adults, may be more prone to emotional aspects inherent to a decision task.

7.1 An Example for Future Research.

The field of decision-making research is characterized by a dual focus, encompassing two distinct yet complementary approaches. On one hand, there is a pursuit of universal principles, which is essential for developing generalizable theories that apply across diverse populations. This approach seeks to identify common cognitive mechanisms and behavioral patterns that underlie human decision-making processes. On the other hand, researchers also strive to elucidate the complexities and idiosyncrasies that arise from individual differences, recognizing that personal experiences, cognitive styles, and personality traits can significantly influence the value we place on alternatives and how we make decisions. The integration of these two approaches is essential for a comprehensive understanding of decision-making phenomena.

Research that illuminates how individual differences impact decision-making processes provides a vital component of the knowledge base, enabling further theoretical development and refinement. By synthesizing insights from research seeking universal principles with findings depicting how individual variations influence decision outcomes, researchers can construct a more comprehensive and robust model of human decision-making. This type of approach should ultimately enhance our ability to predict and explain this fundamental aspect of human behavior.

To highlight the potential of this approach, consider one individual difference trait that has received less attention in decision-making literature: openness to experience. Openness to experience, one of the Big Five personality traits, refers to an individual's willingness to consider new ideas, experiences, and perspectives. Recent research suggests that openness to experience may provide new insights into the decision-making process.

For example, Sobkow et al. [89] investigated the relationship between intuitive abilities, intelligence, and openness to experience. They found that openness to experience was positively associated with intuitive abilities, even after controlling for intelligence. This suggests that individuals high in openness to experience may have a greater capacity for intuitive decision-making.

In another example, McElroy and Dowd [90] examined how openness to experience influences susceptibility to anchoring effects, a common cognitive bias in decision-making. They found that individuals high in openness to experience were more susceptible to anchoring effects, possibly due to their greater willingness to consider and integrate new information, even when the information is irrelevant [89].

As the field of decision-making research moves forward, it is important to continue investigating how individual differences, such as openness to experience, shape the decision-making process. It is equally important to work vigorously to develop more robust universal principles that better apply to human decision-making broadly. By understanding how factors that vary within the individual are associated with decision-making, while developing more comprehensive general models, we can develop better insight into how people make choices.

Future research in the field should work to integrate individual differences into existing theoretical frameworks, such as prospect theory, and develop new models that account for the role of individual variability in decision-making. This will require a multidisciplinary approach, drawing on insights from personality psychology, cognitive psychology, and neuroscience.

Ultimately, by embracing the role of individual differences in decision-making, we can move beyond the assumption that "people are people" and develop a richer, more accurate mathematically-based approach that will allow for better prediction and understanding of how people process both simple and complex choices they face in their lives. This understanding will not only advance the science of decision-making but also inform interventions and policies to help people make better decisions.

References

1. Kahneman D (2011) Thinking fast and slow
2. Evans JStBT, Stanovich KE (2013) Dual-process theories of higher cognition: advancing the debate. Perspec Psychol Sci 8(3):223–241
3. Stanovich KE, West RF (2000) Individual differences in reasoning: implications for the rationality debate? Behav Brain Sci 23(5):645–665
4. Sloman SA (1996) The empirical case for two systems of reasoning. Psychol Bull 119(1):3–22
5. De Neys W (ed) (2017) Dual process theory 2.0. Routledge
6. Strack F, Deutsch R (2004) Reflective and impulsive determinants of social behavior. Pers Soc Psychol Rev 8(3):220–247
7. Kahneman D, Tversky A (1979) Prospect theory: An analysis of decision under risk. Econometrica 47(2):263–291
8. Chaiken S, Trope Y (eds) (1999) Dual-process theories in social psychology
9. Fiske ST, Taylor SE (2017) Social cognition: from brains to culture, 3rd ed. Sage Publications
10. Lieberman MD, Gaunt R, Gilbert DT, Trope Y (2002) Reflexion and reflection: a social cognitive neuroscience approach to attributional inference. Adv Exp Soc Psychol 34:199–249
11. Shiv B, Fedorikhin A (1999) Heart and mind in conflict: the interplay of affect and cognition in consumer decision making. J Cons Res 26(3):278–292

12. Greene JD, Haidt J (2002) How (and where) does moral judgment work? Trends Cogn Sci 6(12):517–523
13. Camerer CF, Loewenstein G, Prelec D (2005) Neuroeconomics: how neuroscience can inform economics. J Econ Literature 43(1):9–64
14. Levin IP, McElroy T, Gaeth GJ, Hedgcock W, Denburg NL (2014) Behavioral and neuroscience methods for studying neuroeconomic processes: what we can learn from framing effects
15. Reyna VF, Brainerd CJ (1991) Fuzzy-trace theory and framing effects in choice: gist extraction, truncation, and conversion. J Behav Decis Mak 4(4):249–262
16. Reyna VF, Brainerd CJ (1995) Fuzzy-trace theory: an interim synthesis. Learn Individ Differ 7(1):1–75
17. Reyna VF, Brainerd CJ (1995) Fuzzy-trace theory: some foundational issues. Learn Individ Differ 7(2):145–162
18. Reyna VF, Brainerd CJ (2008) Numeracy, ratio bias, and denominator neglect in judgments of risk and probability. Learn Individ Differ 18(1):89–107
19. Reyna VF, Brainerd CJ (2011) Dual processes in decision making and developmental neuroscience: a fuzzy-trace model. Dev Rev 31(2–3):180–206
20. Reyna VF (2012) A new intuitionism: Meaning, memory, and development in fuzzy-trace theory. Judgm Decis Mak 7(3):332–359
21. Reyna VF, Brainerd CJ (1998) Fuzzy-trace theory and children's false memories. J Exp Child Psychol 71(2):81–129
22. Reyna VF, Lloyd FJ (2006) Physician decision making and cardiac risk: effects of knowledge, risk perception, risk tolerance, and fuzzy processing. J Exp Psychol Appl 12(3):179–195
23. Reyna VF (2008) A theory of medical decision making and health: fuzzy trace theory. Med Decis Making 28(6):850–865
24. Hans VP, Reyna VF (2011) To dollars from sense: qualitative to quantitative translation in jury damage awards. J Empir Leg Stud 8(S1):120–147
25. Reyna VF, Hans VP, Corbin JC, Yeh R, Lin K, Royer C (2015) The gist of juries: testing a model of damage award decision making. Psychol Public Policy Law 21(3):280–294
26. Helm RK, Hans VP, Reyna VF (2017) Trial by numbers. Cornell J Law Public Policy 27(1):107–143
27. Reyna VF, Corbin JC, Weldon RB, Brainerd CJ (2016) How fuzzy-trace theory predicts true and false memories for words, sentences, and narratives. J Appl Res Mem Cogn 5(1):1–9
28. McElroy T (2024) Variation in valence intensity within frames: testing predictions of prospect theory and fuzzy-trace theory. J Behav Decis Mak 37(4):e2412
29. Glimcher PW, Fehr E (Eds) (2013) Neuroeconomics: decision making and the brain. Academic Press
30. Rangel A, Camerer C, Montague PR (2008) A framework for studying the neurobiology of value-based decision making. Nat Rev Neurosci 9(7):545–556
31. Haber SN, Knutson B (2010) The reward circuit: linking primate anatomy and human imaging. Neuropsychopharmacology 35(1):4–26
32. Schultz W (2015) Neuronal reward and decision signals: from theories to data. Physiol Rev 95(3):853–951
33. Montague PR, Dayan P, Sejnowski TJ (1996) A framework for mesencephalic dopamine systems based on predictive Hebbian learning. J Neurosci 16(5):1936–1947
34. Platt ML, Huettel SA (2008) Risky business: the neuroeconomics of decision making under uncertainty. Nat Neurosci 11(4):398–403
35. Padoa-Schioppa C, Assad JA (2006) Neurons in the orbitofrontal cortex encode economic value. Nature 441(7090):223–226
36. Gold JI, Shadlen MN (2007) The neural basis of decision making. Annu Rev Neurosci 30:535–574
37. Heekeren HR, Marrett S, Ungerleider LG (2008) The neural systems that mediate human perceptual decision making. Nat Rev Neurosci 9(6):467–479
38. Frank MJ (2006) Hold your horses: a dynamic computational role for the subthalamic nucleus in decision making. Neural Netw 19(8):1120–1136

39. LeDoux J (2003) The emotional brain, fear, and the amygdala. Cell Mol Neurobiol 23(4):727–738
40. Miller EK, Cohen JD (2001) An integrative theory of prefrontal cortex function. Annu Rev Neurosci 24(1):167–202
41. Frank MJ, Hutchison K (2009) Genetic contributions to avoidance-based decisions: striatal D2 receptor polymorphisms. Neuroscience 164(1):131–140
42. Jasper JD, Christman SD (2005) A neuropsychological dimension for anchoring effects. J Behav Decis Mak 18(5):343–369
43. Verdejo-García A, Pérez-García M (2007) Profile of executive deficits in cocaine and heroin polysubstance users: common and differential effects on separate executive components. Psychopharmacology 190(4):517–530
44. Killgore WD, Grugle NL, Balkin TJ (2009) Gambling when sleep deprived: don't bet on stimulants. Chronobiol Int 26(6):1285–1301
45. Frank MJ (2005) Dynamic dopamine modulation in the basal ganglia: a neurocomputational account of cognitive deficits in medicated and nonmedicated Parkinsonism. J Cogn Neurosci 17(1):51–72
46. Epstein J, Pan H, Kocsis JH, Yang Y, Butler T, Chusid J, Silbersweig DA (2006) Lack of ventral striatal response to positive stimuli in depressed versus normal subjects. Am J Psychiatry 163(10):1784–1790
47. Dimoka A (2012) How to conduct a functional magnetic resonance (fMRI) study in social science research. MIS Q 36(3):811–840
48. Bogacz R (2007) Optimal decision-making theories: linking neurobiology with behaviour. Trends Cogn Sci 11(3):118–125
49. McElroy T, Dickinson DL (2019) Thinking about complex decisions: how sleep and time-of-day influence complex choices. Conscious Cogn 76:102824
50. Dickinson DL, McElroy T (2019) Bayesian versus Heuristic-based choice under sleep restriction and suboptimal times of day. Games Econom Behav 119:48–59
51. Dickinson DL, McElroy T (2017) Sleep restriction and circadian effects on social decisions. Eur Econ Rev 97:57–71
52. Killgore WD, Killgore DB, Day LM, Li C, Kamimori GH, Balkin TJ (2007) The effects of 53 hours of sleep deprivation on moral judgment. Sleep 30(3):345–352
53. Dickinson DL, McElroy T (2012) Circadian effects on strategic reasoning. Exp Econ 15:444–459
54. Dickinson DL, McElroy T (2010) Rationality around the clock. Sleep and time-of-day effects on guessing game responses. Econ Lett 108:245–248
55. Hastings M, O'Neill JS, Maywood ES (2007) Circadian clocks: regulators of endocrine and metabolic rhythms. J Endocrinol 195(2):187–198
56. McElroy T, Dickinson DL (2010) Thoughtful days and valenced nights: how much will you think about the problem? Judgm Decis Mak 5:516–523
57. Stacey D, Légaré F, Lewis K, Barry MJ, Bennett CL, Eden KB, Trevena L et al (2017) Decision aids for people facing health treatment or screening decisions. Cochrane Database Syst Rev (4)
58. Elwyn G, Stiel M, Durand MA, Boivin J (2011) The design of patient decision support interventions: addressing the theory–practice gap. J Eval Clin Pract 17(4):565–574
59. Jones OD, Marois R, Farah MJ, Greely HT (2013) Law and neuroscience. J Neurosci 33(45):17624–17630
60. Morse SJ (2011) The future of neuroscientific evidence. In: The future of evidence. Routledge, pp 137–164
61. Bodenhausen GV (1990) Stereotypes as judgmental heuristics: evidence of circadian variations in discrimination. Psychol Sci 1(5):319–322
62. Zeelenberg M, Pieters R (2007) A theory of regret regulation 1.0. J Cons Psychol 17(1):3–18
63. Greely HT, Wagner AD (2011) Reference guide on neuroscience. In: Reference manual on scientific evidence. National Academies Press
64. Morse SJ (2006) Brain overclaim syndrome and criminal responsibility: a diagnostic note. Ohio State J Criminal Law 3(2):397–412

65. Farahany NA (2016) Neuroscience and behavioral genetics in US criminal law: an empirical analysis. J Law Biosci 2(3):485–509
66. Schwartz B (2004) The paradox of choice: why more is less. Harper Collins
67. Baumeister RF, Vohs KD, Tice DM (2007) The strength model of self-control. Curr Dir Psychol Sci 16(6):351–355
68. Satel S, Lilienfeld SO (2013) Brainwashed: the seductive appeal of mindless neuroscience. Basic Civitas Books
69. Pustilnik AC (2013) Neurotechnologies at the intersection of criminal procedure and constitutional law. In: The constitution and the future of criminal justice in America, pp 109–136
70. Schwartz B (2016) The paradox of choice: why more is less, revised edition. Ecco Press
71. Markman KD, Medin DL (1995) Similarity and alignment in choice. Organ Behav Hum Decis Process 63(2):117–130
72. Cokely ET, Kelley CM (2009) Cognitive abilities and superior decision making under risk: a protocol analysis and process model evaluation. Judgm Decis Mak 4(1):20–33
73. Dolan P, White MP (2007) How can measures of subjective well-being be used to inform public policy? Perspect Psychol Sci 2(1):71–85
74. Edwards W (1954) The theory of decision making. Psychol Bull 51(4):380–417
75. Bruine de Bruin W, Del Missier F, Levin IP (2012) Individual differences in decision-making competence. J Behav Decis Mak 25(4):329–330
76. Gamliel E, Kreiner H (2019) Applying fuzzy-trace theory to attribute-framing bias: Gist and verbatim representations of quantitative information. J Exp Psychol Learn Mem Cogn 46(3):497–506
77. Levin IP (1999) Relating statistics and experimental design: an introduction. Sage Publications, Inc
78. Mohammed S, Schwall A (2009) Individual differences and decision making: what we know and where we go from here. Int Rev Ind Organ Psychol 24:249–312
79. Stanovich KE, West RF (2008) On the relative independence of thinking biases and cognitive ability. J Pers Soc Psychol 94(4):672
80. Peters E, Västfjäll D, Slovic P, Mertz CK, Mazzocco K, Dickert S (2006) Numeracy and decision making. Psychol Sci 17(5):407–413
81. Liberali JM, Reyna VF, Furlan S, Stein LM, Pardo ST (2012) Individual differences in numeracy and cognitive reflection, with implications for biases and fallacies in probability judgment. J Behav Decis Mak 25(4):361–381
82. Peters E, Levin IP (2008) Dissecting the risky-choice framing effect: numeracy as an individual-difference factor in weighting risky and riskless options. Judgm Decis Mak 3(6):435–448
83. Eberhardt W, Bruine de Bruin W, Strough J (2019) Age differences in financial decision making: the benefits of more experience and less negative emotions. J Behav Decis Mak 32(1):79–93
84. Casey BJ, Giedd JN, Thomas KM (2000) Structural and functional brain development and its relation to cognitive development. Biol Psychol 54(1–3):241–257
85. Gozdas E, Holland SK, Altaye M (2019) Developmental changes in functional brain networks from birth to adolescence. Hum Brain Mapp 40(5):1434–1444
86. Reyna VF, Wilhelms EA, McCormick MJ, Weldon RB (2015) Development of risky decision making: Fuzzy-trace theory and neurobiological perspectives. Child Dev Perspect 9(2):122–127
87. Bruine de Bruin W (2017) Aging and economic decision making. In: Aging and decision making. Academic Press, pp 371–384
88. Bruine de Bruin W, Parker AM, Strough J (2020) Age differences in reported social networks and well-being. Psychol Aging 35(2):159–168
89. Sobkow A, Traczyk J, Kaufman SB, Nosal C (2018) The structure of intuitive abilities and their relationships with intelligence and openness to experience. Intelligence 67:1–10
90. McElroy T, Dowd K (2007) Susceptibility to anchoring effects: How openness-to-experience influences responses to anchoring cues. Judgm Decis Mak 2(1):48–53

91. Stanovich KE, West RF (2008) On the relative independence of thinking biases and cognitive ability. J Pers Soc Psychol 94(4):672
92. Peters E, Västfjäll D, Slovic P, Mertz CK, Mazzocco K, Dickert S (2006) Numeracy and decision making. Psychol Sci 17(5):407–413.
93. McElroy T, Seta JJ (2003) Framing effects: an analytic–holistic perspective. J Exp Soc Psychol 39(6):610–617. https://doi.org/10.1016/S0022-1031(03)00036-2
94. Eberhardt W, Bruine de Bruin W, Strough J (2019) Age differences in financial decision making: The benefits of more experience and less negative emotions. J Behav Decis Making 32(1):79–93. https://doi.org/10.1002/bdm.v32.1https://doi.org/10.1002/bdm.2097
95. McElroy T, Seta JJ, Waring DA (2007) Reflections of the self: how self-esteem determines decision framing and increases risk taking. J Behav Decis Making 20(3):223–240. https://doi.org/10.1002/bdm.v20:3. https://doi.org/10.1002/bdm.551
96. McElroy T, Dickinson DL, Levin IP (2020) Thinking about decisions: an integrative approach of person and task factors. J Behav Decis Making 33(4):538–555. https://doi.org/10.1002/bdm.v33.4. https://doi.org/10.1002/bdm.2175
97. McElroy T, Mascari D (2007) When is it going to happen? How temporal distance influences processing for risky–choice framing tasks. Soc Cogn 25(4):495–517. https://doi.org/10.1521/soco.2007.25.4.495

Todd McElroy is an Associate Professor in the Department of Psychology at Florida Gulf Coast University. His primary areas of research are in decision-making, the psychology of physical activity, and examining factors that influence complex choices.

Todd is interested in how cognitive elements, such as individual differences in thinking style and working memory capacity, as well as social forces like context, influence our decisions. He takes an integrative approach, investigating decision-making through the lens of personality differences, situational factors, biological influences, and their interactions.

In his exploration of the biological basis of decision-making, Todd examines how factors such as sleep, circadian rhythm, and glucose levels shape our choices. His work has uncovered fascinating insights, such as the impact of sleep restriction and circadian mismatch on risk preferences and strategic decision-making.

Throughout his research career, Todd has sought to develop a clearer understanding of the decision-making process. His studies have investigated the roles of regret, cognitive reflection, hemispheric processing, and message framing on our decisions and judgments.

Todd's recent work includes research on the relationship between physical activity and happiness, romantic regret, and a new model of the psychological determinants of physical activity called the Mind-Motion Matrix. In 2024, he published the book "The Science of Laziness: Achieving More by Doing Less" as part of The Self-Enhancement Science Series.

With over 50 peer-reviewed publications to date, Todd McElroy continues to advance our understanding of the complex interplay of factors shaping human judgment and decision-making. His integrative approach and novel findings offer valuable insights for both research and real-world applications.

Replications in Judgment and Decision Making: A Critical Review and Future Pathways

2

Agata Sobkow, Kacper Jurewicz, and Jakub Traczyk

Abstract

Judgment and Decision Making (JDM), an interdisciplinary field integrating psychology, economics, management, and cognitive science, has provided substantial insights into human decision-making processes over the past 70 years. Despite its rich history, the field's credibility relies on replication to validate findings across diverse contexts and enhance methodological rigor. This chapter examines the replication landscape within JDM, highlighting key milestones such as the "replication crisis" and major collaborative replication projects. We assess current replication efforts of JDM effects using comprehensive sources, including academic databases and collaborative networks. Our review of over 100 studies (osf.io/j58t2) reveals a 44% success rate in replications, with an additional 25% showing partial replication. Successfully replicated effects often involve well-established paradigms like economic games and numerous heuristics and biases. However, challenges persist in replicating priming effects and other subtle manipulations, such as incidental anchors and research on the unconscious thought advantage. The chapter underscores the importance of transparency practices and collaborative efforts in addressing replication challenges, ultimately aiming to strengthen the integrity of JDM research.

A. Sobkow (✉) · K. Jurewicz · J. Traczyk
Faculty of Psychology in Wroclaw, Center for Research on Improving Decision Making (CRIDM), SWPS University, Wrocław, Poland
e-mail: asobkow@swps.edu.pl

K. Jurewicz
e-mail: kjurewicz@st.swps.edu.pl

J. Traczyk
e-mail: jtraczyk@swps.edu.pl

© The Author(s), under exclusive license to Springer Nature Switzerland AG 2025

T. McElroy (ed.), *Decision-Making in Life and Work*, Integrated Science 36,
https://doi.org/10.1007/978-3-031-88415-3_2

Keywords

Replication • Judgment and decision making • Heuristics and biases • Review • Open science • Registered reports

1 Introduction

> More ambiguous results would be painful, of course, but they would still protect the reputations of scholars who sincerely believe in their work – even if they are sometimes wrong.
>
> Daniel Kahneman [1]

Judgment and Decision Making (JDM) with a history spanning over 70 years, investigates the intricate processes through which individuals evaluate information and make choices [2]. Rooted in psychology and extending into economics, management, and cognitive science, JDM has evolved to provide profound insights into the mechanisms underlying human judgment. The field covers a wide array of topics, including heuristics and biases—such as anchoring, correspondence bias, framing, and sunk cost—as well as moral decision making, risk preferences, financial and health behaviors. This interdisciplinary approach enables JDM research to shed light on the complexities of human decision making and inform practical applications across various domains.

Despite JDM's extensive history, replication remains crucial for its continued advancement and credibility. Replication is essential for validating findings across different contexts, ensuring the robustness and applicability of conclusions about decision-making processes. It serves as a quality control measure by verifying observed effects, addressing methodological flaws and biases, enhancing generalizability, and building theory [3, 4]. While replication and the open science movement have gained traction in psychology [5], a comprehensive review assessing replicability in JDM is notably absent. This chapter aims to bridge this gap by scrutinizing replication practices, evaluating their impact on the field, and suggesting pathways to enhance the rigor and reliability of future research.

The chapter is structured as follows: We begin with a historical overview of key milestones in replication research within psychology, including discussions on the "replication crisis" and initiatives like the Many Labs project. Next, we review current research involving the replication of JDM paradigms, such as heuristics and biases. We then explore the potential moderators and implications of these replication efforts and propose strategies to improve methodological practices.

2 Origins of Replication Crisis: Tracing the Birth of an Open Science Movement

The replication crisis in psychology, which emerged in the 2010s, was catalyzed by events that shook the field's foundations [6]. Notable failures to replicate effects described in famous studies—particularly in social priming research such as Bargh's work [7, 8]—cast doubt on the robustness of key findings. Concurrently, high-profile cases of fraud, exemplified by Diederik Stapel's fabricated data [9] and controversies surrounding Bem's extrasensory perception studies [10, 11], further eroded trust in psychological research. Additionally, the prevalence of questionable research practices [12, 13], including *p*-hacking and selective reporting, raised concerns about the integrity of study results.

These replication failures have attracted significant attention and prompted questions about the robustness of key findings in psychology. Studies such as those investigating the social priming of the elderly stereotype [7], the facial feedback hypothesis [14], power posing [15], and the ego depletion effect [16] have all faced challenges in replication attempts. Subsequent researchers have often produced inconsistent or null results [8, 17, 18], casting doubt on the robustness and generalizability of these findings.

Moreover, the field was rocked by fraud involving researchers like Diederik Stapel, Dirk Smeesters, and Lawrence Sanna [19, 20]. These instances of misconduct, particularly Stapel's intentional data fabrication, caused a significant retraction of nearly 60 articles attributed to him (https://retractionwatch.com/the-retraction-watch-leaderboard/). Recent revelations in 2023 also highlighted further instances of data manipulation, such as anomalies in data across four papers on moral judgments uncovered by researchers from Data Colada, including Uri Simonsohn, Leif Nelson, and Joe Simmons (https://datacolada.org/112). Another major case involved the work of a neuroscientist, whose faked data on amyloid beta plaques misled the scientific community for over a decade [21]. This misinformation resulted in significant resources being allocated toward the production and clinical trials of Alzheimer's disease drugs based on findings that were later discredited. The implications of this fraud are vast, potentially delaying the development of effective treatments and wasting millions in research funding.

While cases of intentional data manipulation represent extreme misconduct, the prevalence of questionable research practices is a broader issue within the scientific community. Practices like *p*-hacking, which exploit data collection and analysis flexibility, potentially lead to false-positive results [22]. In 2012 [23], many psychologists acknowledged failing to report all dependent measures or collecting additional data post-hoc, both practices inflating the likelihood of significant results. Such questionable practices likely contributed to controversial findings in Bem's [10] extrasensory perception studies, which, despite publication in a prestigious journal, failed to withstand scrutiny and sparked considerable debate.

The replication crisis, with its associated fraud cases and prevalence of questionable practices, has raised significant concerns about the reliability of key findings in psychological research, including JDM phenomena. These challenges have ignited intense debates within the scientific community, prompting broader discussions on research methodology, publication biases, and the replication crisis. A notable response to this crisis was the Nobel Memorial Prize in Economics in 2002 Daniel Kahneman's [1] open letter urging social priming researchers to clear their field by adopting robust replication protocols and transparency practices like preregistration, open materials, and open data. Despite these reform calls, the transition faced resistance, with some researchers criticizing the open science movement for using social media platforms like blogs and Twitter to expose data integrity and replicability issues, labeling them "self-appointed data police" and accusing them of "methodological terrorism" [24].

Nevertheless, the field has progressively embraced remedies to address the replication crisis. These efforts include implementing open science practices such as preregistration, making research materials and data openly accessible, and introducing registered reports [25] where manuscripts are peer-reviewed and accepted based on methodological rigor before data collection. Additionally, guidelines for conducting replications have been developed to ensure replication efforts are valid and convincing [26].

The replication crisis has driven the development of large-scale, collaborative networks that combine resources to tackle specific research questions across diverse cultures, countries, and disciplines. For example, the Many Labs project [27] assessed the replicability of 13 classic and contemporary effects using 36 independent samples totaling 6344 participants. Of these effects, 10 were consistently replicated, one showed weak replicability, and two failed to replicate. Building on this, Many Labs 2 [28] extended the investigation to 28 effects, testing replicability across 125 samples from 36 countries and territories, involving 15,305 participants. Results indicated that 15 (54%) of these replications yielded statistically significant effects in the same direction as the original studies. Similarly, the Reproducibility Project [29] systematically replicated 100 studies from three major psychological journals: *the Journal of Experimental Psychology: Learning, Memory, and Cognition*, *the Journal of Personality and Social Psychology*, and *Psychological Science*. Despite using original materials, many replications showed weaker evidence for the original findings, with only 36% producing significant effects in the same direction as initially reported.

To advance these efforts, another study [30] replicated 21 selected experimental studies in the social sciences published in *Nature* and *Science*. This replication found significant effects in the same direction as the original studies in 13 out of 21 cases (62%). Moreover, the Experimental Economics Replication Project [31] examined 18 studies published in the *American Economic Review* and the *Quarterly Journal of Economics*. This project identified significant effects in the same direction as the original studies for 11 replications (61%). These findings underscore the critical need for rigorous, high-powered replication efforts across various fields to effectively evaluate and strengthen the robustness of research outcomes.

3 Current State of Replications in the Judgment and Decision Making Field

Building on these historical and contemporary insights, the current section explores the replication landscape within the JDM field. This analysis provides a detailed overview of recent replication studies. By assessing the current state of replication practices, this section aims to highlight progress made, ongoing challenges, and potential strategies for enhancing the rigor and credibility of future JDM research.

We adopted a comprehensive, multi-source strategy to select studies replicating JDM effects. Initially, we systematically reviewed replication efforts from collaborative initiatives such as Many Labs [27], Many Labs 2 [28], Many Labs 3 [32], Many Labs 4 [33], Many Labs 5 [34], and the Reproducibility Project [29], the Social Sciences Replication Project [30], the Experimental Economics Replication Project [31], and the Pipeline project [35]. We extended our search by utilizing academic databases such as EBSCO, Scopus, and Google Scholar, employing keywords such as "replication" and "judgment"/"decision making" to identify relevant studies. Furthermore, we consulted the replication database [36] and reviewed the work of the CORE team led by Gilad Feldman [37], given their prominence in conducting replication studies within the JDM field. Our inclusion criteria encompassed studies published or available as preprints (in the case of the Registered Reports—on Stage 2, after the data collection). Our selection also incorporated the collective expertise of our team, who have significant experience in conducting and publishing replication studies. This multifaceted approach ensures a robust and thorough identification of replicated effects in JDM.

The results of our extensive search are summarized in a dedicated OSF project (https://osf.io/j58t2/). The table details the publication information for each identified effect, including the year, main authors, and journal. Additionally, we indicate whether the study was preregistered or not. Each entry also notes the outcome of the replication attempt, categorizing it as either unsuccessful, partially successful, or successful.

Based on our comprehensive review, we analyzed 190 records across topics encompassing heuristics and biases, analytic and intuitive processing, moral judgments, economic games, risk preferences, and numeracy. Our findings reveal that among these studies, 84 (44%) demonstrated successful replication, while 48 (25%) showed partial replication, and 60 (31%) did not replicate. These replication studies were disseminated across various journals in the field, as detailed in Table 1, highlighting the most popular venues for publishing replication studies in JDM topics. Leading the list, *Judgment and Decision Making* featured prominently with 15 of the reviewed studies, followed by *the Journal of Experimental Social Psychology* with 11 publications. Many articles were also published in emerging journals dedicated to Open Science practices, such as *Collabra Psychology* and *Meta-Psychology*.

From a historical perspective (see Fig. 1), before 2014, replication efforts in JDM were scarce, with only six papers explicitly focused on replication. The landscape underwent a significant transformation after 2014, marked by the inception

Table 1 Journals publishing replication studies in Judgment and Decision Making

Journal	Number of identified replication articles (with JDM effects)
Judgment and Decision Making	15
Journal of Experimental Social Psychology	11
Collabra Psychology	10
International Review of Social Psychology	5
Meta-Psychology	5
Journal of Economic Psychology	3
Social Psychological and Personality Science	3
Social Psychology	3

of pioneering large-scale replication projects (see also Table 2 for brief descriptions of these projects and their results). The Open Science Framework (https://osf.io/) was developed during this period, setting new standards such as pre-registration, open data, and materials sharing.

These advancements sparked a resurgence in replication studies, gaining substantial momentum around 2019 and 2020, with approximately 10-20 replication papers published annually. Notably, many of these studies adopted the registered reports format, enabling peer review before data collection commenced. In 2021, a noteworthy development was the establishment of the non-profit initiative Peer Community In Registered Reports (PCI RR; https://rr.peercommunityin.org/). This platform allows scientists to peer-review preprints before data collection. Importantly, an increasing number of PCI RR-friendly journals (e.g., *Advances in Cognitive Psychology*, *Advances in Methods and Practices in Psychological Science*, *Collabra Psychology*, *Experimental Psychology*, *International Review of Social Psychology*, *Journal of Cognition*, *Meta-Psychology*, *Social Psychological Bulletin*) accept manuscripts that receive a positive final recommendation from PCI RR without further peer review. This evolution underscores the JDM community's growing commitment to enhancing research transparency, rigor, and the reliability of findings through systematic replication efforts.

Finally, our review results highlight significant variability in the replicability of different types of effects. Notably, the lowest replicability rates were observed for effects involving priming and other subtle manipulations of thinking modes. For instance, priming effects related to currency [27], with heat [27], and haptic [29] failed to be successfully replicated. Additionally, subtle manipulations intended to evoke a more analytical or intuitive mindset, such as font disfluency [28], observing Rodin's sculpture "The Thinker" [30], or describing situations where intuition versus careful reasoning led to good or bad outcomes [40] also faced replication challenges. Moreover, several studies failed to replicate effects like the unconscious thought advantage [38, 41–43].

It is important to note that replication issues were not limited to research involving intuitive System 1 processing. Other unsuccessful replications included the

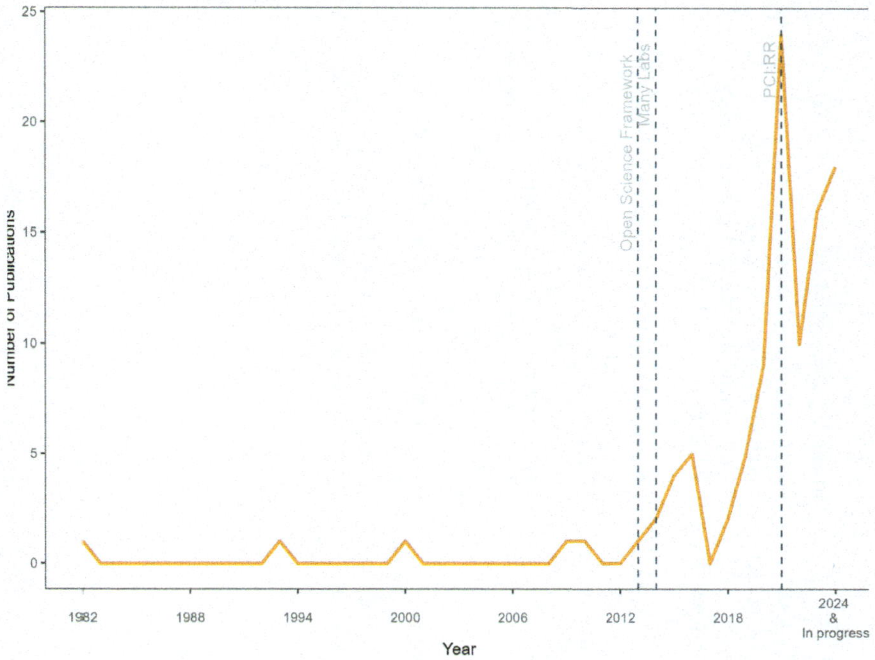

Fig. 1 The number of publications reporting replications of Judgment and Decision Making (JDM) effects across different years. Vertical lines indicate significant milestones in the field: establishing the Open Science Framework, the first Many Labs project, and introducing the PCI Registered Reports initiative.

effects of motivated numeracy [44, 45], subtle forms of anchoring [28, 46, 47], or framing [38, 48]. Nevertheless, for these effects (i.e., anchoring and framing), some replication attempts were at least partially successful [27–29, 38, 49–54], suggesting that procedural details and moderators might influence replication outcomes.

Among the effects with the highest replicability rates, we observed those involving economic games, such as the prisoner's dilemma [31], and moral judgment scenarios, for example, in the Trolley Dilemma [28, 35], the sunk cost effect [27, 38, 55–57], the less-is-better effect [28, 38, 58], and the predictive power of numeracy on decision outcomes [38, 39, 59].

Our review highlights significant variability in the replicability of different psychological effects. Effects involving priming and subtle manipulations of thinking modes had notably low replicability rates, while effects related to economic games and moral judgment showed higher replicability. However, replication attempts were often influenced by procedural details and moderators.

Table 2 Overview of large-scale replication projects investigating JDM effects

Date	Project	Brief description	Results (regarding JDM effects)
2014	The Many Labs project [27]	A pioneering replication study involving 36 independent laboratories. This extensive project assessed 13 classic and contemporary psychological effects, utilizing a total sample size of 6,344 participants	• Successful replication: 4 effects (sunk cost; framing; anchoring; gambler's fallacy) • Unsuccessful replication: 1 effect (currency priming)
2015	The Reproducibility Project [29]	Systematic replication of 100 studies from three leading psychological journals: *Journal of Experimental Psychology: Learning, Memory, and Cognition*, *Journal of Personality and Social Psychology*, and *Psychological Science*	• Successful replication: 1 effect (the ultimate sampling dilemma) • Partial replication: 3 effects (anchoring; cognitive abilities, and rationality; preference for decision-consistent information) • Unsuccessful replication: 5 effects (tempting fate; belief in determinism and cheating; drug use and risk taking; value heuristic)
2016	The Many Labs 3 project [32]	This project analyzed how ten effects, ten individual differences, and data quality indicators varied throughout the academic semester across 20 student participant pools (total N = 2696) and an online sample (N = 737).	• Successful replication: 1 effect (availability heuristic) • Unsuccessful replication: 2 effects (power and perspectives not taken; elaboration likelihood)

(continued)

Table 2 (continued)

Date	Project	Brief description	Results (regarding JDM effects)
2016	The Pipeline project [35]	This project introduced a new collaborative method - effects replicated in independent laboratories before publication. 25 research groups replicated ten moral judgment effects that the last author and their team were preparing for publication	• Successful replication: 8 effects (bigot–misanthrope effect; cold-hearted prosociality effect; bad tipper effect; belief–act inconsistency effect; moral inversion effect; moral cliff effect; intuitive economics effect; burn-in-hell effect) • Unsuccessful replication: 2 effects (presumption of guilt effect; higher standard effect)
2016	The Experimental Economics Replication Project [31]	The project aimed to systematically replicate 18 studies published in *the American Economic Review* and *the Quarterly Journal of Economics*	• Successful replication: 11 effects (e.g., asset markets, prisoner's dilemma, job design) • Unsuccessful replication: 7 effects (e.g., reference point and effort provision; gift exchange vs monetary exchange; happiness and time preference)
2018	The Many Labs 2 project [28]	The project aimed to systematically replicate 28 effects to investigate variations in effect magnitudes across different samples and settings. Each protocol was implemented in roughly half of 125 samples, involving a total of 15,305 participants from 36 countries and territories	• Successful replication: 11 effects (e.g., correspondence bias; trolley dilemma; false consensus effect; less is better effect; framing) • Unsuccessful replication: 6 (e.g., disfluency and analytic processing; incidental anchors; priming "heat")

(continued)

Table 2 (continued)

Date	Project	Brief description	Results (regarding JDM effects)
2018	The Social Sciences Replication Project [30]	The project aimed to replicate 21 experimental studies in the social sciences published in *Nature* and *Science*. The replication efforts adhered to analysis plans that were reviewed by the original authors and pre-registered before the replications began	• Successful replication: 4 effects (avoiding overhead aversion in charity; cooperating with the future; communication, punishment, and earnings; inequality and visibility of wealth) • Unsuccessful replication: 5 effects (haptic priming, priming analytic thinking, washing away post-decisional dissonance, priming giving)
2020	The Many Labs 5 [34]	The project aimed to select replication studies from the Reproducibility Project (Open Science Collaboration, 2015), addressing concerns raised by original authors. New protocols were subsequently revised and underwent formal peer review before being implemented in new replication studies (Registered reports) and conducted across multiple laboratories	• Unsuccessful replication: 2 effects (encouraging a belief in determinism increases cheating; irrational tempting fate behavior and cognitive load)

(continued)

Table 2 (continued)

Date	Project	Brief description	Results (regarding JDM effects)
2022	Replication of 15 JDM effects in a Polish sample [38]	The project aimed to conduct pre registered replications of 15 JDM effects, examining their applicability across diverse contexts. These effects were tested by undergraduate psychology students enrolled in a decision-making course	• Successful replication: 7 effects (procrastination and dependent decision style; sunk cost; openness to experience and anchoring effect; self-control and financial behavior; framing of students' assessment; less is better effect I; anchoring) • Partial replication: 5 effects (priming risk; sex differences in risk; subjective numeracy and risk; framing of moral choices, less in better effect II) • Unsuccessful replication: 3 effects (framing; focusing on emotions; frequency program requirements)

(continued)

Table 2 (continued)

Date	Project	Brief description	Results (regarding JDM effects)
2024	Replicating numeracy project [39]	The project aimed to conduct pre-registered replications of 11 effects demonstrating the significant role of individual differences in numeracy on decision outcomes. These replications were carried out by students enrolled in a decision-making course	• Successful replication: 6 effects (numeracy and consistency in risk perception; decision rules; choices maximizing expected value; medical risk comprehension; financial knowledge; preference for suboptimal options) • Partial replication: 2 effects (numeracy and no-loss bet; visual aids support risk understanding in people with low numeracy) • Unsuccessful replication: 3 effects (numeracy and resistance to sunk cost; numeracy and financial well-being; numeracy and framing effect)
2019–2024	The CORE team [37]	This ongoing project is dedicated to practicing and promoting best practices in open science and meta-research. Under the supervision of Gilad Feldman, a group of early-career researchers and students conducts replications. They also actively develop tools and integrate registered reports into their research methodologies	Because this is an ongoing project, we only report the results of completed studies published or being at Stage 2 Registered Report. Among those: • Successful replication: 16 effects (e.g., omission bias in morality; money illusion effect; better than average effect) • Partial replication: 22 effects (e.g., decoy effect; distinction bias; inaction inertia effect; less is better; numeracy and decision making) • Unsuccessful replication: 6 effects (e.g., conjunction effect; identifiable victim effect; moral feelings)

4 Factors Influencing Replication Success in Psychological Research

Moderators and boundary conditions that influence replication success may include characteristics from both the original and replication studies. A survey of 215 psychological scientists [60] identified 'large sample size' as the most important predictor of an effect's replicability, followed by 'large effect size,' 'low variance in outcome measures,' and 'pre-registered analyses,' among others. In contrast, factors such as experimental versus correlational design, behavioral versus self-report measures, and study setting (field, lab, online) were rated by scientists as having little importance for replicability, with scores below 3 on a 5-point scale.

Nevertheless, empirical data [60] from replication efforts indicate that the original study's sample size is not associated with replication likelihood, whereas effect size is a significant predictor ($r_s = .21$, $p < 0.001$). Moreover, no direct empirical evidence has yet demonstrated that pre-registration alone improves replicability. A recent systematic review [61] found no significant difference between pre-registered and non-preregistered studies in psychology regarding effect sizes, prevalence of positive results, and statistical consistency. The main benefits of pre-registration were the inclusion of power analyses and the use of larger sample sizes compared to non-preregistered studies.

Further evidence on factors influencing replication success comes from Artificial Intelligence (AI) modeling research. Interestingly, AI models trained on data from past replication attempts were found to predict replication success more effectively than reviewers and comparably to prediction markets [62]. An analysis of over 14,000 papers from six leading psychological journals [63] using an AI model showed that study design was a crucial factor: experimental studies had significantly lower replication rates than non-experimental ones in areas such as social psychology, cognitive psychology, and organizational psychology. Another factor linked to replication success was the cumulative publication count and citation impact of the authors, which serves as a proxy for their expertise. Interestingly, a paper's citation count was not associated with replicability, while higher media attention correlated with replication failure, likely due to the media's preference for surprising or counterintuitive findings.

Regarding replication study characteristics, factors such as study setting (e.g., in-lab vs. online) and differences in language or culture can influence replication outcomes. For example, in student replication projects [64], studies that shifted to an online format were less likely to replicate compared to those maintaining the original modality. Nevertheless, the Many Labs 2 project [28] found minimal variability in outcomes based on setting or cultural differences, except for certain effects.

The AI models [63] suggested that the higher expertise of original authors could increase the likelihood of successfully replicating their studies. However, while involving original authors may enhance replication efforts due to their in-depth knowledge and familiarity with specific study details, it does not guarantee success.

This is evident from the failed replications of effects like Ego-Depletion [65] and Mortality Salience [33], even when the original authors were involved.

In summary, replication success in psychology depends on various factors, including characteristics of both original and replication studies, such as effect size, author expertise, and study design. Factors such as replication setting, cultural differences, and author involvement also play roles, underscoring the complexity of replication in psychology. Given these findings, the next section will address strategies to mitigate these challenges and outline future steps for enhancing the reliability and robustness of JDM research.

5 Addressing Limitations and Future Directions in Replication Studies

Despite the valuable insights gained from our comprehensive review, several limitations must be acknowledged. First, the selection of studies for replication was often based on practical considerations, such as feasibility and citation frequency [66, 67], rather than a systematic representation of the broader JDM literature. This approach may bias the results and limit the generalizability of the findings [68].

Second, the studies selected for our review might not represent all conducted replications of JDM effects, largely because researchers conducting replications face limited dissemination opportunities. A recent analysis [69] revealed that although journal policies about publishing replications in social psychology increased from 12% to 25% between 2015 and 2022, replications are still far from mainstream. Among the 100 highest-impact psychology journals screened, only 0.2% of articles (169 out of 84,834) were direct replications, reflecting only a modest increase over time [70]. To enhance the visibility of their replication studies, JDM researchers may refer to Table 1, which lists journals that have published multiple replication articles. They could also consider disseminating their work as preprints (e.g., on https://osf.io/preprints/psyarxiv/) or including it in a replication database [36] to help other researchers find the results of a particular replication attempt.

Third, most replication efforts have concentrated on Western, Educated, Industrialized, Rich, and Democratic (WEIRD) populations, often using online platforms like Prolific or mTurk. This narrow focus limits the generalizability of the findings to more diverse cultural contexts. Future research should prioritize including underrepresented populations to better evaluate the universality of JDM findings. Jarke et al. [66] provide a detailed guide on the workings of various collaborative efforts and essential considerations for conducting multi-country replications, emphasizing the critical importance of diverse samples for enhancing generalizability.

We also encourage researchers to engage younger members of the JDM community in replication projects. Several research teams, including ours, have successfully ran replication studies with undergraduate students [38, 39, 64, 71].

One of the largest and most influential replication initiatives is the CORE team [37], where replication studies are designed as part of a one-semester course. Engaging students in replication projects benefits their education and career development and strengthens the overall research community by promoting a culture of thorough and reliable scientific practices.

Fourth, that single failed replications cannot unambiguously identify fraud or deeply entrenched methodological flaws. Effectively addressing such issues requires replication efforts that involve multiple well-powered studies and thorough data analyses [72–74]. This comprehensive approach is crucial for reliably assessing the robustness of findings and upholding the integrity of psychological research. Therefore, more strategic, representative sampling and robust analytical techniques are needed in replication endeavors to draw meaningful conclusions about the validity and generalizability of psychological theories. Additionally, our review did not distinguish between different types of replication, which may have different aims and consequences [3]. For example, close or direct replications help to rule out false positive effects (Type I errors), while conceptual replications, using different research materials and samples than the original study, help evaluate the general validity of a theoretical prediction or the generalizability of the effect to other contexts.

Lastly, we want to emphasize the importance of improving replications and fostering more credible JDM research, especially in relation to real-world outcomes and policies. For instance, an influential study [75], which tested whether signing a form before completing a self-report increased honesty, was retracted after evidence raised doubts about the data's validity (https://datacolada.org/98 and https://datacolada.org/109). Interestingly, one of the studies in that paper was conducted in collaboration with an insurance company, suggesting potential real-life implications for institutions such as governments and corporations. Policies designed to reduce dishonesty could have been based on this study, much like how fabricated research on amyloid plaques [21] led to significant financial losses.

Conducting credible and trustworthy research—preferably through open, transparent, large-scale collaborative projects, such as those led by the Psychological Science Accelerator—is essential for developing evidence-based policies. For example, one such project tested the effectiveness of loss-framed messages on emotions, policy attitudes, and COVID-19-related behavioral intentions [76]. Another large-scale initiative explored the impact of various interventions (e.g., psychological distancing, negative emotions) on climate change mitigation [77]. These open, collaborative studies, conducted on large, diverse samples from multiple countries, have the potential to significantly influence policy.

However, executing these projects requires enormous effort, and testing the replicability of basic JDM effects can be extremely valuable in identifying the most promising mechanisms for designing interventions. These interventions can then be further tested and eventually implemented in real-world settings.

6 Conclusion

In conclusion, significant progress has been made in the field of JDM. However, we are still navigating the transition from the "replication crisis" to the "credibility revolution" [5], and continuous efforts are necessary to ensure the reliability and validity of JDM research. The field can further strengthen its scientific foundation by embracing more inclusive, transparent, and collaborative practices. As Daniel Kahneman [1] noted, accepting the potential for ambiguous results can safeguard the integrity of scholars and the scientific process, ultimately enhancing our understanding of human judgment and decision making.

Acknowledgments We want to thank Joanna Dixa for her help in searching and selecting articles.

References

1. Kahneman D (2012) A proposal to deal with questions about priming effects
2. Keren G, Wu G (2015) A bird's-eye view of the history of judgement and decision making. In: The Wiley Blackwell handbook of judgement and decision making. Wiley Blackwell
3. Hüffmeier J, Mazei J, Schultze T (2016) Reconceptualizing replication as a sequence of different studies: a replication typology. J Exp Soc Psychol 66:81–92. https://doi.org/10.1016/j.jesp.2015.09.009
4. Simons DJ (2014) The value of direct replication. Perspect Psychol Sci 9:76–80. https://doi.org/10.1177/1745691613514755
5. Vazire S, Nosek B (2023) Introduction to special topic "Is psychology self-correcting? Reflections on the credibility revolution in social and personality psychology." Social Psychol Bull 18:1–4. https://doi.org/10.32872/spb.12927
6. Romero F (2019) Philosophy of science and the replicability crisis. Philos Compass 14:e12633. https://doi.org/10.1111/phc3.12633
7. Bargh JA, Chen M, Burrows L (1996) Automaticity of social behavior: direct effects of trait construct and stereotype activation on action. J Pers Soc Psychol 71:230–244. https://doi.org/10.1037/0022-3514.71.2.230
8. Doyen S, Klein O, Pichon C-L, Cleeremans A (2012) Behavioral priming: it's all in the mind, but whose mind? PLoS ONE 7:e29081. https://doi.org/10.1371/journal.pone.0029081
9. Levelt, WJ, Drenth, P, Noort E (2012) Flawed science: the fraudulent research practices of social psychologist Diederik Stapel
10. Bem DJ (2011) Feeling the future: experimental evidence for anomalous retroactive influences on cognition and affect. J Pers Soc Psychol 100:407–425. https://doi.org/10.1037/a0021524
11. Galak J, LeBoeuf RA, Nelson LD, Simmons JP (2012) Correcting the past: failures to replicate psi. J Pers Soc Psychol 103:933–948. https://doi.org/10.1037/a0029709
12. Banks GC, Rogelberg SG, Woznyj HM, Landis RS, Rupp DE (2016) Editorial: evidence on questionable research practices: the good, the bad, and the ugly. J Bus Psychol 31:323–338. https://doi.org/10.1007/s10869-016-9456-7
13. Ioannidis JPA, Munafò MR, Fusar-Poli P, Nosek BA, David SP (2014) Publication and other reporting biases in cognitive sciences: detection, prevalence, and prevention. Trends Cogn Sci 18:235–241. https://doi.org/10.1016/j.tics.2014.02.010
14. Strack F, Martin LL, Stepper S (1988) Inhibiting and facilitating conditions of the human smile: a nonobtrusive test of the facial feedback hypothesis. J Pers Soc Psychol 54:768–777. https://doi.org/10.1037/0022-3514.54.5.768

15. Carney DR, Cuddy AJC, Yap AJ (2010) Power posing: brief nonverbal displays affect neuroendocrine levels and risk tolerance. Psychol Sci 21:1363–1368. https://doi.org/10.1177/0956797610383437
16. Sripada C, Kessler D, Jonides J (2014) Methylphenidate blocks effort-induced depletion of regulatory control in healthy volunteers. Psychol Sci 25:1227–1234. https://doi.org/10.1177/0956797614526415
17. Hagger MS, Chatzisarantis NLD, Alberts H, Anggono CO, Batailler C, Birt AR, Brand R, Brandt MJ, Brewer G, Bruyneel S, Calvillo DP, Campbell WK, Cannon PR, Carlucci M, Carruth NP, Cheung T, Crowell A, De Ridder DTD, Dewitte S, Elson M, Evans JR, Fay BA, Fennis BM, Finley A, Francis Z, Heise E, Hoemann H, Inzlicht M, Koole SL, Koppel L, Kroese F, Lange F, Lau K, Lynch BP, Martijn C, Merckelbach H, Mills NV, Michirev A, Miyake A, Mosser AE, Muise M, Muller D, Muzi M, Nalis D, Nurwanti R, Otgaar H, Philipp MC, Primoceri P, Rentzsch K, Ringos L, Schlinkert C, Schmeichel BJ, Schoch SF, Schrama M, Schütz A, Stamos A, Tinghög G, Ullrich J, vanDellen M, Wimbarti S, Wolff W, Yusainy C, Zerhouni O, Zwienenberg M (2016) A multilab preregistered replication of the ego-depletion effect. Perspect Psychol Sci 11:546–573. https://doi.org/10.1177/1745691616652873
18. Wagenmakers E-J, Beek T, Dijkhoff L, Gronau QF, Acosta A, Adams RB, Albohn DN, Allard ES, Benning SD, Blouin-Hudon E-M, Bulnes LC, Caldwell TL, Calin-Jageman RJ, Capaldi CA, Carfagno NS, Chasten KT, Cleeremans A, Connell L, DeCicco JM, Dijkstra K, Fischer AH, Foroni F, Hess U, Holmes KJ, Jones JLH, Klein O, Koch C, Korb S, Lewinski P, Liao JD, Lund S, Lupianez J, Lynott D, Nance CN, Oosterwijk S, Ozdoğru AA, Pacheco-Unguetti AP, Pearson B, Powis C, Riding S, Roberts T-A, Rumiati RI, Senden M, Shea-Shumsky NB, Sobocko K, Soto JA, Steiner TG, Talarico JM, van Allen ZM, Vandekerckhove M, Wainwright B, Wayand JF, Zeelenberg R, Zetzer EE, Zwaan RA (2016) Registered replication report: Strack, Martin, & Stepper (1988). Perspect Psychol Sci 11:917–928. https://doi.org/10.1177/1745691616674458
19. Cooper J (2013) On fraud, deceit and ethics. J Exp Soc Psychol 49:314. https://doi.org/10.1016/j.jesp.2012.12.006
20. Simonsohn U (2013) Just post it: the lesson from two cases of fabricated data detected by statistics alone. Psychol Sci 10:1875–88. https://doi.org/10.1177/0956797613480366
21. Piller C (2022) BLOTS ON A FIELD? A neuroscience image sleuth finds signs of fabrication in scores of Alzheimer's articles, threatening a reigning theory of the disease. Science 377. https://doi.org/10.1126/science.ade0209
22. Simmons JP, Nelson LD, Simonsohn U (2011) False-positive psychology: undisclosed flexibility in data collection and analysis allows presenting anything as significant. Psychol Sci 22:1359–1366. https://doi.org/10.1177/0956797611417632
23. John LK, Loewenstein G, Prelec D (2012) Measuring the prevalence of questionable research practices with incentives for truth telling. Psychol Sci 23:524–532. https://doi.org/10.1177/0956797611430953
24. Fiske S (2016) Mob rule or wisdom of crowds? http://datacolada.org/wp-content/uploads/2016/09/Fiske-presidential-guest-column_APS-Observer_copy-edited.pdf
25. Nosek BA, Lakens D (2014) Registered reports: a method to increase the credibility of published results. Social Psychol 45:137–141. https://doi.org/10.1027/1864-9335/a000192
26. Brandt MJ, IJzerman H, Dijksterhuis A, Farach FJ, Geller J, Giner-Sorolla R, Grange JA, Perugini M, Spies JR, van't Veer A (2014) The replication recipe: what makes for a convincing replication? J Exp Social Psychol 50:217–224. https://doi.org/10.1016/j.jesp.2013.10.005
27. Klein RA, Ratliff KA, Vianello M, Adams RB, Bahník Š, Bernstein MJ, Bocian K, Brandt MJ, Brooks B, Brumbaugh CC, Cemalcilar Z, Chandler J, Cheong W, Davis WE, Devos T, Eisner M, Frankowska N, Furrow D, Galliani EM, Hasselman F, Hicks JA, Hovermale JF, Hunt SJ, Huntsinger JR, Ijzerman H, John MS, Joy-Gaba JA, Kappes HB, Krueger LE, Kurtz J, Levitan CA, Mallett RK, Morris WL, Nelson AJ, Nier JA, Packard G, Pilati R, Rutchick AM, Schmidt K, Skorinko JL, Smith R, Steiner TG, Storbeck J, Van Swol LM, Thompson D, Van'T Veer AE,

Vaughn LA, Vranka M, Wichman AL, Woodzicka JA, Nosek BA (2014) Investigating variation in replicability: a "many labs" replication project. Social Psychol 45:142–152. https://doi.org/10.1027/1864-9335/a000178

28. Klein RA, Vianello M, Hasselman F, Adams BG, Adams RB, Alper S, Aveyard M, Axt JR, Babalola MT, Bahník Š, Batra R, Berkics M, Bernstein MJ, Berry DR, Bialobrzeska O, Binan ED, Bocian K, Brandt MJ, Busching R, Rédei AC, Cai H, Cambier F, Cantarero K, Carmichael CL, Ceric F, Chandler J, Chang J-H, Chatard A, Chen EE, Cheong W, Cicero DC, Coen S, Coleman JA, Collisson B, Conway MA, Corker KS, Curran PG, Cushman F, Dagona ZK, Dalgar I, Dalla Rosa A, Davis WE, de Bruijn M, De Schutter L, Devos T, de Vries M, Doğulu C, Dozo N, Dukes KN, Dunham Y, Durrheim K, Ebersole CR, Edlund JE, Eller A, English AS, Finck C, Frankowska N, Freyre M-Á, Friedman M, Galliani EM, Gandi JC, Ghoshal T, Giessner SR, Gill T, Gnambs T, Gómez Á, González R, Graham J, Grahe JE, Grahek I, Green EGT, Hai K, Haigh M, Haines EL, Hall MP, Heffernan ME, Hicks JA, Houdek P, Huntsinger JR, Huynh HP, IJzerman H, Inbar Y, Innes-Ker ÅH, Jiménez-Leal W, John M-S, Joy-Gaba JA, Kamiloğlu RG, Kappes HB, Karabati S, Karick H, Keller VN, Kende A, Kervyn N, Knežević G, Kovacs C, Krueger LE, Kurapov G, Kurtz J, Lakens D, Lazarević LB, Levitan CA, Lewis NA, Lins S, Lipsey NP, Losee JE, Maassen E, Maitner AT, Malingumu W, Mallett RK, Marotta SA, Međedović J, Mena-Pacheco F, Milfont TL, Morris WL, Murphy SC, Myachykov A, Neave N, Neijenhuijs K, Nelson AJ, Neto F, Lee Nichols A, Ocampo A, O'Donnell SL, Oikawa H, Oikawa M, Ong E, Orosz G, Osowiecka M, Packard G, Pérez-Sánchez R, Petrović B, Pilati R, Pinter B, Podesta L, Pogge G, Pollmann MMH, Rutchick AM, Saavedra P, Saeri AK, Salomon E, Schmidt K, Schönbrodt FD, Sekerdej MB, Sirlopú D, Skorinko JLM, Smith MA, Smith-Castro V, Smolders KCHJ, Sobkow A, Sowden W, Spachtholz P, Srivastava M, Steiner TG, Stouten J, Street CNH, Sundfelt OK, Szeto S, Szumowska E, Tang ACW, Tanzer N, Tear MJ, Theriault J, Thomae M, Torres D, Traczyk J, Tybur JM, Ujhelyi A, van Aert RCM, van Assen MALM, van der Hulst M, van Lange PAM, van't Veer AE, Vásquez-Echeverría A, Ann Vaughn L, Vázquez A, Vega LD, Verniers C, Verschoor M, Voermans IPJ, Vranka MA, Welch C, Wichman AL, Williams LA, Wood M, Woodzicka JA, Wronska MK, Young L, Zelenski JM, Zhijia Z, Nosek BA (2018) Many Labs 2: investigating variation in replicability across samples and settings. Adv Methods Pract Psychol Sci 1:443–490. https://doi.org/10.1177/2515245918810225
29. Open Science Collaboration (2015) Estimating the reproducibility of psychological science. Science 349:aac4716–aac4716. https://doi.org/10.1126/science.aac4716
30. Camerer CF, Dreber A, Holzmeister F, Ho T-H, Huber J, Johannesson M, Kirchler M, Nave G, Nosek BA, Pfeiffer T, Altmejd A, Buttrick N, Chan T, Chen Y, Forsell E, Gampa A, Heikensten E, Hummer L, Imai T, Isaksson S, Manfredi D, Rose J, Wagenmakers E-J, Wu H (2018) Evaluating the replicability of social science experiments in Nature and Science between 2010 and 2015. Nat Hum Behav 2:637–644. https://doi.org/10.1038/s41562-018-0399-z
31. Camerer CF, Dreber A, Forsell E, Ho T-H, Huber J, Johannesson M, Kirchler M, Almenberg J, Altmejd A, Chan T, Heikensten E, Holzmeister F, Imai T, Isaksson S, Nave G, Pfeiffer T, Razen M, Wu H (2016) Evaluating replicability of laboratory experiments in economics. Science 351:1433–1436. https://doi.org/10.1126/science.aaf0918
32. Ebersole CR, Atherton OE, Belanger AL, Skulborstad HM, Allen JM, Banks JB, Baranski E, Bernstein MJ, Bonfiglio DBV, Boucher L, Brown ER, Budiman NI, Cairo AH, Capaldi CA, Chartier CR, Chung JM, Cicero DC, Coleman JA, Conway JG, Davis WE, Devos T, Fletcher MM, German K, Grahe JE, Hermann AD, Hicks JA, Honeycutt N, Humphrey B, Janus M, Johnson DJ, Joy-Gaba JA, Juzeler H, Keres A, Kinney D, Kirshenbaum J, Klein RA, Lucas RE, Lustgraaf CJN, Martin D, Menon M, Metzger M, Moloney JM, Morse PJ, Prislin R, Razza T, Re DE, Rule NO, Sacco DF, Sauerberger K, Shrider E, Shultz M, Siemsen C, Sobocko K, Weylin Sternglanz R, Summerville A, Tskhay KO, van Allen Z, Vaughn LA, Walker RJ, Weinberg A, Wilson JP, Wirth JH, Wortman J, Nosek BA (2016) Many Labs 3: evaluating participant pool quality across the academic semester via replication. J Exp Soc Psychol 67:68–82. https://doi.org/10.1016/j.jesp.2015.10.012

33. Klein RA, Cook CL, Ebersole CR, Vitiello C, Nosek BA, Hilgard J, Ahn PH, Brady AJ, Chartier CR, Christopherson CD, Clay S, Collisson B, Crawford JT, Cromar R, Gardiner G, Gosnell CL, Grahe J, Hall C, Howard I, Joy-Gaba JA, Kolb M, Legg AM, Levitan CA, Mancini AD, Manfredi D, Miller J, Nave G, Redford L, Schlitz I, Schmidt K, Skorinko JLM, Storage D, Swanson T, Van Swol LM, Vaughn LA, Vidamuerte D, Wiggins B, Ratliff KA (2022) Many Labs 4: failure to replicate mortality salience effect with and without original author involvement. Collabra: Psychology 8:35271. https://doi.org/10.1525/collabra.35271
34. Ebersole CR, Mathur MB, Baranski E, Bart-Plange D-J, Buttrick NR, Chartier CR, Corker KS, Corley M, Hartshorne JK, IJzerman H, Lazarević LB, Rabagliati H, Ropovik I, Aczel B, Aeschbach LF, Andrighetto L, Arnal JD, Arrow H, Babincak P, Bakos BE, Baník G, Baskin E, Belopavlović R, Bernstein MH, Białek M, Bloxsom NG, Bodroža B, Bonfiglio DBV, Boucher L, Brühlmann F, Brumbaugh CC, Casini E, Chen Y, Chiorri C, Chopik WJ, Christ O, Ciunci AM, Claypool HM, Coary S, Čolić MV, Collins WM, Curran PG, Day CR, Dering B, Dreber A, Edlund JE, Falcão F, Fedor A, Feinberg L, Ferguson IR, Ford M, Frank MC, Fryberger E, Garinther A, Gawryluk K, Ashbaugh K, Giacomantonio M, Giessner SR, Grahe JE, Guadagno RE, Hałasa E, Hancock PJB, Hilliard RA, Hüffmeier J, Hughes S, Idzikowska K, Inzlicht M, Jern A, Jiménez-Leal W, Johannesson M, Joy-Gaba JA, Kauff M, Kellier DJ, Kessinger G, Kidwell MC, Kimbrough AM, King JPJ, Kolb VS, Kołodziej S, Kovacs M, Krasuska K, Kraus S, Krueger LE, Kuchno K, Lage CA, Langford EV, Levitan CA, de Lima TJS, Lin H, Lins S, Loy JE, Manfredi D, Markiewicz Ł, Menon M, Mercier B, Metzger M, Meyet V, Millen AE, Miller JK, Montealegre A, Moore DA, Muda R, Nave G, Nichols AL, Novak SA, Nunnally C, Orlić A, Palinkas A, Panno A, Parks KP, Pedović I, Pękala E, Penner MR, Pessers S, Petrović B, Pfeiffer T, Pieńkosz D, Preti E, Purić D, Ramos T, Ravid J, Razza TS, Rentzsch K, Richetin J, Rife SC, Rosa AD, Rudy KH, Salamon J, Saunders B, Sawicki P, Schmidt K, Schuepfer K, Schultze T, Schulz-Hardt S, Schütz A, Shabazian AN, Shubella RL, Siegel A, Silva R, Sioma B, Skorb L, de Souza LEC, Steegen S, Stein LAR, Sternglanz RW, Stojilović D, Storage D, Sullivan GB, Szaszi B, Szecsi P, Szöke O, Szuts A, Thomae M, Tidwell ND, Tocco C, Torka A-K, Tuerlinckx F, Vanpaemel W, Vaughn LA, Vianello M, Viganola D, Vlachou M, Walker RJ, Weissgerber SC, Wichman AL, Wiggins BJ, Wolf D, Wood MJ, Zealley D, Žeželj I, Zrubka M, Nosek BA (2020) Many Labs 5: testing pre-data-collection peer review as an intervention to increase replicability. Adv Methods Pract Psychol Sci 3:309–331. https://doi.org/10.1177/2515245920958687
35. Schweinsberg M, Madan N, Vianello M, Sommer SA, Jordan J, Tierney W, Awtrey E, Zhu LL, Diermeier D, Heinze JE, Srinivasan M, Tannenbaum D, Bivolaru E, Dana J, Davis-Stober CP, du Plessis C, Gronau QF, Hafenbrack AC, Liao EY, Ly A, Marsman M, Murase T, Qureshi I, Schaerer M, Thornley N, Tworek CM, Wagenmakers E-J, Wong L, Anderson T, Bauman CW, Bedwell WL, Brescoll V, Canavan A, Chandler JJ, Cheries E, Cheryan S, Cheung F, Cimpian A, Clark MA, Cordon D, Cushman F, Ditto PH, Donahue T, Frick SE, Gamez-Djokic M, Grady RH, Graham J, Gu J, Hahn A, Hanson BE, Hartwich NJ, Hein K, Inbar Y, Jiang L, Kellogg T, Kennedy DM, Legate N, Luoma TP, Maibuecher H, Meindl P, Miles J, Mislin A, Molden DC, Motyl M, Newman G, Ngo HH, Packham H, Ramsay PS, Ray JL, Sackett AM, Sellier A-L, Sokolova T, Sowden W, Storage D, Sun X, Van Bavel JJ, Washburn AN, Wei C, Wetter E, Wilson CT, Darroux S-C, Uhlmann EL (2016) The pipeline project: pre-publication independent replications of a single laboratory's research pipeline. J Exp Soc Psychol 66:55–67. https://doi.org/10.1016/j.jesp.2015.10.001
36. Röseler L, Kaiser L, Doetsch C, Klett N, Seida C, Schütz A, Aczel B, Adelina N, Agostini V, Alarie S, Albayrak-Aydemir N, Aldoh A, Al-Hoorie AH, Azevedo F, Baker BJ, Barth CL, Beitner J, Brick C, Brohmer H, Chandrashekar SP, Chung KL, Cockcroft JP, Cummins J, Diveica V, Dumbalska T, Efendic E, Elsherif M, Evans T, Feldman G, Fillon A, Förster N, Frese J, Genschow O, Giannouli V, Gjoneska B, Gnambs T, Gourdon-Kanhukamwe A, Graham CJ, Hartmann H, Haviva C, Herderich A, Hilbert LP, Holgado D, Hussey I, Ilchovska ZG, Kalandadze T, Karhulahti V-M, Kasseckert L, Klingelhöfer-Jens M, Koppold A, Korbmacher M, Kulke L, Kuper N, LaPlume A, Leech G, Lohkamp F, Lou NM, Lynott D, Maier M, Meier M, Montefinese M, Moreau D, Mrkva K, Nemcova M, Oomen D, Packheiser J, Pandey S,

Papenmeier F, Paruzel-Czachura M, Pavlov YG, Pavlović Z, Pennington CR, Pittelkow M-M, Plomp W, Plonski PE, Pronizius E, Pua AA, Pypno-Blajda K, Rausch M, Rebholz TR, Richert E, Röer JP, Ross R, Schmidt K, Skvortsova A, Sperl MFJ, Tan AWM, Thürmer JL, Tołopiło A, Vanpaemel W, Vaughn LA, Verheyen S, Wallrich L, Weber L, Wolska JK, Zaneva M, Zhang Y (2024) The replication database: documenting the replicability of psychological science. J Open Psychol Data 12. https://doi.org/10.5334/jopd.101

37. CORE Team (2023) Collaborative Open-science and meta REsearch. Retrieved from http://osf.io/5z4a8 and https://mgto.org/core-team/
38. Sobkow A, Surowski M, Olszewska A, Antoniewska N, Barcik K, Bartkiewicz U, Brzeska A, Brzozowska A, Budrewicz O, Choja J, Choma K, Chorbotowicz P, Filimoniak M, Filip Ł, Gambuś P, Gierlik W, Gonczar T, Goryczka K, Góra M, Haczek M, Hetmańczuk W, Holka Z, Janosz A, Kikowska N, Kołcun J, Kozłowska Z, Kujawińska M, Kuleszczyk M, Lach-Galińska A, Latacz K, Ławniczak A, Majewska K, Makowska K, Mamzer M, Marciniszyn I, Masternak A, Matuszek M, Mehr J, Miela E, Mleczko M, Morga P, Niemczyk M, Ostrowski D, Pełdiak J, Piotrowicz K, Płuciennik A, Ryśkiewicz O, Sekuła W, Sikora M, Sikora N, Sitko D, Sobczak A, Sosenko J, Stando S, Starek K, Ślak Ł, Świtała J, Świtniewska N, Tyc A, Urban O, Wcisło N, Wiśniewska K, Wodzińska J, Zabiełło A, Żygadło M, Zaleskiewicz T, Traczyk J (2022) Conceptual replication study of fifteen JDM effects: insights from the Polish sample. Polish Psychol Bull:138–151. https://doi.org/10.24425/ppb.2022.141862
39. Sobkow A, Jurewicz K, Zaleskiewicz T, Pronczuk A, Mondal S, Badecka J, Bartkiewicz U, Bobowiec W, Bochniak M, Chwalibog K, Czech L, Daniluk-Malcherek J, Domanska M, Dziwusz N, Falborska P, Gazda Z, Gruszczynski L, Grzejdziak S, Kalus A, Knotz K, Kocik A, Krzysiek A, Kuriata T, Kuriata W, Kusiak K, Litwinczuk P, Daniel, Marek J, Mnich L, Mrowiec A, Myslikowska I, Ossowska A, Panek P, Pawelec A, Pozaroszczyk J, Rogowska A, Seredyuk A, Seweryn M, Shakhraichuk I, Sielicki M, Siemieniec M, Stec A, Szczepaniak A, Trufanova O, Zgajewska A, Traczyk J (2025) The role of numeracy in judgment and decision making: the replication of eleven effects using several numeracy scales. Collabra: Psychology 11:128514. https://doi.org/10.1525/collabra.128514
40. Saribay SA, Yilmaz O, Körpe GG (2020) Does intuitive mindset influence belief in God? A registered replication of. Judgm Decis Mak 15:193–202. https://doi.org/10.1017/S1930297500007348
41. Nieuwenstein MR, Wierenga T, Morey RD, Wicherts JM, Blom TN, Wagenmakers E-J, van Rijn H (2015) On making the right choice: a meta-analysis and large-scale replication attempt of the unconscious thought advantage. Judgm Decis Mak 10:1–17. https://doi.org/10.1017/S1930297500003144
42. Nieuwenstein M, van Rijn H (2012) The unconscious thought advantage: further replication failures from a search for confirmatory evidence. Judgm Decis Mak 7:779–798. https://doi.org/10.1017/S1930297500003338
43. Calvillo DP, Penaloza A (2009) Are complex decisions better left to the unconscious? Further failed replications of the deliberation-without-attention effect. Judgm Decis Mak 4:509–517. https://doi.org/10.1017/S1930297500004046
44. Persson E, Andersson D, Koppel L, Västfjäll D, Tinghög G (2021) A preregistered replication of motivated numeracy. Cognition 214:104768. https://doi.org/10.1016/j.cognition.2021.104768
45. Stagnaro MN, Tappin BM, Rand DG (2023) No association between numerical ability and politically motivated reasoning in a large US probability sample. Proc Natl Acad Sci USA 120:e2301491120. https://doi.org/10.1073/pnas.2301491120
46. Röseler L, Schütz A, Blank PA, Dück M, Fels S, Kupfer J, Scheelje L, Seida C (2021) Evidence against subliminal anchoring: two close, highly powered, preregistered, and failed replication attempts. J Exp Soc Psychol 92:104066. https://doi.org/10.1016/j.jesp.2020.104066
47. Shanks DR, Barbieri-Hermitte P, Vadillo MA (2020) Do incidental environmental anchors bias consumers' price estimations? Collabra: Psychology 6:19. https://doi.org/10.1525/collabra.310

48. Bogdanov B, Corbin J, Dobreva S, McElroy T, Rachev NR (2023) Working memory capacity and the risky-choice framing effect: a preregistered replication and extension of Corbin et al. (2010). Judgm decis mak 18:e39. https://doi.org/10.1017/jdm.2023.40
49. Lee C-Y, Morewedge CK (2022) Noise increases anchoring effects. Psychol Sci 33:60–75. https://doi.org/10.1177/09567976211024254
50. Ball S, Katz B, Li F, Smith A (2023) The effect of cognitive load on economic decision-making: a replication attempt. J Econ Behav Organ 210:226–242. https://doi.org/10.1016/j.jebo.2023.03.018
51. Bergman O, Ellingsen T, Johannesson M, Svensson C (2010) Anchoring and cognitive ability. Econ Lett 107:66–68. https://doi.org/10.1016/j.econlet.2009.12.028
52. Yoon S, Fong NM, Dimoka A (2019) The robustness of anchoring effects on preferential judgments. Judgm Decis Mak 14:470–487. https://doi.org/10.1017/S1930297500006148
53. Zhou L, Liu N, Liao Y-Q, Li A-M (2021) Risky choice framing with various problem descriptions: a replication and extension study. Judgm Decis Mak 16:394–421. https://doi.org/10.1017/S1930297500008615
54. Chandrashekar SP, Adelina N, Zeng S, Chiu YYE, Leung GYS, Henne P, Cheng BL, Feldman G (2023) Defaults versus framing: revisiting default effect and framing effect with replications and extensions of Johnson and Goldstein (2003) and Johnson, Bellman, and Lohse (2002). Meta-Psychology 7. https://doi.org/10.15626/MP.2022.3108
55. Macaskill AC, Hackenberg TD (2013) Optimal and nonoptimal choice in a laboratory-based sunk cost task with humans: a cross-species replication. J Exper Analysis Behavior 100:301–315. https://doi.org/10.1002/jeab.52
56. Petrov N, Chan M, Lau C, Kwok TH, Chow L, Lo W, Song W, Feldman G (2023) Sunk cost effects for time versus money: replication and extensions registered report of Soman (2001). Int Rev Social Psychol 36:17. https://doi.org/10.5334/irsp.883
57. Bolinger TC, Ostermiller MA, Martin SD (2023) Replication of the interpersonal sunk cost effect. PsiChiJournal 28:79–86. https://doi.org/10.24839/2325-7342.JN28.1.79
58. Vonasch AJ, Hung WY, Leung WY, Nguyen ATB, Chan S, Cheng BL, Feldman G (2023) "Less is better" in separate evaluations versus "More is better" in joint evaluations: mostly successful close replication and extension of Hsee (1998). Collabra: Psychology 9:77859. https://doi.org/10.1525/collabra.77859
59. Zhu M, Feldman G (2023) Revisiting the links between numeracy and decision making: replication registered report of Peters et al. (2006) with an extension examining confidence. Collabra: Psychology 9:77608. https://doi.org/10.1525/collabra.77608
60. Li X, Liu J, Gao W, Cohen GL (2024) Challenging the N-Heuristic: effect size, not sample size, predicts the replicability of psychological science. PLoS ONE 19:e0306911. https://doi.org/10.1371/journal.pone.0306911
61. van den Akker OR, van Assen MALM, Bakker M, Elsherif M, Wong TK, Wicherts JM (2024) Preregistration in practice: a comparison of preregistered and non-preregistered studies in psychology. Behav Res 56:5424–5433. https://doi.org/10.3758/s13428-023-02277-0
62. Yang Y, Youyou W, Uzzi B (2020) Estimating the deep replicability of scientific findings using human and artificial intelligence. Proc Natl Acad Sci 117:10762–10768. https://doi.org/10.1073/pnas.1909046117
63. Youyou W, Yang Y, Uzzi B (2023) A discipline-wide investigation of the replicability of Psychology papers over the past two decades. Proc Natl Acad Sci 120:e2208863120. https://doi.org/10.1073/pnas.2208863120
64. Boyce V, Mathur M, Frank MC (2023) Eleven years of student replication projects provide evidence on the correlates of replicability in psychology. R Soc Open Sci. https://doi.org/10.1098/rsos.231240
65. Vohs KD, Schmeichel BJ, Lohmann S, Gronau QF, Finley AJ, Ainsworth SE, Alquist JL, Baker MD, Brizi A, Bunyi A, Butschek GJ, Campbell C, Capaldi J, Cau C, Chambers H, Chatzisarantis NLD, Christensen WJ, Clay SL, Curtis J, De Cristofaro V, del Rosario K, Diel K, Doğruol Y, Doi M, Donaldson TL, Eder AB, Ersoff M, Eyink JR, Falkenstein A, Fennis

BM, Findley MB, Finkel EJ, Forgea V, Friese M, Fuglestad P, Garcia-Willingham NE, Geraedts LF, Gervais WM, Giacomantonio M, Gibson B, Gieseler K, Gineikiene J, Gloger EM, Gobes CM, Grande M, Hagger MS, Hartsell B, Hermann AD, Hidding JJ, Hirt ER, Hodge J, Hofmann W, Howell JL, Hutton RD, Inzlicht M, James L, Johnson E, Johnson HL, Joyce SM, Joye Y, Kaben JH, Kammrath LK, Kelly CN, Kissell BL, Koole SL, Krishna A, Lam C, Lee KT, Lee N, Leighton DC, Loschelder DD, Maranges HM, Masicampo EJ, Mazara K, McCarthy S, McGregor I, Mead NL, Mendes WB, Meslot C, Michalak NM, Milyavskaya M, Miyake A, Moeini-Jazani M, Muraven M, Nakahara E, Patel K, Petrocelli JV, Pollak KM, Price MM, Ramsey HJ, Rath M, Robertson JA, Rockwell R, Russ IF, Salvati M, Saunders B, Scherer A, Schütz A, Schmitt KN, Segerstrom SC, Serenka B, Sharpinskyi K, Shaw M, Sherman J, Song Y, Sosa N, Spillane K, Stapels J, Stinnett AJ, Strawser HR, Sweeny K, Theodore D, Tonnu K, van Oldenbeuving Y, vanDellen MR, Vergara RC, Walker JS, Waugh CE, Weise F, Werner KM, Wheeler C, White RA, Wichman AL, Wiggins BJ, Wills JA, Wilson JH, Wagenmakers E-J, Albarracín D (2021) A multisite preregistered paradigmatic test of the ego-depletion effect. Psychol Sci 32:1566–1581. https://doi.org/10.1177/0956797621989733

66. Jarke H, Anand-Vembar S, Alzahawi S, Andersen TL, Bojanić L, Carstensen A, Feldman G, Garcia-Garzon E, Kapoor H, Lewis S, Todsen AL, Većkalov B, Zickfeld JH, Geiger SJ (2022) A roadmap to large-scale multi-country replications in psychology. Collabra: Psychology 8:57538. https://doi.org/10.1525/collabra.57538
67. Isager PM, Giner-Sorolla R, Krueger JI, Perugini M, Ropovik I, van AE, Vranka M, Lakens D (2023) Deciding what to replicate: a decision model for replication study selection under resource and knowledge constraints. Psychol Methods 28:438–451. https://doi.org/10.1037/met0000438
68. Stroebe W (2019) What can we learn from many labs replications? Basic Appl Soc Psychol 41:91–103. https://doi.org/10.1080/01973533.2019.1577736
69. Torka A-K, Mazei J, Hüffmeier J (2023) Are replications mainstream now? A comparison of support for replications expressed in the policies of social psychology journals in 2015 and 2022. Soc Psychol Bull 18:e9695. https://doi.org/10.32872/spb.9695
70. Clarke B, Lee PY, Schiavone SR, Rhemtulla M, Vazire S (2023) The prevalence of direct replication articles in top-ranking psychology journals
71. Jekel M, Fiedler S, Allstadt Torras R, Mischkowski D, Dorrough AR, Glöckner A (2020) How to teach open science principles in the undergraduate curriculum—the Hagen cumulative science project. Psychol Learn Teach 19:91–106. https://doi.org/10.1177/1475725719868149
72. Maxwell SE, Lau MY, Howard GS (2015) Is psychology suffering from a replication crisis? What does "failure to replicate" really mean? Am Psychol 70:487–498. https://doi.org/10.1037/a0039400
73. Shrout PE, Rodgers JL (2018) Psychology, science, and knowledge construction: broadening perspectives from the replication crisis. Annu Rev Psychol 69:487–510. https://doi.org/10.1146/annurev-psych-122216-011845
74. Field SM, Hoekstra R, Bringmann L, van Ravenzwaaij D (2019) When and why to replicate: as easy as 1, 2, 3? Collabra: Psychology 5:46. https://doi.org/10.1525/collabra.218
75. Shu LL, Mazar N, Gino F, Ariely D, Bazerman MH (2012) RETRACTED: Signing at the beginning makes ethics salient and decreases dishonest self-reports in comparison to signing at the end. Proc Natl Acad Sci 109:15197–15200. https://doi.org/10.1073/pnas.1209746109
76. Dorison CA, Lerner JS, Heller BH, Rothman AJ, Kawachi II, Wang K, Rees VW, Gill BP, Gibbs N, Ebersole CR, Vally Z, Tajchman Z, Zsido AN, Zrimsek M, Chen Z, Ziano I, Gialitaki Z, Ceary CD, Lin Y, Kunisato Y, Yamada Y, Xiao Q, Jiang X, Du X, Yao E, Wilson JP, Cyrus-Lai W, Jimenez-Leal W, Law W, Collins WM, Richard KL, Vranka M, Ankushev V, Schei V, Križanić V, Kadreva VH, Adoric VC, Tran US, Yeung SK, Hassan W, Houston R, Lima TJS, Ostermann T, Frizzo T, Sverdrup TE, House T, Gill T, Fedotov M, Paltrow T, Jernsäther T, Koptjevskaja-Tamm M, Hostler TJ, Ishii T, Szaszi B, Adamus S, Suter L, Habib S, Studzinska A, Stojanovska D, Janssen SMJ, Stieger S, Schulenberg SE, Tatachari S, Azouaghe S, Sorokowski P, Sorokowska A, Song X, Lewis SC, Sinkolova S, Grigoryev D, Drexler SM,

Daches S, Geniole SN, Vračar S, Massoni S, Zorjan S, Sarıoğuz E, Izquierdo SM, Alves SG, Pöntinen S, Solas SÁ, Ordoñez-Riaño S, Očovaj SB, Onie S, Lins S, Çoksan S, Sacakli A, Ruiz-Fernández S, Geiger SJ, FatahModares S, Walczak RB, Betlehem R, Vilar R, Cárcamo R, Ross RM, McCarthy R, Ballantyne T, Westgate EC, Afhami R, Ren D, Monteiro RP, Reips U-D, Reggev N, Calin-Jageman RJ, Pourafshari R, Oliveira R, Nedelcheva-Datsova M, Rahal R-M, Ribeiro RR, Radtke T, Searston R, Habte R, Zdybek P, Chen S-C, Maturan PLG, Perillo JT, Isager PM, Kačmár P, Macapagal PM, Szwed P, Hanel PHP, Forbes PAG, Arriaga P, Paris B, Papachristopoulos K, Correa PS, Kácha O, Bernardo M, Campos O, Bravo ON, Galindo-Caballero OJ, Ogbonnaya CE, Bialobrzeska O, Kiselnikova N, Simonovic N, Cohen N, Nock NL, Johannes N, Albayrak-Aydemir N, Say N, Torunsky N, Van Doren N, Sunami N, Rachev NR, Majeed NM, Schmidt N-D, Nadif K, Corral-Frías NS, Ouherrou N, Pantazi M, Lucas MY, Vasilev MR, Ortiz MV, Butt MM, Kabir M, Muda R, Tejada Rivera M del CM, Sirota M, Seehuus M, Parzuchowski M, Toro M, Hricova M, Maldonado MA, Marszalek M, Karekla M, Mioni G, Bosma MJ, Westerlund M, Vdovic M, Bialek M, Silan MA, Anne M, Misiak M, Grinberg M, Capizzi M, Espinoza Barría MF, Kurfali MA, Mensink MC, Harutyunyan M, Khosla M, Korbmacher M, Adamkovič M, Ribeiro MFF, Terskova M, Hruška M, Martončik M, Voracek M, Čadek M, Frías-Armenta M, Kowal M, Topor M, Roczniewska M, Oosterlinck M, Kohlová MB, Paruzel-Czachura M, Romanova M, Papadatou-Pastou M, Lund ML, Antoniadi M, Jones MV, Ortiz MS, Manavalan M, Muminov A, Kossowska M, Friedemann M, Wielgus M, Varella MAC, Colloff MF, Bradford M, Vaughn LA, Eudave L, Vieira L, Pineda LMS, Pérez LC, Lazarevic LB, Jaremka LM, Kushnir E, Anton-Boicuk L, de Holanda Coelho GL, Ahlgren L, Levitan CA, Micheli L, Volz L, Stojanovska M, Boucher L, Samojlenko L, Delgado LGJ, Kaliska L, Warmelink L, Rojas-Berscia LM, Yu K, Wachowicz J, Desai K, Barzykowski K, Kozma L, Evans K, Kirgizova K, Agesin BE, Koehn MA, Wolfe K, Korobova T, Klevjer K, van Schie K, Vezirian K, Damnjanović K, Thommesen KK, Schmidt K, Filip K, Grzech K, Hoyer K, Moon K, Rana K, Janjić K, Suchow JW, Kielińska J, Cruz Vásquez JE, Beitner J, Vargas-Nieto JC, Roxas JCT, Taber J, Urriago-Rayo J, Pavlacic JM, Bavolar J, Soto JA, Olofsson JK, Vilsmeier JK, Messerschmidt J, Czamanski-Cohen J, Boudesseul J, Lee JM, Kamburidis J, Zickfeld J, Miranda JF, Verharen JPH, Hristova E, Beshears JE, Đorđević JM, Bosch J, Valentova JV, Antfolk J, Berkessel JB, Schrötter J, Urban J, Röer JP, Norton JO, Silva JR, Pickering JS, Vintr J, Uttley J, Kunst JR, Ndukaihe ILG, Iyer A, Vilares I, Ivanov A, Ropovik I, Sula I, Sarieva I, Metin-Orta I, Prusova I, Pinto I (2022) In COVID-19 health messaging, loss framing increases anxiety with little-to-no concomitant benefits: experimental evidence from 84 countries. Affec Sci 3:577–602. https://doi.org/10.1007/s42761-022-00128-3

77. Vlasceanu M, Doell KC, Bak-Coleman JB, Todorova B, Berkebile-Weinberg MM, Grayson SJ, Patel Y, Goldwert D, Pei Y, Chakroff A, Pronizius E, Van Den Broek KL, Vlasceanu D, Constantino S, Morais MJ, Schumann P, Rathje S, Fang K, Aglioti SM, Alfano M, Alvarado-Yepez AJ, Andersen A, Anseel F, Apps MAJ, Asadli C, Awuor FJ, Azevedo F, Basaglia P, Bélanger JJ, Berger S, Bertin P, Białek M, Bialobrzeska O, Blaya-Burgo M, Bleize DNM, Bø S, Boecker L, Boggio PS, Borau S, Bos B, Bouguettaya A, Brauer M, Brick C, Brik T, Briker R, Brosch T, Buchel O, Buonauro D, Butalia R, Carvacho H, Chamberlain SAE, Chan H-Y, Chow D, Chung D, Cian L, Cohen-Eick N, Contreras-Huerta LS, Contu D, Cristea V, Cutler J, D'Ottone S, De Keersmaecker J, Delcourt S, Delouvée S, Diel K, Douglas BD, Drupp MA, Dubey S, Ekmanis J, Elbaek CT, Elsherif M, Engelhard IM, Escher YA, Etienne TW, Farage L, Farias AR, Feuerriegel S, Findor A, Freira L, Friese M, Gains NP, Gallyamova A, Geiger SJ, Genschow O, Gjoneska B, Gkinopoulos T, Goldberg B, Goldenberg A, Gradidge S, Grassini S, Gray K, Grelle S, Griffin SM, Grigoryan L, Grigoryan A, Grigoryev D, Gruber J, Guilaran J, Hadar B, Hahnel UJJ, Halperin E, Harvey AJ, Haugestad CAP, Herman AM, Hershfield HE, Himichi T, Hine DW, Hofmann W, Howe L, Huaman-Chulluncuy ET, Huang G, Ishii T, Ito A, Jia F, Jost JT, Jovanović V, Jurgiel D, Kácha O, Kankaanpää R, Kantorowicz J, Kantorowicz-Reznichenko E, Kaplan Mintz K, Kaya I, Kaya O, Khachatryan N, Klas A, Klein C, Klöckner CA, Koppel L, Kosachenko AI, Kothe EJ, Krebs R, Krosch AR, Krouwel APM, Kyrychenko Y, Lagomarsino M, Lamm C, Lange F, Lee Cunningham J, Lees

J, Leung TY, Levy N, Lockwood PL, Longoni C, López Ortega A, Loschelder DD, Lu JG, Luo Y, Luomba J, Lutz AE, Majer JM, Markowitz E, Marsh AA, Mascarenhas KL, Mbilingi B, Mbungu W, McHugh C, Meijers MHC, Mercier H, Mhagama FL, Michalakis K, Mikus N, Milliron S, Mitkidis P, Monge-Rodríguez FS, Mora YL, Moreau D, Motoki K, Moyano M, Mus M, Navajas J, Nguyen TL, Nguyen DM, Nguyen T, Niemi L, Nijssen SRR, Nilsonne G, Nitschke JP, Nockur L, Okura R, Öner S, Özdoğru AA, Palumbo H, Panagopoulos C, Panasiti MS, Pärnamets P, Paruzel-Czachura M, Pavlov YG, Payán-Gómez C, Pearson AR, Pereira Da Costa L, Petrowsky HM, Pfattheicher S, Pham NT, Ponizovskiy V, Pretus C, Rêgo GG, Reimann R, Rhoads SA, Riano-Moreno J, Richter I, Röer JP, Rosa-Sullivan J, Ross RM, Sabherwal A, Saito T, Sarrasin O, Say N, Schmid K, Schmitt MT, Schoenegger P, Scholz C, Schug MG, Schulreich S, Shreedhar G, Shuman E, Sivan S, Sjåstad H, Soliman M, Soud K, Spampatti T, Sparkman G, Spasovski O, Stanley SK, Stern JA, Strahm N, Suko Y, Sul S, Syropoulos S, Taylor NC, Tedaldi E, Tinghög G, Huynh LDT, Travaglino GA, Tsakiris M, Tüter İ, Tyrala M, Uluğ ÖM, Urbanek A, Valko D, Van Der Linden S, Van Schie K, Van Stekelenburg A, Vanags E, Västfjäll D, Vesely S, Vintr J, Vranka M, Wanguche PO, Willer R, Wojcik AD, Xu R, Yadav A, Zawisza M, Zhao X, Zhao J, Żuk D, Van Bavel JJ (2024) Addressing climate change with behavioral science: a global intervention tournament in 63 countries. Sci Adv 10:eadj5778. https://doi.org/10.1126/sciadv.adj5778

Agata Sobkow is an Associate Professor at the Center for Improving Decision Making (CRIDM) at SWPS University, Faculty of Psychology in Wroclaw, Poland. Her research focuses on intuitive processing, decision making, numeracyand (ir)rationality. She is particularly interested in developing and testing innovative methods, such as cognitive training and visual aids, to help individuals with low numeracy make better decisions. Additionally, she is dedicated to promoting Open Science practices and leads several replication projects.

Kacper Jurewicz is a researcher and student at SWPS University, Faculty of Psychology in Wroclaw, Poland specializing in cognitive processes behind decision making, with a particular focus on autism. His work examines the mechanisms driving differences in decision making between autistic and neurotypical individuals, aiming to support autistic individuals in navigating decision-making challenges more effectively. He is involved in research on decision making and motivation, investigating cognitive factors influencing risk perception, financial decisions, and the dynamics of yielding to temptation in the pursuit of self-defining goals.

Jakub Traczyk is an Associate Professor and the head of the Center for Improving Decision Making (CRIDM) at SWPS University, Faculty of Psychology in Wroclaw, Poland. His research investigates the mechanisms underlying risky decision making, with a focus on the impact of numeracy on decision quality. He explores the dynamics of adaptive decision making, emphasizing the roles of contextual factors, feedback, metacognition, and emotional influences in optimizing decision-making processes.

Fuzzy-Trace Theory, Decision Making, and the Law

3

Jordan E. Roue, Sarah M. Edelson, Aadya Singh, and Valerie F. Reyna

Abstract

In this chapter we explain key components of fuzzy-trace theory, a dual-process theory of memory, judgment, and decision making. We use two examples from legal decision-making—how people make decisions about plea bargains in criminal law and how jurors award civil damage awards—to illustrate these core concepts. In particular, people independently form multiple representations of information in parallel: verbatim (surface level, precise details) and gist (bottom line, meaning based). These mental representations influence judgment and decision-making in important and different ways. Reliance on gist emphasizes key categorical differences between decision options, whereas reliance on verbatim emphasizes precise differences between decision options and facilitates trading off probability and outcomes. People also tend to rely on the simplest representation of information that distinguishes between two options to make their decision; this principle is known as the fuzzy-processing preference. The tendency to rely on gist vs. verbatim also differs developmentally and among individuals. For example, adolescents tend to rely more on verbatim information. Additionally, reliance on gist representations tends to cue relevant values that are stored in memory in gist form, thus incorporating such values into the decision-making process. Both of our examples from the legal field illustrate how gist and verbatim representations explain important features of decision-making in life and work.

Keywords

Fuzzy-trace theory • Gist • Verbatim • Plea bargaining • Civil damage awards • Fuzzy-processing preference

J. E. Roue (✉) · S. M. Edelson · A. Singh · V. F. Reyna
Department of Psychology, Cornell University, New York, USA
e-mail: jr976@cornell.edu

© The Author(s), under exclusive license to Springer Nature Switzerland AG 2025

T. McElroy (ed.), *Decision-Making in Life and Work*, Integrated Science 36,
https://doi.org/10.1007/978-3-031-88415-3_3

In this chapter, we summarize core ideas of fuzzy-trace theory (FTT). FTT explains how memory, judgment, and decision-making work in the human mind and brain (for an overview and neural framework, see Reyna et al. [1]). By memory, we mean the "picture" in someone's mind as they encounter information whether during the time they see the information or later as they recall that information. One of the take-home messages of this chapter is that people form multiple representations of information, and these representations then influence the tasks people perform in life and work, such as making judgments or decisions, in different ways. In particular, as we explain below, people often rely on fuzzy gist representations of information rather than precise verbatim representations. We use decision-making in law as an illustration of these core ideas, focusing on how people make decisions about plea bargains in criminal law (defendants' pleading guilty to a lesser charge or sentence versus taking a risk by going to trial) and damage awards in civil law (jurors' judgments of dollars to be paid to redress economic injuries as well as pain and suffering). FTT applies to both kinds of decisions because it offers an explanation of how people translate information into meaningful forms that make contact with their basic values; for example, more severe injuries should be worth more dollars in damages—even when they do not know exactly how many dollars should be awarded. Below, we introduce the core ideas of FTT first, then discuss each application of the theory in turn, followed by implications for future work.

1 Background

Fuzzy-trace theory (FTT) is a dual-process theory that explains how people encode and use information when making decisions [2, 3]. FTT has been applied to many areas across life and work where decision-making is relevant, such as vaccination behaviors, using protection during sex, reporting a concussion, taking financial risks, getting screened for cancer, and comparing risky decisions by humans and artificial intelligence [4–7]. According to FTT, people encode information in two forms: verbatim or exact surface-level representations, and gist or bottom-line, meaning-based representations. These independent verbatim and gist mental representations are formed in parallel as people encounter information in everyday life, though adults primarily tend to rely on gist when making decisions [2, 8]. These multiple representations are encoded simultaneously and used in memory and decision-making processes. Verbatim representations encode surface-level information, while gist representations involve extracting and synthesizing the bottom-line meaning of the information in a way that is not captured by just considering the precise details. Although these processes occur simultaneously, humans tend to primarily rely on gist information in decision-making, often giving categorical distinctions (e.g., yes or no, something or nothing) the most weight.

Decision-making emphasizes the least precise gist representation that has been encoded, which then cues simple core values which are also stored in memory in a gisty way. That is, people usually do not store core values in long-term memory in terms of exact numerical values or precise memorized forms but in terms of simple, categorical, good/bad principles: saving lives is good, money is good, natural is good, family is important and so on [9, 10]. These forms of values are then more likely to be cued when the question is also formed and considered in a gisty way. This cuing occurs due to the encoding specificity principle, which says that information stored in the same form as the retrieval cue is more likely to be remembered.

The encoding specificity principle and FTT together explain how people process information to arrive at various conclusions. First, people encounter a decision between a certain number of options; they process both gist and verbatim information for each option. The two forms of information are then encoded in their mind as gist and verbatim representations. It is these representations that an individual will rely on when making a decision. Specifically, in risky decision-making, verbatim stimuli (information inputs, such as decision options) are converted to both verbatim and gist representations, meaning that both influence decisions but gist influences them more. As described in FTT, typical adults tend to rely on gisty information, so they will preferentially apply the gist representations of their options when making their decision. This gist-based information will then be likely to bring other gisty core values into play in the decision-making process, as explained by the encoding specificity principle. Alternatively, if the person making the decision applies their verbatim representations of the options, the gisty core values will not be cued. In that case, decision-making is more likely to be focused on comparing the costs and benefits of each option and encourage the direct trading off between risks and rewards.

Despite the general fuzzy-processing preference, there are situations where the simplest gist cannot differentiate between two options and then decision makers rely on the next least precise representation. For example, an individual deciding what to order at a restaurant may end up trying to decide between two green salads. The two salads would have the same gist representation (healthy meal) which does not distinguish between options. At this point, the decision might involve considering which salad is healthier, which involves ordinal gist representations such as being low versus high in calories. If those ordinal representations do not discriminate between options (e.g., both are low in calories), one might then turn to counting how many calories each dish has. If they then decided based on number of calories but did not consider if that amount of calories were "a lot" or "a little," they would be relying on verbatim processing. Preference for reliance on gist vs. verbatim also varies developmentally: Children and adolescents tend to rely more on verbatim representations, and reliance on gist increases with age and experience (for a review, see [11]).

We now turn to two examples of legal decision-making to illustrate some of the core theoretical predictions and explanations of FTT described above. Specifically, we first illustrate how FTT principles explain decision-making where numbers are transformed into a gist in plea bargaining. We then turn to FTT's explanation of the transformation of gist into numbers through decisions about civil damage awards. These topics provide illustrations of the important role of gist and verbatim mental representations in decision-making, along with other important concepts from FTT such as how gist representations of information can cue relevant values, the independent nature of gist and verbatim, the fuzzy-processing preference, and the ways that FTT's predictions vary developmentally. In addition, these specific types of legal decisions are made by everyday people (e.g., those chosen to serve on juries and defendants without special legal training or knowledge), demonstrating the importance of such decisions as part of everyday life.

2 Plea Bargaining: Turning Numbers into Gist

We begin by examining a form of decision making that, according to FTT, involves the transformation of numerical information (e.g., probability of conviction, sentence length) into gist interpretations that can then be relied on to reach a decision. Plea bargaining refers to a process in criminal cases whereby the prosecutor offers a defendant a deal that would resolve the case without going to trial, typically requiring the defendant to plead guilty in exchange for a more favorable ultimate outcome (e.g., reduced sentence or fewer charges; [12]). This practice is very common in the United States legal system, with almost 90% of defendants in all federal criminal cases (and 99% of defendants whose cases were not dismissed) taking plea deals in 2022 [13]. In this section, we discuss decision-making theories more generally before turning to the specific application of FTT to plea bargaining.

According to classic economic theory, probabilities and outcomes are crucial building blocks of decision-making: People compare their options based on these features and pick the most favorable best alternative based on this process (see Kahneman and Tversky [14]; Tversky and Kahneman [15]) [16]. This basic conception of outcomes being weighted by their probabilities in determining the overall value of an option is maintained in what is called "prospect theory". However, as research involving the famous dread disease problem and similar "framing" problems illustrate, people's actual behavior often differs from these expectations in predictable ways [8, 16]. The dread disease problem involves two versions of a mathematically identical question, presented in a gain frame and a loss frame [15]. The prelude to the problem (for both frames) states: "Imagine that the United States is preparing for the outbreak of an unusual disease, which is expected to kill 600 people. Please indicate which option you prefer." We first address the gain frame, where participants must choose between a sure option (200 people *saved* for sure) and a risky option (1/3 probability 600 *saved* and 2/3 probability no one *saved*). Importantly, the expected value of the sure and risky options is equal (200 saved = 1/3 * 600 saved). Therefore, on that basis, decision makers should be indifferent. However, people are not indifferent, and tend to choose the sure option, in the gain frame [15]. Prior decision theories explained the preference for the sure option in terms of perceiving outcomes (number of lives saved) and probabilities (1/3 probability) in slightly distorted ways. However, more recent research shows that those perceptual distortions do not explain many aspects of risk preferences [1, 2, 17, 18]. FTT predicts and explains risk preferences through gist representations and the fuzzy-processing preference. In particular, the simplest gist of these options boils down to save some people for sure versus a risky option in which some people are saved or no one is saved, and saving some is better than saving none. Most people prefer saving some people for sure over a chance of saving no one. Thus, FTT's prediction is that most people will choose the sure option in gain frame problems, which is what has been observed for the dread disease problem and similar problems [8]. The fuzzy-processing preference is demonstrated by the fact that the gist representation is given more weight than the verbatim representation of expected value that the options are equal.

In contrast to the gain frame, the options in a loss frame are as follows: a sure option (400 people *die* for sure) and a risky option (2/3 probability 600 people *die* and 1/3 probability no one *dies*). While the verbatim information conveyed in these choices is identical to the gain frame (400 out of 600 die means 200 out of 600 saved and vice versa), the gist of the options in the loss frame is different from those in the gain frame. The loss frame problem in its simplest form boils down to a choice between some people dying for sure or a chance that some die or no one dies. When presented with these loss options, most people prefer the option that has a chance that no one dies. People tend to be risk-seeking in the loss frame. Thus, per FTT, reliance on gist and the important ways that the gist differs between the gain and the loss frame of such problems, explains why people show these kinds of preference reversals.

Plea bargains resemble the dread disease problem described above in that they involve a choice between a sure option and a risky option: taking a plea deal (a sure option) or going to trial (a risky option) with a chance of being found not guilty and a chance of being found guilty and having a longer sentence [19] (See Table 1 for a more in-depth comparison). In the plea-bargaining context, the shadow of trial model (SOT) is similar to classic economic theories mentioned above in that it proposes that defendants trade off between sentence lengths and certainty of the outcomes to determine if they should accept a plea deal or not [20, 21]. According to SOT, a risk-neutral defendant would discount the expected sentence length they expect to get at trial by the likelihood of conviction in order to weigh this product against the length of the sentence offered in the plea deal (which is certain). This approach is very focused on the exact details of the two outcomes, or the verbatim information.

However, when people are deciding to take plea deals or go to trial, not all of the information required for the SOT calculation is precisely knowable (e.g., exact sentence length if convicted, although there are sentencing guidelines and expectations of experienced attorneys). Using statistical modeling to estimate such implicit or ambiguous information, the general form of the model tended to hold on average, suggesting that the expected value of the potential sentence at trial and the plea bargain are roughly similar (but see [19]). In other words, outcomes and probabilities roughly balanced out as expected value/utility theories suggest [20, 21], but see [22].

Although there is some evidence to support the general form of SOT, it is also important to evaluate whether the model holds at an individual level [20–23].

Table 1 Comparison between the dread disease problem and a plea bargain decision

	Premise	Sure option	Risky option	Conclusion
Dread disease (Gain Frame)	Imagine that the United States is preparing for the outbreak of an unusual disease, which is expected to kill 600 people. Please indicate which option you prefer	200 people will be saved	1/3 probability that 600 people will be saved and 2/3 probability that no one will be saved	The sure option, where 200 people will be saved, is preferable to the risky option where there is a chance that no one will be saved
Plea bargaining (Gain Frame, without numbers)	Imagine that you have been arrested and charged with a felony. Please indicate which option you prefer	Accept a plea deal where you get to stay out of prison for sure	Go to trial where you either will stay out of prison (acquittal) or won't stay out of prison (conviction)	The sure option, where you get to stay out of prison is preferable to the risky option where there is a chance that you won't stay out of prison
Plea bargaining (Gain Frame, with numbers)	Imagine your attorney tells you that you are facing a sentence of 20 years for a felony charge if convicted at trial. This is a typical sentence, and what you should expect to receive if convicted	Accept a plea deal, gaining 8 years of freedom compared to the expected sentence	Go to trial and maybe gain 20 years of freedom (acquittal) or maybe gain no years of freedom (conviction)	The sure option, where you gain 8 years of freedom for sure, is preferable to the risky option where there is a chance you won't gain any years of freedom

Note As with framing problems such as the dread disease problem, plea bargains can be framed as gains or losses. Here we illustrate the similarities between the dread disease problem and two plea bargains presented with and without specific numerical information in a gain frame

Using a dataset of almost 3400 case files, Bushway and Redlich [20] found that 16% of people took a plea deal when they should have expected a better outcome at trial, suggesting that SOT model's optimal predictions substantially depart from real-world behavior for some individuals. However, as we know from research on gain–loss framing effects and other research discussed above, risky choices are not fully explained by expected value/utility as SOT assumes.

Plea bargaining preferences, and differences from expected value/utility theory's predictions, can be easily explained using FTT. The SOT model's reliance on the specific details of the plea bargain—what FTT would call verbatim details—means that the SOT model does not account for the meaning-based distinctions inherent in plea deals that drive decision-making according to FTT (e.g., [24]).

For example, the qualitative distinction between a misdemeanor and a felony is not fully reducible to units of utilities; there is a categorical gist difference that embraces the ambiguity surrounding these two sentences: Having a felony on one's record is worse than a misdemeanor in many ways that cannot be precisely enumerated.

To wit, when deciding between two options, according to FTT, people tend to consider multiple levels of precision at once, but if simpler levels do not distinguish options, they revert to considering more precise distinctions: Categorical gist differences matter if they distinguish options (e.g., jail/prison time vs. no jail/prison time), as do ordinal gist (e.g., shorter sentence vs. longer sentence), and specific verbatim differences (4 year sentence vs. 5 year sentence) [25]. For example, researchers manipulated the probability of conviction and possible sentence length in hypothetical scenarios given to participants. The participants were then asked whether they would plead guilty to an offered deal and asked to name the longest sentence they would accept in a plea deal relative to the potential sentence at trial. Instead of being affected by details about exact probability of conviction or potential sentence length (e.g., 20% chance of conviction vs. 30%), participants treated the probabilities and sentence lengths as belonging to meaningful categories (e.g., small or large) and made their decision accordingly [26]. Both rates of plea acceptance and maximum plea deal sentence length were similar within and different between the categorical 'groupings' of the variables. These meaningful categories fall directly out of FTT's predictions about gist representations of information [1, 2, 8]. The conclusion that people are primarily relying on gist-based meaningful categorizations is supported by the fact that the verbatim differences used in expected value calculations (or expected utilities) do not seem to drive defendants' decision-making as anticipated by the SOT model when the decision options have different gists [26]. Research shows that decisions of whether to take a plea are not related to probability of conviction and sentence length at trial as SOT would predict, but instead vary only across 'gist-based bands' (e.g., low and high probability of conviction), which is consistent with FTT's predictions [26].

In the context of plea bargaining, the decision typically involves several important categorical differences such as type of charge (felony vs. misdemeanor), custodial status (being held in jail until trial v. being released on bail), and factual guilt or innocence. We first examine the categorical distinction between misdemeanor and felony convictions in plea bargaining and how that affects decision-making according to FTT. Felony convictions are categorically and meaningfully different from misdemeanor convictions, as they can result in consequences like losing one's right to vote and severely limiting employment opportunities [27–29]. Knowledge becomes essential here, as first-time defendants may not fully understand the collateral consequences that come with a felony conviction, which affects their gist representation. Attorneys are better positioned to understand the meaningful differences between a felony and misdemeanor, and thus have a responsibility to communicate information to their client in a way that highlights the gist.

To illustrate, consider a plea deal offering the defendant a one-year prison term and a misdemeanor conviction on their record. Assume further that if defendants go to trial, they know they would risk a one-year prison sentence for a felony charge and that the chance of conviction is unknown, but probably on the order of 80% [9]. A verbatim analysis of these options (as SOT model proposes) might predict that the defendant would choose to go to trial: 1.0 * 1 year = 1 years of a misdemeanor but the trial is 0.8 * 1 year = 0.8 years of a felony. In an SOT approach, the number of years could compensate for the difference in outcomes, misdemeanor or felony. However, this verbatim analysis does not acknowledge an entirely different, qualitative level of analysis in which having a felony conviction on one's criminal record is categorically worse for the defendant's life than having a misdemeanor conviction, which is a gist-based distinction. FTT would predict that a defendant who was a gist-based thinker would likely take the plea deal in this situation because the sentences are categorically similar and a felony is categorically bad, but a verbatim-based thinker might trade off the years, probabilities, and severity of charges against one another and go to trial.

Next, we examine the categorical distinction between custodial sentences (e.g., jail or prison time) and non-custodial sentences (e.g., probation or community service). To illustrate, consider an innocent defendant who has been offered a plea deal where their options have the following gists: (1) a misdemeanor conviction with no prison time for sure vs. (2) a gamble between either a felony conviction with at least 2 years in prison or acquittal (no conviction) with no prison time. Although the just outcome for this innocent person may be going to trial and getting acquitted, the costs associated with going to trial can be prohibitively high (e.g., financial costs of hiring an attorney or opportunity costs of time spent in jail waiting for trial; [9]) and the outcome at trial is risky—acquittal is far from guaranteed. The plea deal is the better option based on the categorical gist representations because it offers the possibility of no prison time (among other categorical differences). Ordinally, the plea deal is also superior (e.g., a misdemeanor conviction is less severe than a felony conviction). Even in the case of a naïve defendant who does not understand all the exact collateral consequences that may come with a felony conviction, if they understand the categorical differences between the two types of charges, they will be able to extract the relevant gist of felonies being much less desirable. Values matter, too. FTT predicts and explains why reliance on gist favors taking the plea deal because staying out of jail or prison is an important value to most defendants. Gist representations that are used in this decision-making process—no prison for sure versus the possibility of prison—also cue core values—no prison is better than some prison—which affect decision-making according to FTT [11, 25].

Gist-based decision making is more likely to allow one's closely held values to come into play in decision-making because values are stored in memory in a similar form to the way gist representations are stored and are thus cued when gist representations are relied upon [25, 30]. Additional examples of strong values that may be cued during plea bargaining are ones of not confessing to a crime one did not commit or providing for one's family [9, 25, 31]. A person's gist representation of a situation depends on their knowledge and background, which then influences which values are cued and how strongly someone relies on them. The ordering of conflicting values like 'I will not plead guilty to something I am innocent of' and 'I have to stay out of jail to continue providing for my family' depends on the individual, their beliefs, and their gist representations of the specific situation. For example, in a situation where the difference in possible jail time between the plea and trial sentence is very large, the gist distinction between options will be much stronger, and prioritizing minimizing jail time will likely be the most salient and relevant value. Developmental considerations will also affect how these values are incorporated or ignored in decision-making. Adolescents, who rely more on verbatim thinking, and those with high presence of psychopathic traits, who are less likely to hold strong moral values, will be less influenced by value-based considerations like guilt or innocence in their decision-making.

Situational factors, such as whether the defendant is in jail or out on bail might also influence the gist representation of the situation. For example, a plea deal offering a sentence of time served means a return to freedom if the defendant is already in jail, but no change in freedom status if the defendant is out on bail. If a defendant is in jail, viewing a plea deal as a categorical return to freedom (vs. taking a chance on freedom or no freedom by going to trial) might make the defendant be relatively insensitive to the probability of conviction, the magnitude of the trial outcome, or the length of time served. Roughly consistent with this interpretation, in one study, being detained pre-trial (as opposed to being released on bail) made participants 3.67 times more likely to take a plea deal [32]. This situational factor is especially likely to influence defendants who are the sole breadwinner or caretaker for their family and/or children. Research on scope insensitivity (insensitivity to quantitative differences) in this context has not yet been conducted.

FTT predicts that those relying on gist would be affected by guilt status and charge difference (felony vs. misdemeanor) more than those relying on verbatim, since verbatim representations of information do not cue core values as effectively as gist representations do [11, 25]. In one study, people were grouped as either verbatim or gist thinkers based on their performance on typical framing problems like the dread disease question discussed earlier, and each group was asked to rate the importance of values (e.g., not pleading guilty to a crime I did not commit) to them [25]. People were then asked to make four decisions about plea deals and were randomly assigned guilt status for the crimes (i.e., told to assume they were innocent or guilty). Consistent with FTT's predictions, the more an individual was a gist thinker, the more likely that their plea decision reflected their value of not falsely admitting guilt. Conversely, although verbatim thinkers endorsed the value of not falsely admitting guilt as much as gist thinkers did on average, their plea decisions were not associated with their endorsement of this value. Since verbatim and gist thinking groups did not differ in their ratings of how important each value was to them, these differences likely resulted from reliance on gist versus verbatim representations (Fig. 1).

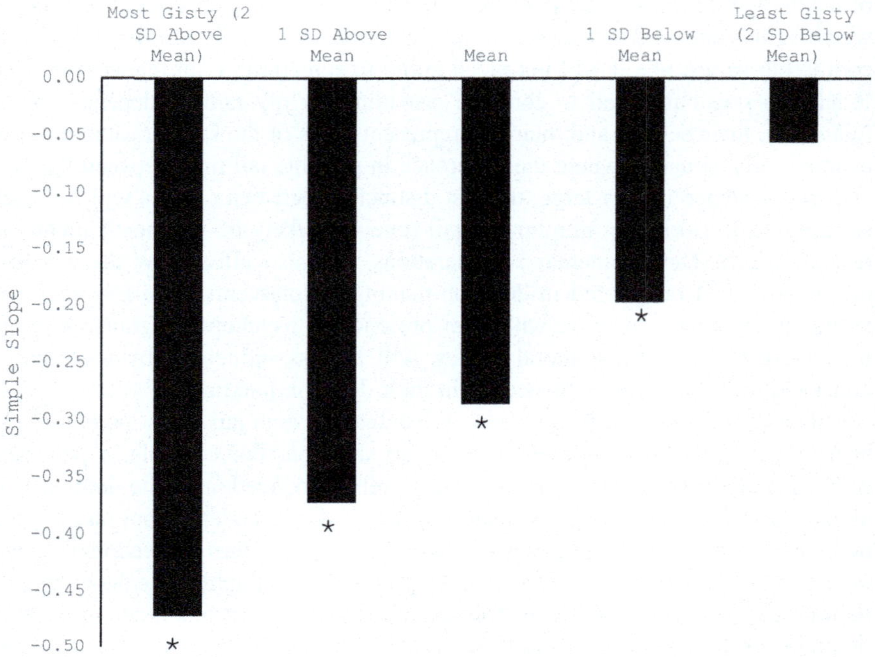

Fig. 1 Relationship between the importance of the principle "I would not plead guilty to a crime I did not commit" and Accepting a Plea Deal Grouped By Reliance on Gist. *Note.* Reliance on Gist is based on the extent to which individuals demonstrated standard framing, based on a measure called the framing index [33]. "SD" means standard deviation. Having a framing index above the mean suggests higher reliance on gist, whereas having a framing index below the mean suggests lower reliance on gist. This figure shows that when people rely most heavily on gist, they are least inclined to accept a plea deal when they endorse the principle "I would not plead guilty to a crime I did not commit." This relationship between endorsement of this value and accepting a plea deal becomes weaker as reliance on gist decreases, as FTT predicts. In other words, endorsement of this value became less relevant to the decision to accept a plea deal the less participants relied on gist, despite the fact that this value was endorsed similarly among those relying more and less on gist. The data from this figure are from Helm et al. [33]

Similarly, FTT predicts that adolescents rely more on verbatim representations and less on gist than adults do [10, 11]. Thus, the theory predicts that adolescents would be more sensitive to specific outcomes of taking a plea or going to trial (i.e., the difference in number of years in prison) even when outcomes fall into equivalent categories and less sensitive to their value of not falsely admitting guilt. Research confirmed these predictions: Innocent adolescents took plea deals more than innocent adults did based on quantitative differences between options and relied less on their qualitative values [28, 29]. Importantly, the values were endorsed equally strongly across age groups, but were not implicated equally in their decision-making.

These developmental differences have implications for adolescents' competency to accept plea deals [28, 29, 34, 35]. For example, the requirements for a plea deal to be valid and non-coercive are for the defendant to accept knowingly, intelligently, and voluntarily [34]. The first two criteria most directly address the definition of competence and require the person to have an opportunity to both discuss the plea with their lawyer and consider the consequences of the deal (*Boykin v. Alabama* [36]; *People v. Harris* [37]). Defendants also must be mentally capable of understanding the terms of the plea agreement. If adolescents process and evaluate risks in a different way than typical adults, this would raise the question of if they are able to fulfill these standards and if plea bargaining practices are problematic within this age group. To examine this question through an FTT lens we need to compare adolescents' and adults' reasoning, both using the standard definition and considering how our understanding of the fuzzy-processing preference might be implicated in competency.

On one level, adolescents and adults have very similar levels of logical skills and the two groups tend to evaluate risks as similarly harmful [38]. This would seem to suggest that adolescents and adults have equal opportunity to consider the consequences of a plea deal. Assuming the two groups also have equal opportunity to discuss the plea with an attorney, they would both be considered competent under the current legal definition (although developmental differences in impulsivity have been acknowledged; see Roper v. Simmons [39]). A long line of research has shown that the verbatim risks and consequences are encoded very similarly by adults and adolescents [5, 38, 40, 41]. However, because adults tend to rely on gist representations more than verbatim, they are better able to connect the dots integrating the big picture implications of accepting the plea. For example, adults might not realize that they give up at least 40 rights (e.g., the right to have your attorney cross-examine witnesses, the right to appeal) by accepting a plea bargain and not going to trial. However, they are more likely to grasp the gist that pleading guilty to a felony rather than a misdemeanor would change their lives in unforeseeable ways [42]. Adolescents will typically not rely on that sort of gist-based meaningful understanding, instead being captured by the minutiae and details of what is being offered and relinquished. For example, adolescents might process differences between 7 months of probation and 9 months of probation but be less likely to appreciate the categorical distinction between probation and incarceration.

This processing difference highlights the contrast between literally being able to list the potential outcomes of the risk and really understanding the impact of them. By engaging in verbatim analysis, adolescents are likely to trade off risks and rewards to get the best outcome without considering the qualitative nature of how their life will be affected going forward even when innocent [7]. In this view, an adolescent may be fully capable of knowingly, intelligently, and voluntarily accepting a plea deal to the same standard as an adult by the current legal definition and still not fully, deeply appreciating in an intuitive sense what the consequences really mean, which may point to a deficiency in the definition of competence.

Per FTT, plea bargaining illustrates the important role of gist and verbatim mental representations in decision-making, how categorical gist guides decision-making, and how values are or are not integrated into the decision-making process [26, 28, 29]. Viewing plea bargaining through an FTT lens gives a more complete picture of how defendants make decisions, pinpointing the important factors that people are using to make these decisions in real life. In this context, it is very difficult to generally operationalize "improving" a defendant's decision given the individual differences and case-level factors discussed above. However, an understanding of how multiple representations of information and values play into decision-making will allow legal professionals to get the correct gist to their client. This does not mean just using plain language or exact statistics, but accurately conveying the meaning of the information (e.g., going to jail for a misdemeanor might be undesirable for the length of the sentence, but being convicted of a felony will ruin your life forever, even after the sentence is over). Specifically, attorneys should be asking their clients to summarize their gist representations of their options to ensure the most complete understanding of the possible outcomes.

3 Civil Damage Awards: From Gist to Numbers

We now turn to civil damage awards to evaluate decision-making that involves the transformation of gist representations into numbers. The outcome of civil litigation cases is typically monetary compensation in the form of damages, unlike criminal cases. There are multiple type of damages, including economic compensatory damages (e.g., medical bills, lost work), non-economic compensatory damages (e.g., pain and suffering) and punitive damages. Punitive damages are those imposed upon the defendant by the jury as punishment for wrongdoing and research has shown the amount awarded tends to correlate strongly with compensatory damages [43, 44]. Punitive damages tend to be based in a sense of just punishment, which may mean this decision is especially likely to cue moral values (e.g., hurting people is wrong) and thus be based on gist. In this chapter, we focus on compensatory damages, which are awarded to a plaintiff (the party bringing the lawsuit) based on harm they experienced as a result of a defendant's actions [45, 46]. Although the different kinds of damages correlate, verbatim economic damages are not necessarily equivalent to non-economic gist-based damages. Consider a stay-at-home mother who sprains her back due to another party's negligence. Despite the fact that the economic damages will be small, as medical bills will be minimal and she was not deprived of any employment, she may be unable to pick up her children or play with them to the same extent she used to, which would likely be categorized as large amounts of pain and suffering.

In civil trials, juries (which are made up of lay people) are typically tasked with determining the appropriate monetary amount the plaintiff should be awarded. While economic damages typically involve a relatively straightforward calculation, non-economic damages are based on abstract factors such as emotional damage or pain and suffering. The usual assumption is that factors such as numeracy (the ability to use and understand numbers) and focusing on the precision and reliability of dollar judgments are key. Yet, jurors report having a great deal of difficulty putting a precise dollar value on non-economic damages even when liability is clear, a challenge that goes beyond being innumerate [24].

FTT predicts and explains that a conversion takes place from a qualitative, gist-based determination of severity of harm to a verbatim monetary amount owed. The former gist representation of damages is intuitive and a natural way of thinking. However, translating these fuzzy intuitions into verbatim numbers is very difficult for jurors [47], and results in a large amount of variability in damage award outcomes [43, 44]. For example, the abstract nature of severity of harm judgments (e.g., unnecessarily losing a lung in surgery is really bad, [47] lends itself to gist-based meaningful categorizations (e.g., small or large amount of harm). However, jurors are asked to take that judgment (according to FTT) and decide on a specific amount of money to award the plaintiff [33, 43–45]. Converting gist information (e.g., deservedness, harm) to verbatim (e.g., amount of money to give) is also required when people are asked to donate money to charity. Specifically, decisions like how much money to donate to hurricane relief efforts requires the same exercise. People encode verbatim and gist representations about information like the severity of the hurricane (e.g. 150 mph winds and very bad) and amount of damage done (e.g. $50,000 and devastating amounts). They will also use value-based judgments like determining how in need the harmed population is of their donation or other contributions. Rather than simply trying to arrive at a number that is difficult to estimate, the gisty concepts of neediness and harm guide a person in determining if they will donate, and if so, how much to give. In this section, we explore how gist and verbatim representations affect decision-making in the context of civil damages, specifically by examining the Hans-Reyna model and its formalization of the fuzzy-processing preference in this context [43].

Mirroring FTT's fuzzy-processing preference, the Hans-Reyna model of civil damage award decision-making assumes parallel processing of different representations of damages. The only difference is that decisions of liability (categorical gist: liable or not) often precede judgments of award amounts, which then are a matter of ordinal gist: less or more (Table 1; [43]). Thus, jurors first decide based on their categorical gist whether damages are owed to the plaintiff or not. Then, based on ordinal gist, they decide if the deserved damages are low or high (and if neither, then are medium). Finally, they decide on the verbatim monetary amount that is in line with their previous gist determinations [43]. FTT posits that much of the variation observed in both real life and mock juror damage awards can then be explained by the challenges of determining a precise verbatim number that matches the jurors' gists [24, 48], as many people lack "number sense" (i.e., ability to contextualize numbers and give them meaning, [49, 50]. FTT predicts that awards will tend to make sense ordinally, but lack internal coherence *within* categories of 'small', 'medium', and 'large', which fits with the observed trends in courtroom data [43, 44].

As described above, gist and verbatim are parallel but are independent representations of information [8, 10]. This is supported by differences in the observed variation in award amounts (a verbatim judgment) being much larger than variation in the underlying judgments of severity of harm (a gist judgment, [51]). This also suggests that people have less difficulty making the gist-based determination of harm than identifying a specific, verbatim monetary amount to award, which is consistent with the FTT tenet that most adults prefer to rely primarily on their gist representations of information. However, despite the challenge of identifying one number to match an individual's gist of the severity of harm, there is still ordinal coherence in award amounts (i.e., as the severity of harm increases, so too does the monetary amount jurors decide the plaintiff is owed [44, 45, 52]).

For example, Reyna et al. [53] found that jurors' gist-based judgements of the size of award they gave (e.g., small, medium, or large) fully mediated the relationship between the damage award and evaluations of injury severity, supporting the idea that the ordinal gist categorization of the injury was being translated into the quantitative award amount. This finding is consistent with FTT's predictions and suggests that meaningful category distinctions (e.g., mild or severe) are coherent, while the verbatim details (i.e., exact award amount) vary widely from case to case. In real life, the exact correct damage award that would make the plaintiff whole is unknowable; the only knowable "correct" answer is the answer the jury gives in the award. Humans are more able to identify a correct or coherent gist of the award, so it is possible to be more certain about the reliability of the gist coherence than the verbatim coherence.

According to FTT, one reason for the difficulty of identifying an exact number to match their gist representations is that there is an absence of a bounded scale and meaningful anchor points (i.e., information that would provide context or meaning to numbers). Numbers themselves do not hold an independent gist, and their meaning and relative size changes in different contexts (e.g., $1000 might be a lot of money to a high school student, but a small amount to a recent college graduate with student loans; [54]). In essence, this means that civil jurors are being asked to identify an award amount that is gist consistent (i.e., has the same meaningful representation) with their severity of harm judgement, while numbers do not have a gist to compare with. Consistent with FTT, the evidence shows that people tend to cohere in their gist-based categorical judgments of a case, such as determining that the amount of harm the plaintiff underwent was large as opposed to small [26, 43, 53]. However, despite this gist-based ordinal coherence, there is a predictable breakdown in the process when jurors attempt to choose a verbatim amount of money to award based on this harm.

Consistent with this account, Reed et al. [55] found that a gist-based classification of severity of harm significantly predicted the amount awarded in a multiple regression analysis, while controlling for all other reasons given for award determination (e.g., wanting the award to be fair, considering medical or legal bills). Additionally, Hans and Reyna [43] found that most jurors awarded round numbers (e.g., $1000 or $15,000) rather than more precise amounts, supporting the FTT-based notion that award decisions are based on gist distinctions converted into monetary amounts.

Anchoring and adjustment as a process has been observed in generating damage awards, though typically individuals adjust insufficiently away from the anchor [56], see also Epley and Gilovich [57]. This insufficient adjustment is the driving force behind most theoretical explanations of how people decide on specific numerical judgments in the legal field. Specifically, research has examined the effect of meaningless anchor numbers on damage judgements, which resulted in a bias towards the provided anchor number [56]. Both external anchors provided by researchers and internal anchors generated by the person themselves have similar effects; people first process the original anchor and then adjust away from the anchor to arrive at a final number that is within an implicitly generated plausible range [57]. The fact that the adjustment is typically insufficient suggests that adjusting is an effortful process of which people do only as much as necessary. In the context of civil damage awards, this adjustment based on meaningless anchors results in increased ordinal coherence among individual award amounts, but the increase does not necessarily signal an increase in the quality of judgments, just the illusion of correctness in an arbitrary verbatim way.

In contrast, FTT-based predictions suggest that providing meaningful anchor numbers (e.g., average yearly salary) to jurors before deliberation will result in a different type of adjustment [43, 53]. Meaningless anchors by nature do not convey a particular gist, so the number is adjusted arbitrarily towards the target compensation category. In contrast, meaningful anchors will convey information about what gist the anchor number has, which allows jurors to orient themselves and make purposeful adjustments. This is one of the key differences between the standard view and the FTT account, that while the former provides an anchor, the latter makes sense of an anchor. In fact, one study found that jurors who were provided with meaningful anchors had greater coherence within the gist categories of award amount than those who had meaningless anchors [53]. Predictions grounded in FTT imply that meaning-based manipulations will make it easier to understand how the gist of the award amount relates to that of the anchor amount [43, 53].

This prediction has been empirically validated, as Reyna and colleagues [53] examined the effect of meaningful vs. meaningless anchor numbers on mock juror damage award decisions, varying anchor size (small or large). Meaningful anchors were given in terms of income, a relevant consideration for damage awards—described as large compared to average annual income or as small compared to average lifetime income—while meaningless anchors were numerically the same but phrased in terms of irrelevant courtroom construction costs. When a meaningful anchor was provided to jurors, the award amounts were pulled more towards the anchor number (either increasing or decreasing average awards depending on the size of the anchor) compared to meaningless anchors, not the usual meaningless "anchoring effect". The fact that a meaningful anchor had an effect while the meaningless anchor had little or no effect specifically supports the Hans-Reyna model's tenet (undergirded by FTT) that jurors struggle mostly with trying to identify a verbatim amount of money that fits their gist of severity because the gist of the severity of the injury is straightforward [43, 53]. When the meaningful anchors were placed within a frame of reference, jurors were able to pick up on if anchor was large or small and then connect their gist representation of the injury to the gist of the anchor number. Also as predicted, awards given by jurors who received a meaningless anchor had significantly larger variability than those given by those who saw a meaningful one [43, 53]. These results show how providing a meaningful context to numbers (i.e., giving them a gist) can help increase the validity and reliability of damage award amounts, which supports FTT's predictions.

Another study by Hans and colleagues [49] demonstrates that providing numerical guidance (specific number amounts like $25,000 or $1,000,000) along with verbal guidance (very small amount or great deal of money) can help jurors with challenges of connecting specific damage awards to their gist of the severity of the plaintiff's harm. This process mimics the meaningful anchor intervention in that the guidance is aimed at conveying a gist of the severity to jurors and then giving them a number that matches that gist. In this study, the researchers manipulated the presence of this numerical and verbal guidance in closing statements. When mock jurors were provided with both numerical and verbal guidance, their overall awards were less variable and they reported having less difficulty deciding on an award amount. Importantly, these effects were not observed for verbal guidance alone, suggesting that the difference is specifically due to the provision of a gist-consistent number that participants could then use to make their decision [49].

Similar results were also observed by Reed and colleagues [58], who found that providing numbers alone without context did not result in any improvements (as predicted by FTT), but providing numbers in such a way so as to be gist-consistent with the facts of the trial did result in anchoring effects. This study also revealed that providing a specific number to the jury seemed to act as a 'cap' on the awards, where jurors almost never ended up awarding an amount that exceeded the anchor amount mentioned in the closing statement. As discussed earlier, FTT proposes that a large part of the difficulty of the qualitative to quantitative conversion necessary here is the lack of a bounded scale that would imply categorical boundaries. Thus, it makes sense that providing jurors with an upper bound resulted in increased reliability of awards both in individual and group award judgments [43, 58]. These effects are consistent with FTT and Hans-Reyna model predictions, and demonstrate the positive effects that meaningful anchors could have on the current system of civil damage award decision-making.

According to other theories of decision-making, numeracy, or a person's skill and ability to use and understand numbers, reduces or eliminates biases [24, 59]. However, FTT is unique in that it holds that numeracy in the rote sense is not enough [24]. Though high numeracy may help jurors evaluate scientific evidence presented (e.g., how likely it is to have a fingerprint match or the inherent risk of a negative outcome of surgery), jurors still have to attach bottom-line meaning to numbers. Thus, according to FTT, the civil damage award decision-making process instead involves gist to verbatim conversions, which technical expertise with numbers may not help [24, 33, 52, 59]. Some research suggests that awards given by highly numerate people tend to be slightly less variable and better aligned with ordinal gist judgments compared to less numerate people [53, 60]. However, meaningful anchors (as discussed above) tend to produce the same effects, suggesting that even if numeracy does help jurors, providing anchor numbers may provide enough benefit so as to render individual differences in numeracy moot [58, 60]. For example, $100,000 is a large amount of money for the average adult, but a small amount for a large corporation; understanding this gist-based difference does not necessarily involve numeracy [33].

Overall, how jurors make decisions about damage award amounts in civil cases illustrates core tenets of FTT such as independent gist and verbatim processing and the fuzzy-processing preference. The three stages of the Hans-Reyna model exemplify (and are modeled after) the kinds of mental representations put forth in FTT, and the theoretically-based intervention of meaningful anchors addresses the challenges of the qualitative to quantitative conversion when numbers themselves lack gist. FTT also has unique explanations for how different kinds of numeracy will be (and sometimes will not be) implicated in this type of decision-making that are supported above other models' explanations [24].

4 Conclusion

Application of FTT's framework of gist and verbatim mental representations predicts and explains important aspects of legal decision-making in plea bargaining and civil damage awards. These two specific examples of legal decision-making are excellent scenarios with which to examine core principles of FTT such as the independent nature of gist and verbatim representations. Understanding how gist and verbatim thinking differ and are associated with different types of decision-making has important implications for understanding real world behaviors and everyday choices that people make. Plea-bargaining decisions are choices involving a sure option and a risky one that draw on mental processes identified in choices in framing problems [19], whereas deciding on dollars in damage awards is finding a verbatim match for gist criteria that draws on mental processes identified in other numerical judgment problems [24]. These two examples of plea bargaining and damage awards highlight the mental processes of converting verbatim information to gist and converting gist to verbatim judgments, respectively. Without FTT's concept of gist, categorical distinctions among decision options might be regarded as non-influential, though our examples above illustrate robust empirical support for how categorical distinctions drive decisions more than fine-grained verbatim differences [9, 26, 32, 33, 43, 53]. Overall, FTT's concept of both gist and verbatim representations, their independent nature, the fuzzy-processing preference, and the role of gist in cuing relevant values allows the model to account for important considerations in the decision-making process that other models (e.g., SOT as discussed above) do not account for. This examination of FTT in two separate legal contexts illuminates these decision-making processes, both providing an excellent setting for demonstration of FTT principles and an opportunity to explain how people make these kinds of important life decisions. Using this analysis, legal professionals will be better able to understand what factors are most important in their client's decision-making about plea deals and can offer useful guidance to juries in civil cases where damages are awarded.

References

1. Reyna VF, Müller SM, Edelson SM (2023) Critical tests of fuzzy trace theory in brain and behavior: Uncertainty across time, probability, and development. Cogn Affect Behav Neurosci 23(3):746–772. https://doi.org/10.3758/s13415-022-01058-0
2. Reyna VF (2012) A new intuitionism: meaning, memory, and development in fuzzy-trace theory. Judgm Decis Mak 7(3):332–359
3. Reyna VF, Edelson S, Hayes B, Garavito D (2022) Supporting health and medical decision-making: Findings and insights from fuzzy-trace theory. Med Decis Making 42(6):741–754. https://doi.org/10.1177/0272989X221105473
4. Garavito DMN, Reyna VF, DeTello JE, Landow BR, Tarpinian LM (2021) Intentions to report concussion symptoms in nonprofessional athletes: a fuzzy-trace theory approach. Appl Cogn Psychol 35(1):26–38. https://doi.org/10.1002/acp.3734
5. Reyna VF, Mills BA (2014) Theoretically motivated interventions for reducing sexual risk taking in adolescence: a randomized controlled experiment applying fuzzy-trace theory. J Exp Psychol Gen 143(4):1627–1648. https://doi.org/10.1037/a0036717
6. Blalock SJ, Reyna VF (2016) Using fuzzy-trace theory to understand and improve health judgments, decisions, and behaviors: a literature review. Health Psychol 35(8):781–792. https://doi.org/10.1037/hea0000384
7. Edelson SM, Roue JE, Singh A, Reyna VF (2023) How decision making develops: adolescents, irrational adults, and should AI be trusted with the car keys? Policy Insights Behav Brain Sci 11(1):11–18. https://doi-org.proxy.library.cornell.edu/10.1177/23727322231220423 (Original work published 2024)
8. Broniatowski DA, Reyna VF (2018) A formal model of fuzzy-trace theory: variations on framing effects and the Allais paradox. Decision 5(4):205–252. https://doi.org/10.1037/dec0000083
9. Reed K, Franz A, Calderon V, Meschkow A, Reyna V (2022) Reported experiences with plea bargaining: a theoretical analysis of the legal standard. West Virginia Law Review 124(2):421
10. Reyna VF, Estrada SM, DeMarinis JA, Myers RM, Stanisz JM, Mills BA (2011) Neurobiological and memory models of risky decision making in adolescents versus young adults. J Exp Psychol Learn Mem Cogn 37(5):1125–1142. https://doi.org/10.1037/a0023943
11. Edelson SM, Reyna VF (2021) How fuzzy-trace theory predicts development of risky decision making, with novel extensions to culture and reward sensitivity. Dev Rev 62. https://doi.org/10.1016/j.dr.2021.100986
12. Garrett B (2016) Why plea bargains are not confessions. William and Mary Law Review 57(4):1415
13. Gramlich J (2023) Fewer than 1% of federal criminal defendants were acquitted in 2022. Pew Research Center. https://www.pewresearch.org/short-reads/2023/06/14/fewer-than-1-of-defendants-in-federal-criminal-cases-were-acquitted-in-2022/
14. Kahneman D, Tversky A (1979) Prospect theory: an analysis of decision under risk. Econometrica 47(2):263–291. https://doi.org/10.2307/1914185
15. Tversky A, Kahneman D (1981) The framing of decisions and the psychology of choice. Science 211(4481):453–458. https://doi.org/10.1126/science.7455683
16. Kühberger A (1998) The influence of framing on risky decisions: a meta-analysis. Organ Behav Hum Decis Process 75(1):23–55. https://doi.org/10.1006/obhd.1998.2781
17. Reyna VF, Brainerd CJ, Chen Z, Bookbinder SH (2021) Explaining risky choices with judgments: framing, the zero effect, and the contextual relativity of gist. J Exp Psychol Learn Mem Cogn 47(7):1037–1053. https://doi.org/10.1037/xlm0001016

18. Kühberger A, Tanner C (2010) Risky choice framing: Task versions and a comparison of prospect theory and fuzzy-trace theory. J Behav Decis Making 23(3):314–329. https://doi.org/10.1002/bdm.656
19. Reyna VF, Reed K, Meschkow A, Calderon V, Helm RK (in press) Framing biases in plea bargaining decisions in those with and without criminal involvement: tests of theoretical assumptions. J Behav Decision Making.
20. Bushway SD, Redlich AD (2012) Is plea bargaining in the "shadow of the trial" a mirage? J Quant Criminol 28(3):437–454. https://doi.org/10.1007/s10940-011-9147-5
21. Bushway SD, Redlich AD, Norris RJ (2014) An explicit test of plea bargaining in the "Shadow of the Trial." Criminology 52(4):723–754. https://doi.org/10.1111/1745-9125.12054
22. Abrams DS (2011) Is pleading really a bargain? J Empir Leg Stud 8(s1):200–221. https://doi.org/10.1111/j.1740-1461.2011.01234.x
23. Bushway SD (2019) Defendant decision-making in plea bargains. In: Edkins V, Redlich A (eds) A system of pleas: social science's contributions to the real legal system, pp 24–36
24. Reyna VF, Brainerd CJ (2023) Numeracy, gist, literal thinking and the value of nothing in decision making. Nat Rev Psychol:1–19. Advance online publication. https://doi.org/10.1038/s44159-023-00188-7
25. Helm RK, Reyna VF (2017) Logical but incompetent plea decisions: a new approach to plea bargaining grounded in cognitive theory. Psychol Public Policy Law 23(3):367–380. https://doi.org/10.1037/law0000125
26. Zottoli TM, Helm RK, Edkins VA, Bixter MT (2023) Developing a model of guilty plea decision-making: Fuzzy-trace theory, gist, and categorical boundaries. Law Hum Behav 47(3):403–421. https://doi.org/10.1037/lhb0000532
27. Collateral Consequences Inventory. (n.d.). National inventory of collateral consequences of criminal conviction. Retrieved 6 April 2024 from https://niccc.nationalreentryresourcecenter.org/consequences
28. Helm RK, Reyna VF, Franz AA, Novick RZ (2018a) Too young to plead? Risk, rationality, and plea bargaining's innocence problem in adolescents. Psychol Public Policy Law 24(2):180–191. https://doi.org/10.1037/law0000156
29. Helm RK, Reyna VF, Franz AA, Novick RZ, Dincin S, Cort AE (2018b) Limitations on the ability to negotiate justice: attorney perspectives on guilt, innocence, and legal advice in the current plea system. Psychology, Crime and Law 24(9):915–934. https://doi.org/10.1080/1068316X.2018.1457672
30. Fujita K, Han H (2009) Moving beyond deliberative control of impulses: the effect of construal levels on evaluative associations in self-control conflicts. Psychol Sci 20:799–804. https://doi.org/10.1111/j.1467-9280.2009.02372.x
31. Helm RK (2018) Cognitive theory and plea-bargaining. Policy Insights Behav Brain Sci 5(2):195–201. https://doi.org/10.1177/2372732218786974
32. Edkins VA, Dervan LE (2018) Freedom now or a future later: pitting the lasting implications of collateral consequences against pretrial detention in decisions to plead guilty. Psychol Public Policy Law 24(2):204–215. https://doi.org/10.1037/law0000159
33. Helm RK, Hans V, Reyna V (2017) Trial by numbers. Cornell J Law Public Policy 27(1):107–143
34. Cauffman E, Steinberg L (2012) Emerging findings from research on adolescent development and juvenile justice. Vict Offenders 7(4):428–449. https://doi-org.proxy.library.cornell.edu/10.1080/15564886.2012.713901
35. Redlich AD, Shteynberg RV (2016) To plead or not to plead: a comparison of juvenile and adult true and false plea decisions. Law Hum Behav 40(6):611–625. https://doi.org/10.1037/lhb0000205
36. Boykin v Alabama (1969) 395 US 238
37. People v Harris (1985) 109 AD2d 351 (2d Dept)
38. Reyna VF, Farley F (2006) Risk and rationality in adolescent decision making: implications for theory, practice, and public policy. Psycholog Sci Publ Interest 7(1):1–44. https://doi.org/10.1111/j.1529-1006.2006.00026.x

39. Roper v. Simmons (2005) (Supreme Court of the United States)
40. Brainerd CJ, Reyna VF (2015) Fuzzy-trace theory and lifespan cognitive development. Dev Rev 38:89–121. https://doi-org.proxy.library.cornell.edu/10.1016/j.dr.2015.07.006
41. Defoe IN, Dubas JS, Figner B, van Aken MA (2015) A meta-analysis on age differences in risky decision making: adolescents versus children and adults. Psychol Bull 141(1):48–84. https://doi.org/10.1037/a0038088
42. Redlich AD, Bonventre CL (2015) Content and comprehensibility of juvenile and adult tender-of-plea forms: implications for knowing, intelligent, and voluntary guilty pleas. Law Hum Behav 39(2):162–176. https://doi.org/10.1037/lhb0000118
43. Hans VP, Reyna VF (2011) To dollars from sense: qualitative to quantitative translation in jury damage awards (SSRN Scholarly Paper 1876667). https://doi.org/10.2139/ssrn.1876667
44. Eisenberg T, Rachlinski J, Wells M (2002) Reconciling experimental incoherence with real-world coherence in punitive damages. Cornell Law Faculty Publications. https://scholarship.law.cornell.edu/facpub/377
45. Greene E, Bornstein BH (2003) Determining damages: the psychology of jury awards. American Psychological Association, pp xviii, 238. https://doi.org/10.1037/10464-000
46. Robbennolt JK, Hans VP (2016) The psychology of tort law. In: Advances in psychology and law. Springer International Publishing, pp 249–274. https://doi.org/10.1007/978-3-319-29406-3_8
47. Mott N, Hans V, Simpson L (2000) What's half a lung worth? Civil jurors' accounts of their award decision making. Cornell Law Faculty Publications. https://scholarship.law.cornell.edu/facpub/389
48. Reyna VF, Nelson WL, Han PK, Dieckmann NF (2009) How numeracy influences risk comprehension and medical decision making. Psychol Bull 135(6):943–973. https://doi.org/10.1037/a0017327
49. Hans VP, Reed K, Reyna VF, Garavito D, Helm RK (2022) Guiding jurors' damage award decisions: experimental investigations of approaches based on theory and practice. Psychol Public Policy Law 28(2):188–212
50. Laski EV, Siegler RS (2007) Is 27 a big number? Correlational and causal connections among numerical categorization, number line estimation, and numerical magnitude comparison. Child Dev 78(6):1723–1743. https://doi.org/10.1111/j.1467-8624.2007.01087.x
51. Wissler R, Hart A, Saks M (1999) Decisionmaking about general damages: a comparison of jurors, judges, and lawyers. Mich Law Rev 98(3):751–826
52. Hans VP, Helm RK, Reyna VF (2018) From meaning to money: translating injury into dollars. Law Hum Behav 42(2):95–109. https://doi.org/10.1037/lhb0000282
53. Reyna VF, Hans VP, Corbin JC, Yeh R, Lin K, Royer C (2015) The gist of juries: testing a model of damage award decision making. Psychol Public Policy Law 21(3):280–294. https://doi.org/10.1037/law0000048
54. Reyna VF (2013) Intuition, reasoning and development: a fuzzy-trace theory approach. In: The development of thinking and reasoning. Psychology Press, pp 193–220
55. Reed K, Hans V, Reyna V (2019) Accounting for awards: An examination of juror reasoning behind pain and suffering damage award decisions. Cornell Law Faculty Publications. https://scholarship.law.cornell.edu/facpub/1724
56. Bystranowski P, Janik B, Próchnicki M, Skórska P (2021) Anchoring effect in legal decision-making: a meta-analysis. Law Hum Behav 45(1):1–23. https://doi.org/10.1037/lhb0000438
57. Epley N, Gilovich T (2006) The anchoring-and-adjustment heuristic: why the adjustments are insufficient. Psycholog Sci 17(4):311–318. https://doi-org.proxy.library.cornell.edu/10.1111/j.1467-9280.2006.01704.x
58. Reed K, Hans VP, Rotenstein V, Helm RK, Rodriguez A, McKendall P, Reyna VF (2024) The power of meaningful numbers: attorney guidance and jury deliberation improve the reliability and gist validity of damage awards. Law and Human Behav 48(2):83–103. https://doi.org/10.1037/lhb0000559
59. Peters E (2020) Innumeracy in the wild: misunderstanding and misusing numbers. Oxford University Press, pp ix, 302

60. Helm RK, Hans VP, Reyna VF, Reed K (2020) Numeracy in the jury box: numerical ability, meaningful anchors, and damage award decision making. Appl Cogn Psychol 34(2):434–448. https://doi.org/10.1002/acp.3629

Jordan E. Roue is a joint degree JD/PhD student at Cornell University working with Dr. Valerie Reyna and Dr. Charles Brainerd on studying risky decision making in the context of the law. More specifically, she is researching the interrogation techniques that may lead to obtaining false confessions and false guilty pleas as well as how different neurotypes affect risky decision making and framing behavior. She is especially interested in the Reid technique and their usage of moral minimization to elicit confessions from suspects. Jordan's most recent publication, which was on artificial intelligence and decision making, appeared in Policy Insights from the Behavioral and Brain Sciences.

Valerie F. Reyna is the Lois and Melvin Tukman Professor of Psychology, Director of the Human Neuroscience Institute, and Co-director of the Center for Behavioral Economics and Decision Research at Cornell University. Her research integrates brain and behavioral approaches to understand and improve judgment, decision making, and memory across the life span. She has developed a basic theory of decision making that explains and predicts new phenomena involving framing effects, value allocation tasks, and the Allais paradox. Her work also encompasses psychological mechanisms underlying the believability and spread of misinformation through social media; explainable and interpretable AI; applications of cognitive models and AI to improve understanding of genetics (e.g., in breast cancer); and other medical and legal decisions (e.g., about jury awards, medication decisions, and adolescent risk-taking). Recently, many of her activities have involved improving risk communication (e.g., related to the COVID-19 pandemic) and the dissemination of scientifically valid information to vulnerable communities, including advising the American Psychological Association, National Academies, and World Health Organization. She is a developer of fuzzy-trace theory, a model of the relation between mental representations and decision making that has been widely applied in law, medicine, and public health. Dr. Reyna was awarded the 2025 lifetime achievement award for fundamental contributions to experimental psychology by the Society for Experimental Psychology and Cognitive Science; she is also a Fellow of the Society of Experimental Psychologists and has been elected to the National Academy of Medicine.

The Neuroscience of Choice: How Our Brain Processes Decisions

4

Michael McCormick

"The brain is wider than the sky."

Emily Dickinson

Abstract

The current chapter reviews the neuroscientific literature on decision making. Evidence for reward versus control processes are reviewed along with Prospect Theory, Fuzzy Trace Theory and several other factors that affect choice. Throughout, the evidence for regional versus network-based processing is considered and recommendations are made for future research. While many important advances have been made, significant challenges persist due to the complexity of human decision making, diverse levels of analysis and various types of data being collected. Notwithstanding, the future of the field looks bright and the ability of decision neuroscientists to inform public policy and improve health interventions has never been greater. Additional research should build on the work reviewed herein, seeking to identify the critical units of analysis (regions or networks) that provide the greatest predictive power to realize real-world benefits.

Graphical Abstract/Art Performance

M. McCormick (✉)
Department of Psychology, Florida Gulf Coast University, Fort Myers, Florida, USA
e-mail: mjmccormick@fgcu.edu

© The Author(s), under exclusive license to Springer Nature Switzerland AG 2025

T. McElroy (ed.), *Decision-Making in Life and Work*, Integrated Science 36,
https://doi.org/10.1007/978-3-031-88415-3_4

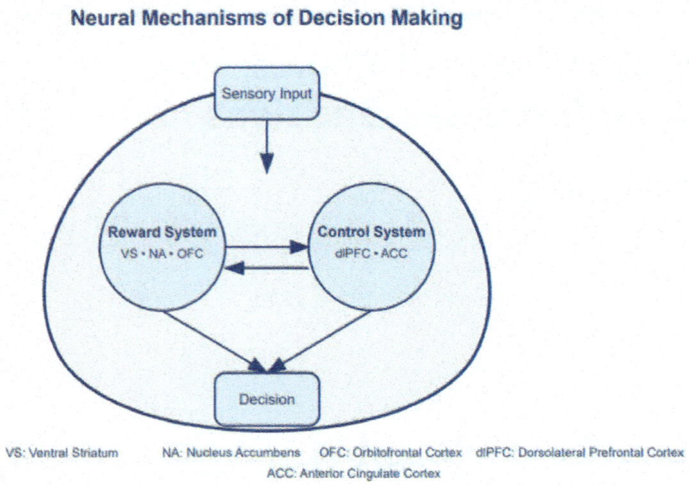

Keywords

Decision making • Neuroscience • Reward • Control • Framing • Prospect theory • Fuzzy trace theory

1 Introduction

The neuroscientific literature on decision making is similarly vast and wide ranging, including studies focused on the personal choices of individuals to how groups make decisions that affect public policy; from general processes such as aging that affect us all to those that are individual in nature. Perhaps not surprisingly then, the specific brain regions and networks that have been reported in decision making studies are also wide-ranging and sometimes inconsistent, presenting a challenge to the field and our ability to inform mental health treatments and shape public policy. Notwithstanding, the last several years have seen significant advancements in our understanding of the neural regions and networks that underlie choice.

In general, the empirical literature examining the neural underpinnings of decision making can be organized into three areas: (1) investigations that study the theoretical or fundamental processes underlying decision making, (2) those that examine additional factors related to decision making (e.g., conflict monitoring, prediction error) and (3) studies that focus on the role of general individual differences such as aging, personality and health. This last category is particularly general, also including studies that examine how social and experiential factors (e.g., ostracism, racism, group size) impact neurological processing when making decisions. Alternative taxonomies are possible, but we feel that this categorization is helpful in capturing a basic division between studies in a diverse body of literature. For the purposes of the current chapter, however, we will focus primarily on research conducted in areas 1 and 2, as these deal more specifically with the basic processes affecting choice for all people.

Notwithstanding, across all areas, a question of increasing importance is whether regions primarily subserve specific but mostly independent functioning, or whether neurological functioning is primarily network based. That is, is it more appropriate to focus on specific areas of activation when explaining the neurological underpinnings of decision making (and brain functioning more generally), or should empirical investigations seek to uncover the larger network systems to which individual regions belong because brain functioning and decision making are primarily network oriented? Numerous reviews have been published regarding the basic neurological processes underlying decision making [1–4], but few have focused directly on the question of regional versus network-oriented activation. For the field to advance, however, it is critical that the fundamental units involved in decision making—either regions or networks (or both)—are identified and studied further to gain a better understanding of what is actually going on in the brain, which will greatly increase our ability to predict important outcomes, improve mental health treatments and guide public policy. Thus, the current chapter reviews the empirical literature with a specific focus on evidence for regional versus network-based functioning. Throughout, areas of need for future research and potential implications for public policy will be discussed.

2 Theoretical/Fundamental Process Studies

Studies examining the fundamental or theoretical processes underlying decision making have been prominent in the literature. Despite differences, these accounts generally agree that reward/valuation processes are active and underly decision bias, while control related processes are sometimes active and resist or inhibit bias [5]. Numerous investigations have supported this view but few accounts have stimulated more research interest than Prospect Theory (PT) [6] and Fuzzy-Trace Theory (FTT) [7]. Both PT and FTT have significantly impacted the field of decision making and public policy more broadly, and each has its enthusiastic supporters. Considerable neuroscientific evidence has also been reported to support each account, but the results are sometimes inconsistent or inconclusive, and studies often do not lend themselves to a critical test between theories. Thus, for ease of exposition, the literature focusing on reward versus control related processes is presented first and separately from the literature on specific theoretical accounts (i.e. PT, and FTT) (Table 1).

2.1 Reward Processing and the Ventral Striatum (vS)

The VS, particularly the nucleus accumbens (NA), plays a crucial role in reward anticipation and processing [8]. This region is part of the mesolimbic dopamine system, which is fundamental to motivation and reward-seeking behavior. Diederen

Table 1 Key brain regions and networks reported in fundamental process studies

Brain region/network	Role in decision making	References
Ventral striatum (VS)	Reward anticipation and processing; involved in motivation and reward-seeking behavior	Diederen et al. [8]; Schultz [9]; Delgado et al. [10]; Setsompop et al. [11]
Orbitofrontal cortex (OFC)	Encodes subjective value of options; integrates aspects of reward for decision-making	Schultz [9]; Padoa-Schioppa and Assad [12]; Howard et al. [13]; Stalnaker et al. [14]; Rolls [15]; Zimmermann et al. [16]
Dorsolateral prefrontal cortex (dlPFC)	Maintains goal representations; modulates reward-driven behaviors and self-control	Wichary & Smolen [4]; Miller & Cohen [17]; Hare et al. [18]
Anterior cingulate cortex (ACC)	Conflict monitoring and error detection; signals need for cognitive control	Shenhav et al. [19]; Botvinick et al. [20]
Insula	Integrates risk and reward information; involved in emotional processing	Basten et al. [21]; Uddin [22]
Ventromedial prefrontal cortex (vmPFC)	Integrates value information to guide decision-making	Berkman [23]
Amygdala	Emotional processing and response to framing effects; involved in loss aversion	De Martino et al. [24, 25]; Tom et al. [26]; Sokol-Hessner et al. [27]
Thalamus	Part of network involved in loss aversion	Canessa et al. [28]

et al. demonstrated that the ventral striatum shows increased activation during the anticipation of rewards, with the magnitude of activation correlating with the expected value of the reward [8]. The NA receives dopaminergic input from the ventral tegmental area (VTA) and is sensitive to both primary rewards (e.g., food, water) and secondary rewards (e.g., money, social approval). Schultz proposed that dopamine neurons in this circuit encode reward prediction errors—the difference between expected and received rewards [9]. This mechanism is crucial for learning and updating value representations, which in turn influence future decision-making. Recent studies have also highlighted the role of the VS in processing aversive outcomes. For instance, Delgado et al. found that the ventral striatum responds to both monetary gains and losses, suggesting a more general role in outcome evaluation rather than just reward processing [10]. In other work, using high-resolution fMRI, Setsompop et al. revealed fine-grained functional organization within the striatum, showing distinct subregions involved in different aspects of reward processing and decision making [11].

2.2 The Orbitofrontal Cortex and Value Representation

The orbitofrontal cortex (OFC) works in concert with the ventral striatum to encode the subjective value of options and is involved in updating value representations based on prediction errors [9]. The OFC is particularly important for integrating various aspects of a reward, including its magnitude, probability, and delay, into a common value signal.

Padoa-Schioppa and Assad demonstrated that neurons in the OFC encode the economic value of offered and chosen goods in a common neural currency [12]. This allows for the comparison of disparate options, such as choosing between an apple and an orange. More recent work by Howard et al. has shown that the OFC also represents the value of imagined outcomes, suggesting its role in prospective decision-making [13]. Additionally, Stalnaker et al. showed that OFC neurons encode both chosen and unchosen option values, supporting its role in credit assignment and learning [14]. The OFC's role extends beyond simple value representation. It is also involved in reversal learning and flexible behavior. When the contingencies between actions and outcomes change, the OFC rapidly updates its representations, allowing for adaptive decision-making [15]. Consistent with this view, Zimmermann et al. showed that the hippocampus and OFC work together to integrate reward information with contextual cues, enabling flexible decision making in changing environments [16].

2.3 Cognitive Control and the Prefrontal Cortex

Cognitive control processes, primarily mediated by the prefrontal cortex (PFC), modulate reward-driven behaviors. The dorsolateral PFC (dlPFC) is involved in maintaining goal representations and exerting top-down control over decision processes [4]. This region is crucial for working memory and the implementation of task rules, allowing for the pursuit of long-term goals in the face of immediate temptations.

Miller and Cohen proposed that the dlPFC biases processing in other brain regions to guide behavior in accordance with internal goals or states [17]. This top-down control is particularly important in situations where automatic, habitual responses are insufficient or inappropriate. Recent work has also highlighted the role of the dlPFC in self-control during decision-making. Hare et al. found that successful dieters showed increased dlPFC activation when choosing healthy foods over tasty but unhealthy alternatives [18]. This suggests that the dlPFC can modulate value signals in regions like the VS and OFC to promote goal-directed behavior.

2.4 The Anterior Cingulate Cortex and Conflict Monitoring

Conflict monitoring will be discussed in greater length below, but as it is central to some accounts of reward versus control processes in decision making, we briefly review its role here. The anterior cingulate cortex (ACC) plays a vital role in conflict monitoring and error detection, signaling when additional cognitive control is needed [19]. The ACC is part of the salience network and is thought to monitor ongoing behavior and environmental feedback to detect situations that require increased cognitive control. Botvinick et al. proposed that the ACC detects conflicts in information processing, such as when multiple competing responses are activated simultaneously (as proposed by Fuzzy Trace Theory; see below) [20]. This conflict detection then triggers the recruitment of cognitive control processes, often implemented by the dlPFC, to resolve the conflict and optimize behavior.

2.5 Integration of Reward and Control Processes

While reward processing and cognitive control have often been studied separately, recent research has focused on how these systems interact to guide decision-making [29]. Taken together, a model of cognitive effort-based decision-making that integrates reward valuation, conflict monitoring, and cognitive control begins to emerge. In this model, the ACC monitors for conflicts and signals the need for control, the dlPFC implements that control, and regions like the ventral striatum, insula and OFC evaluate the rewards and costs associated with exerting control. Value representations are continuously modulated by the dlPFC and integrated by regions such as the ACC and vmPFC in a functionally connected feedback loop until a decision has been made (see Fig. 1).

Neuroimaging studies have provided evidence for this integration. For example, Basten et al. found that during a cost–benefit decision-making task, the VS and OFC encoded benefits, the amygdala and insula encoded costs, and the vmPFC integrated this information to guide choices [21]. More recently, Berkman proposed an integrative model where the dorsolateral PFC modulates the subjective value of options represented in the vmPFC, influencing choice behavior [23]. This interaction is particularly relevant in situations requiring self-control.

Network-based approaches have also gained prominence in recent years. Shine et al. demonstrated how flexible network reconfiguration supports adaptive control during decision making, with the frontoparietal network playing a key role in this flexibility [30]. Uddin reviewed the salience network's role in detecting behaviorally relevant stimuli and initiating cognitive control signals [22]. This network, anchored by the anterior insula and anterior cingulate cortex, plays a crucial role in switching between reward and control networks during decision making.

In conclusion, the interplay between reward processing and cognitive control in decision-making involves a complex network of brain regions, including the VS, insula, OFC, dlPFC, and ACC. These regions work together to evaluate options, anticipate outcomes, detect conflicts, and implement control processes. Notably,

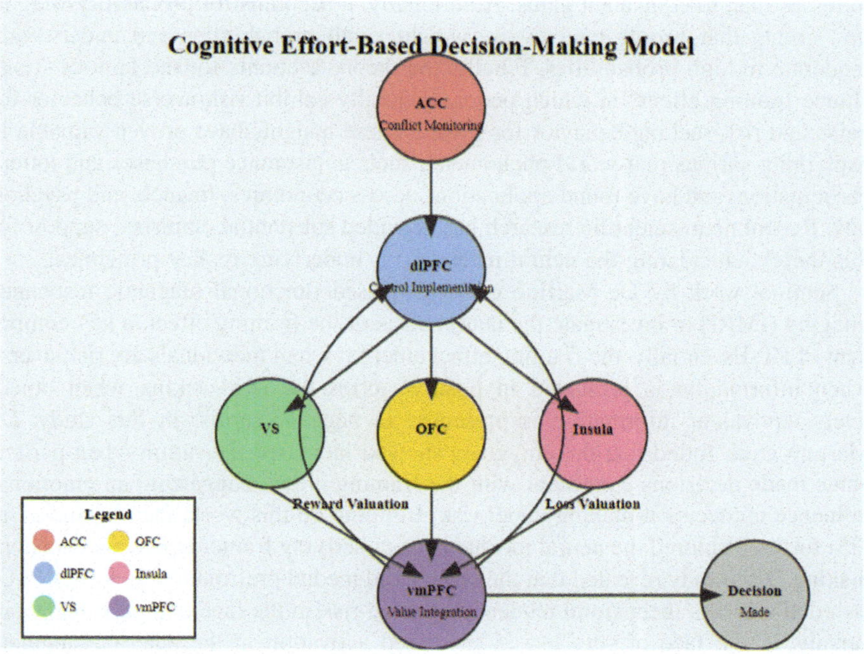

Fig. 1 Conceptual model of effort-based decision making in the brain. *Note* ACC = Anterior Cingulate Cortex; OFC = Orbitofrontal Corex; dlPFC = Dorsolateral Prefrontal Cortex; VS = Ventral Striatum; vmPFC = Ventromedial Prefrontal Cortex. In this model, the ACC monitors for conflicts and signals the need for control, the dlPFC implements that control, and regions like the ventral striatum, insula and OFC evaluate the rewards and costs associated with exerting control. Value representations are continuously modulated by the dlPFC and integrated by regions such as the ACC and vmPFC in a functionally connected feedback loop until a decision has been made

however, region specific findings are still often reported in this literature with little or no reference to the larger networks to which these regions belong. Future work should seek to distinguish whether the functioning of a given region is largely independent of other regions or a part of a larger (and more critical) neural network. Functional and structural connectivity analyses may play a critical role in meeting this goal.

3 Neuroscientific Support for Prospect Theory

Prospect Theory (PT), developed by Kahneman and Tversky [6], has been a cornerstone in understanding decision-making under risk. Unlike traditional expected utility theory, prospect theory incorporates psychological factors to account for observed decision-making behaviors. The theory posits that people evaluate outcomes as gains or losses relative to a reference point, rather than in absolute terms. It also highlights the concept of loss aversion, where individuals are more sensitive to losses than to equivalent gains. Additionally, it accounts for probability weighting, noting that people tend to overweight small probabilities and underweight moderate to high probabilities. Finally, the theory accounts for the famous "risky choice framing effect" in which people generally exhibit risk-averse behavior for gains and risk-seeking behavior for losses. These insights have proven valuable in explaining various real-world phenomena, such as insurance purchases and lottery participation, and have found applications across economics, finance, and psychology. Recent neuroscientific research has provided substantial empirical support for this theory, elucidating the neural mechanisms underlying its key principles.

Seminal work by De Martino et al. [24] used functional magnetic resonance imaging (fMRI) to investigate the neural basis of the framing effect, a key component of PT. Essentially, the framing effect emerges when individuals are risk-averse when information is presented in positive terms but risk-seeking when objectively equivalent information is presented in negative terms. In this study, De Martino et al. found that the amygdala showed increased activation when participants made decisions consistent with the framing effect, suggesting an emotional influence on decision-making under risk. Building on this work, De Martino et al. [25] further explored the neural mechanisms underlying frame-dependent decision-making. This study revealed that the orbital and medial prefrontal cortex (OMPFC) encoded subjects' behavioral tendency to avoid risk in the face of sure gains. Conversely, in the face of sure losses, increased activation in the anterior cingulate cortex (ACC) was associated with subjects' tendency to overcome this behavioral bias and choose risky options. More recently, Xu et al. [31] found that amygdala-vmPFC coupling was associated with frame consistent behavior, while dorsal ACC-vmPFC coupling was associated with frame inconsistent behavior, suggesting distinct neural mechanisms for succumbing to or overcoming framing biases.

The nonlinear probability weighting function proposed by PT has also been investigated using neuroimaging techniques. Employing fMRI, Hsu et al. [32] presented participants with a series of two-part gambles that ranged in probability and reward for gains only. They found that neural responses in the striatum reflect nonlinear probability weighting, with overweighting of low probabilities and underweighting of high probabilities. This nonlinear encoding of probabilities is consistent with the pattern predicted by PT and helps explain why people tend to overvalue unlikely events (e.g., winning the lottery) and undervalue highly probable outcomes (e.g., higher SES with increased education).

Several studies have also examined the neural basis of loss aversion. Tom et al. [26] used fMRI to examine neural responses to potential gains and losses and found that several brain regions, including the ventral striatum and ventromedial prefrontal cortex (vmPFC) showed increased activation for increasing potential gains and decreased activation for increasing potential losses. Sokol-Hessner et al. [27] found that individuals with greater loss aversion showed increased amygdala activation during decision-making, while those with reduced loss aversion exhibited greater dorsolateral prefrontal cortex (dlPFC) activity. Finally, taking a network-based perspective, Canessa et al. [28] identified a network including the amygdala, thalamus, and striatum that showed increased activation correlating with individual differences in loss aversion. This research provides additional support for the involvement of emotion-related brain regions in decision-making under risk.

Lastly for PT, the neural basis of reference point dependence has been the focus of recent research. Louie et al. [33] showed that neurons in the orbitofrontal cortex (OFC) encode value relative to the available choice set, providing a neural substrate for reference-dependent valuation. Furthermore, Brooks et al. [34] found that the VS encodes relative values during risky choices involving electric shocks, supporting the idea that subjective value is inherently comparative.

In conclusion, recent neuroscientific research has provided substantial empirical support for key principles of PT. These studies have elucidated the neural mechanisms underlying framing effects, probability weighting, loss aversion, and reference dependence. As with other areas in the decision neuroscience literature, however, region specific results are often reported with little or no reference to larger network systems. How the interactions of these systems provide support for PT (or not) remains a largely empirical question, therefore, and future research should include greater emphasis on neural network-based analyses.

4 Neuroscientific Support for Fuzzy Trace Theory

Fuzzy Trace Theory (FTT), developed by Valerie F. Reyna and Charles Brainerd, posits that decision-making relies on two types of mental representations: gist-based (essential meaning) and verbatim (exact details) [35]. According to FTT, framing effects emerge when individuals rely on gist processing and are diminished or eliminated with reliance on verbatim processing. This dual-process model has been extensively applied to understand how individuals make decisions in a variety of contexts, and critical behavioral tests have generally provided greater supported for FTT over PT [36–39]. Although not as extensive as the literature supporting PT, recent neuroscientific research has provided valuable insights into the neural correlates of FTT, further elucidating how gist and verbatim processing contribute to decision-making.

In one study, participants completed a series of 60 variations on the classic risky choice framing problem while undergoing fMRI scanning [39]. Increased dlPFC activation covaried with self-reported criminal risk-taking when participants made choices consistent with the framing effect, reflecting gist processing. When choices ran counter to the framing effect, however, (reflecting verbatim processing) risk-taking covaried with amygdala and striatal activation (for noncriminal risk-taking) or covaried with an area around the temporal-parietal junction (for criminal risk-taking). These findings provide neurological support for the distinction between gist and verbatim processing in the brain.

In other work by Venkatraman et al. [40], participants also completed a series of monetary choice trials while undergoing fMRI scanning, but here the focus was on choice strategy and the accompanying neurological mechanisms. Gist-based choice strategies (i.e. Pmax) were associated with increased activation in parietal and lateral prefrontal areas, as well as increased functional connectivity between the dmPFC, dlPFC and PPC. Meanwhile, choice strategies consistent with verbatim processing (Lmin and Pmax) were associated with increased activation in the vmPFC and anterior insular, and greater functional connectivity between the dmPFC and the anterior insula. As before, different neural mechanisms were associated with gist and verbatim processing, supporting their distinction in theoretical accounts.

Earlier work by De Martino [24] that is typically thought of as supporting PT can also be interpreted as providing support for the distinction between gist and verbatim processing. As noted, increased amygdala activation was observed for framing consistent choices whereas dlPFC activation was found when participants were resistant to the framing effect. According to the FTT account of this result, amygdala activation reflects emotional, gist-based processing that results in framing effects, whereas dlPFC activation reflects compensatory, verbatim processing that resists framing bias.

Moreover, the developmental aspects of FTT have also received neuroscientific support. Chick et al. [41] investigated age-related differences in gist-based decision making using fMRI. They found that adolescents showed greater activation in the dorsolateral prefrontal cortex (dlPFC) during gist-based decisions compared to adults, suggesting a developmental shift in reliance on gist versus verbatim processing.

Finally, FTT's applications to medical decision making have also received neuroscientific support. Blalock and Reyna [7] used fMRI to study how patients process health risk information. They found that presenting risk information in a gist-based format led to increased activation in areas associated with emotional processing, such as the amygdala and insula, consistent with FTT's emphasis on the role of affect in gist-based decision making.

In conclusion, recent neuroscientific research has provided substantial empirical support for key principles of Fuzzy Trace Theory. Brain regions involved in gist and verbatim processing have been identified, including the dmPFC, amygdala and dlPFC, and some network specific results have also been reported. Research has also supported FTT predictions about developmental increases in reliance on gist processing, and the role of gist in risky decision making. Future research should continue to investigate how these neural processes interact to produce the complex decision-making behaviors described by FTT.

5 Section Summary

Most studies consistently report the involvement of the PFC, striatum, and ACC in decision-making processes. However, the specific roles attributed to these regions can vary across studies. For example, while some studies emphasize the role of the OFC in representing subjective value [42], others highlight its role in representing task states [43]. The integration of reward and control processes is also a common theme across studies, with regions like the dlPFC and ACC often implicated in this integration [23, 44]. However, the exact mechanisms of this integration and the relative contributions of different regions remain areas of ongoing research.

Some discrepancies also exist in the reported functions of the VS. For instance, while some studies emphasize the role of the VS in encoding reward prediction errors, more recent work suggests it may also encode moment-to-moment fluctuations in motivation [45]. Finally, recent studies have increasingly focused on network-level analyses, highlighting the importance of interactions between brain regions and large-scale network dynamics in decision making [22, 30]. This network-based approach may help reconcile some of the apparent discrepancies in the functions attributed to individual brain regions. Future research should continue to integrate findings across different levels of analysis to develop more comprehensive models of decision-making processes.

6 Additional Factors Related to Decision Making

6.1 Conflict Monitoring

Conflict monitoring plays a crucial role in decision-making processes, particularly when individuals face choices between competing options. Recent neuroimaging studies have shed light on the neural mechanisms underlying conflict detection and resolution during decision-making tasks (Table 2). The ACC has been consistently implicated as a key region in conflict monitoring [19]. Functional magnetic resonance imaging (fMRI) studies have shown increased ACC activation during high-conflict decisions compared to low-conflict ones [46]. This activation is observed not only in perceptual tasks but also in more complex decision-making scenarios involving the integration of beliefs and preferences [44, 47].

Consistent with these results, EEG based studies have found that midfrontal theta oscillations, originating primarily from the ACC, have emerged as a reliable neural marker of conflict strength [48]. These oscillations show higher power during avoidance-avoidance conflicts compared to approach-approach conflicts, indicating that they can differentiate between various types of motivational conflicts. Importantly, this neural signature of conflict is detectable seconds before the actual motor response, suggesting an early conflict detection mechanism.

The role of the ACC in conflict monitoring extends beyond simple perceptual decisions to more complex, value-based choices. Evidence suggests that the ACC is involved in tracking the volatility of the environment [51] and in integrating multiple decision variables, including reward magnitude, probability, and effort costs [52]. Zysset et al. [46] demonstrated that the ACC is involved in multi-attribute decision making, showing increased activation when participants had to integrate multiple attributes to make a choice. Other work has expanded on this view. Shenhav et al. [19] proposed an "expected value of control" model, suggesting that the ACC not only detects conflict but also estimates the expected value of exerting

Table 2 Key brain regions and networks reported for conflict monitoring and prediction error

Brain region/network	Role in decision making	References
Anterior cingulate cortex (ACC)	Key role in conflict monitoring; signals conflict and integrates decision variables	Shenhav et al. [19]; Pochon et al. [47]
Dorsolateral prefrontal cortex (dlPFC)	Implements control mechanisms in response to detected conflicts	Shenhav et al. [19]
Midfrontal theta oscillations	Neural marker of conflict strength; indicates early conflict detection	Levy et al. [48]
Ventral tegmental area (VTA)	Signals reward prediction errors (RPEs); involved in reinforcement learning	Schultz [9]
Substantia Nigra	Part of the dopaminergic system signaling RPEs	Schultz [9]
Ventral striatum	Shows activity patterns consistent with RPE signaling; involved in value updating	Diederen et al. [8]
Orbitofrontal cortex (OFC)	Encodes state prediction errors; important for model-based learning	Wikenheiser and Schoenbaum [49]
Dorsomedial prefrontal cortex (dmPFC)	Encodes social prediction errors; relates to violations of social expectations	Joiner et al. [50]

cognitive control. This model accounts for situations where conflict is present, but the benefits of exerting control do not outweigh the costs.

Recent research has also highlighted the dynamic nature of conflict processing in the brain. Lin et al. [53] proposed that midfrontal theta might serve as a sensitive marker of continuous degrees of conflict, rather than a binary indicator. This suggests that the brain's conflict monitoring system can track subtle variations in decision difficulty.

The conflict monitoring role of the ACC is part of a broader network involved in cognitive control. According to Shenhav et al. [19], upon detecting conflict, the ACC signals other brain regions such as the dorsolateral prefrontal cortex (DLPFC) to implement control mechanisms and resolve the conflict. This interaction between conflict detection and resolution systems allows for adaptive decision-making in complex environments.

Perhaps surprisingly, the relationship between conflict-related neural activity and behavior is not always straightforward. Using EEG, Levy et al. [48] found that higher levels of midfrontal theta were associated with shorter decision times on a single-trial basis, particularly in high-conflict situations. This suggests that successful recruitment of control mechanisms, as indexed by increased theta power, can facilitate faster decision-making despite the presence of conflict.

In conclusion, recent neuroimaging research has significantly advanced our understanding of the neural underpinnings of conflict monitoring in decision making. The ACC, along with its associated midfrontal theta oscillations, plays a central role in detecting and signaling conflict. This system appears to operate across various types of decisions, from simple perceptual choices to complex value-based judgments, and interacts with other brain regions (e.g., the dlPFC) to implement adaptive control mechanisms.

6.2 Prediction Error

Prediction error, the discrepancy between expected and actual outcomes, plays a crucial role in decision-making processes. Recent neuroscientific research has significantly advanced our understanding of how the brain encodes and utilizes prediction errors to guide future decisions. The dopaminergic system, particularly the ventral tegmental area (VTA) and substantia nigra, has been consistently implicated in signaling reward prediction errors (RPEs) [9]. These midbrain dopamine neurons encode the difference between received and expected rewards, with increased firing for positive RPEs and decreased firing for negative RPEs. This dopaminergic signaling is crucial for reinforcement learning and adaptive decision-making.

In separate work, Functional magnetic resonance imaging (fMRI) studies have revealed that the ventral striatum, a key target of dopaminergic projections, shows activity patterns consistent with RPE signaling during decision-making tasks [8]. The magnitude of striatal activation correlates with the size of the prediction error, suggesting a role in value updating and learning.

Additional research has expanded our understanding beyond simple RPEs to include more complex forms of prediction errors. For instance, the ACC has been implicated in signaling unsigned prediction errors, which reflect the absolute deviation from expectations regardless of valence [54]. This ACC activity may contribute to attentional allocation and behavioral adjustment in uncertain environments.

The OFC has emerged as a critical region for representing and updating cognitive maps that guide decision-making [51]. The OFC appears to encode state prediction errors, which occur when the experienced state differs from the expected state, even in the absence of reward discrepancies. This suggests a role for the OFC in model-based learning and flexible decision-making.

Recent studies have also investigated the neural underpinnings of prediction errors in social decision-making contexts. The dorsomedial prefrontal cortex (dmPFC) has been shown to encode social prediction errors, which arise from violations of social expectations [50]. These findings suggest that similar neural mechanisms may underlie both individual and social learning processes in decision-making.

Finally, computational models, particularly reinforcement learning models, have been instrumental in formalizing how prediction errors drive learning and decision-making [43]. These models have been successfully applied to neuroimaging data, allowing researchers to identify neural correlates of specific computational processes involved in prediction error-driven learning.

In conclusion, prediction errors serve as a fundamental mechanism for learning and adaptive decision-making. The brain's ability to compute and respond to various types of prediction errors, involving a network of regions including the midbrain, striatum, ACC, OFC, and prefrontal cortex, allows for flexible behavior in complex and changing environments. Future research will likely continue to refine our understanding of how these neural systems interact to support prediction error-based decision-making across various contexts.

7 Section Summary

The ACC is consistently reported as playing a crucial role in conflict monitoring and detection across studies. However, its function extends beyond simple conflict detection to encoding unsigned prediction errors, suggesting a broader role in signaling deviations from expectations [54]. The PFC, particularly the dlPFC, is implicated in conflict resolution for decision-making. This suggests a general role for the dlPFC in high-level cognitive control and information integration processes.

The dopaminergic system and ventral striatum are consistently associated with reward prediction error signaling. However, the literature also highlights the existence of other types of prediction errors (e.g., state prediction errors in the OFC and social prediction errors in the dmPFC), suggesting a distributed network for different types of prediction error processing. The PPC's role in spatial processing and attention during decision-making adds another dimension to the decision-making network, highlighting the importance of spatial cognition in certain types of decisions.

One potential discrepancy is the level of analysis: while some studies focus on specific regions (e.g., ACC, DLPFC), others emphasize network-level interactions (e.g., fronto-parietal control network), and others focused on neurotransmitter systems (e.g., dopaminergic). This suggests a need for future research to integrate levels of analysis along with region-specific and network-level approaches to better understand decision-making processes.

In conclusion, the studies reviewed highlight the complex and distributed nature of decision-making processes in the brain. While specific regions play crucial roles in aspects like conflict monitoring and prediction error signaling, the interaction between these regions within larger networks appears to be critical for integrating various types of information to guide decisions.

8 Conclusion

The neuroscience of decision-making has made significant strides in recent years, elucidating the complex interplay of neural mechanisms underlying choice behavior. This chapter has reviewed and integrated findings from studies on fundamental decision-making processes, theoretical frameworks, and additional factors influencing choice. Several key themes have emerged from this synthesis.

First, the debate between region-specific and network-based approaches to understanding brain function in decision-making remains ongoing. While many studies report region-specific findings, there is a growing emphasis on network-level analyses. The interaction between reward processing and cognitive control, for instance, involves a complex network of brain regions, including the VS, OFC, dlPFC, and ACC. Future research should aim to directly compare these approaches to determine which provides better predictive power in explaining decision-making behaviors. Presumably, if a neural network is the primary unit underlying a certain cognitive function, then the correlation between network activation (or connectivity) and that function should be greater than the correlation between any one region and that function; if not, and one or more regions is found to correlate to a greater degree with the given function, then the cognitive structures that underly that function should be presumed to be primarily region and not network specific.

Second, despite differences and historically being studied by different groups, there is considerable convergence across theoretical frameworks such as Prospect Theory and Fuzzy Trace Theory in terms of the neural correlates underlying decision making. Both theories implicate regions like the prefrontal cortex and amygdala in decision-making processes, suggesting common neural substrates underlying different theoretical conceptualizations of choice behavior.

Third, the multifaceted nature of decision-making is evident from the diverse cognitive processes involved, including reward processing, cognitive control, conflict monitoring, and prediction error signaling. These processes are presumably supported by a distributed network of brain regions that interact dynamically to produce adaptive decision-making behavior, but as noted earlier, future work is needed to integrate levels of analysis and determine whether network-based measures provide greater predictive power.

Emerging trends in the field include an increased focus on understanding individual differences in decision-making processes and their neural correlates. Additionally, the integration of computational models, particularly reinforcement learning models, with neuroimaging data is providing more mechanistic explanations of decision-making processes.

Despite these advances, several challenges remain. Human decision making is tremendously complex and occurs in such diverse domains that building a cohesive picture of decision making in the brain that applies to most or all people remains elusive. Diverse methods of data collection (e.g., fMRI, EEG) and types of data collected (e.g., activation, functional connectivity, structural connectivity, structure size, neurotransmitter levels) add to this challenge but the potential benefits are tremendous. Future research should continue to integrate findings across different levels of analysis, from specific regions to large-scale networks. A greater number of longitudinal studies examining how decision-making processes and their neural correlates change across the lifespan could provide additional valuable insights. Ultimately, translational research applying these neuroscientific insights to real-world decision-making contexts are needed. By addressing these challenges and building on the solid foundation of research reviewed in this chapter, the field of decision neuroscience is poised to make further significant contributions to our understanding of human choice behavior. These advances have the potential to inform interventions aimed at improving decision-making in various domains, from individual health choices to public policy decisions.

References

1. Genevsky A, Yoon C (2022) Neural basis of consumer decision making and neuroforecasting. In: Kahle LR, Lowrey TM, Huber J (eds) APA handbook of consumer psychology. American Psychological Association, pp 621–635
2. Glimcher PW, Fehr E (eds) (2013) Neuroeconomics: decision making and the brain, 2nd edn. Academic Press
3. Gold JI, Shadlen MN (2007) The neural basis of decision making. Annu Rev Neurosci 30:535–574
4. Wichary S, Smolen T (2016) Neural underpinnings of decision strategy selection: a review and a theoretical model. Front Neurosci 10:500
5. Kuhberger A (1998) The influence of framing on risky decisions: a meta-analysis. Organ Behav Hum Decis Processes 75(1):23–55
6. Tversky A, Kahneman D (1979) Prospect theory: an analysis of decision under risk. Econometrica 47(2):263–291
7. Blalock SJ, Reyna VF (2016) Using fuzzy-trace theory to understand and improve health judgments, decisions, and behaviors: a literature review. Health Psychol 35(8):781–792
8. Diederen KM, Ziauddeen H, Vestergaard MD, Spencer T, Schultz W, Fletcher PC (2017) Dopamine modulates adaptive prediction error coding in the human midbrain and striatum. J Neurosci 37(7):1708–1720
9. Schultz W (2016) Dopamine reward prediction-error signalling: a two-component response. Nat Rev Neurosci 17(3):183–195
10. Delgado MR, Jou RL, Phelps EA (2011) Neural systems underlying aversive conditioning in humans with primary and secondary reinforcers. Front Neurosci 5:71
11. Setsompop K et al (2019) Pushing the limits of in vivo diffusion MRI for the human connectome project. Neuroimage 80:220–233
12. Padoa-Schioppa C, Assad JA (2006) Neurons in the orbitofrontal cortex encode economic value. Nature 441(7090):223–226
13. Howard JD, Gottfried JA, Tobler PN, Kahnt T (2015) Identity-specific coding of future rewards in the human orbitofrontal cortex. Proc Natl Acad Sci 112(16):5195–5200
14. Stalnaker TA, Cooch NK, Schoenbaum G (2015) What the orbitofrontal cortex does not do. Nat Neurosci 18(5):620–627
15. Rolls ET (2019) The orbitofrontal cortex and emotion in health and disease, including depression. Neuropsychologia 128:14–43
16. Zimmermann J, Glimcher PW, Louie K (2018) Multiple timescales of normalized value coding underlie adaptive choice behavior. Nat Commun 9(1):3206
17. Miller EK, Cohen JD (2001) An integrative theory of prefrontal cortex function. Annu Rev Neurosci 24(1):167–202

18. Hare TA, Camerer CF, Rangel A (2009) Self-control in decision-making involves modulation of the vmPFC valuation system. Science 324(5927):646–648
19. Shenhav A, Botvinick MM, Cohen JD (2013) The expected value of control: an integrative theory of anterior cingulate cortex function. Neuron 79(2):217–240
20. Botvinick MM, Braver TS, Barch DM, Carter CS, Cohen JD (2001) Conflict monitoring and cognitive control. Psychol Rev 108(3):624–652
21. Basten U, Biele G, Heekeren HR, Fiebach CJ (2010) How the brain integrates costs and benefits during decision making. Proc Natl Acad Sci 107(50):21767–21772
22. Uddin LQ (2021) Cognitive and behavioural flexibility: neural mechanisms and clinical considerations. Nat Rev Neurosci 22(3):167–179
23. Berkman ET (2018) The neuroscience of goals and behavior change. Consult Psychol J Pract Res 70(1):28–44
24. De Martino B, Kumaran D, Seymour B, Dolan RJ (2006) Frames, biases, and rational decision-making in the human brain. Science 313(5787):684–687
25. De Martino B, Camerer CF, Adolphs R (2008) Amygdala damage eliminates monetary loss aversion. Proc Natl Acad Sci 105(8):3216–3220
26. Tom SM, Fox CR, Trepel C, Poldrack RA (2007) The neural basis of loss aversion in decision-making under risk. Science 315(5811):515–518
27. Sokol-Hessner P, Camerer CF, Phelps EA (2015) Emotion regulation reduces loss aversion and decreases amygdala responses to losses. Soc Cogn Affect Neurosci 10(5):695–704
28. Canessa N et al (2017) The functional and structural neural basis of individual differences in loss aversion. J Neurosci 37(29):6745–6754
29. Dennison JB, Sazhin DB, Smith DV (2022) Integrating reward and cognitive control. Trends Cogn Sci 26(8):690–702
30. Shine JM et al (2019) Human cognition involves the dynamic integration of neural activity and neuromodulatory systems. Nat Neurosci 22(2):289–296
31. Xu P et al (2013) Neural basis of emotional decision making in trait anxiety. J Neurosci 33(47):18641–18653
32. Hsu M, Krajbich I, Zhao C, Camerer CF (2009) Neural response to reward anticipation under risk is nonlinear in probabilities. J Neurosci 29(7):2231–2237
33. Louie K, Glimcher PW, Webb R (2015) Adaptive neural coding: from biological to behavioral decision-making. Curr Opin Behav Sci 5:91–99
34. Brooks AM et al (2014) From bad to worse: striatal coding of the relative value of painful decisions. Front Neurosci 8:375
35. Reyna VF, Brainerd CJ (1995) Fuzzy-trace theory: an interim synthesis. Learn Individ Differ 7(1):1–75
36. Reyna VF, Ellis SC (1994) Fuzzy-trace theory and framing effects in children's risky decision making. Psychol Sci 5(5):275–279
37. Reyna VF et al (2011) Developmental reversals in risky decision making: Intelligence agents show larger decision biases than college students. Psychol Sci 22(4):430–434
38. Reyna VF, Weldon RB, McCormick M (2015) Educating intuition: reducing risky decisions using fuzzy-trace theory. Curr Dir Psychol Sci 24(5):392–398
39. Reyna VF et al (2018) Brain activation covaries with reported criminal behavior when making risky choices: a fuzzy-trace theory approach. J Exp Psychol Gen 147(7):1094–1109
40. Venkatraman V, Payne JW, Bettman JR, Luce MF, Huettel SA (2009) Separate neural mechanisms underlie choices and strategic preferences in risky decision making. Neuron 62(4):593–602
41. Chick CF, Reyna VF, Corbin JC (2016) Framing effects are robust to linguistic disambiguation: a critical test of contemporary theory. J Exp Psychol Learn Mem Cogn 42(2):238–256
42. Wilson RC, Takahashi YK, Schoenbaum G, Niv Y (2014) Orbitofrontal cortex as a cognitive map of task space. Neuron 81(2):267–279
43. Niv Y (2019) Learning task-state representations. Nat Neurosci 22(10):1544–1553
44. Biderman N, Bakkour A, Shohamy D (2020) What are memories for? The hippocampus bridges past experience with future decisions. Trends Cogn Sci 24(7):542–556

45. Mohebi A et al (2019) Dissociable dopamine dynamics for learning and motivation. Nature 570(7759):65–70
46. Zysset S et al (2006) The neural implementation of multi-attribute decision making: a parametric fMRI study with human subjects. Neuroimage 31(3):1380–1388
47. Pochon JB, Riis J, Sanfey AG, Nystrom LE, Cohen JD (2008) Functional imaging of decision conflict. J Neurosci 28(13):3468–3473. https://doi.org/10.1523/JNEUROSCI.4195-07.2008
48. Levy A, Enisman M, Perry A, Kleiman T (2023) Midfrontal theta as an index of conflict strength in approach--approach vs avoidance--avoidance conflicts. Soc Cogn Affect Neurosci 18(1):nsad038
49. Wikenheiser AM, Schoenbaum G (2016) Over the river, through the woods: cognitive maps in the hippocampus and orbitofrontal cortex. Nat Rev Neurosci 17(8):513–523
50. Joiner J, Piva M, Turrin C, Chang SW (2017) Social learning through prediction error in the brain. npj Sci Learn 2(1):1–9
51. Behrens TE, Woolrich MW, Walton ME, Rushworth MF (2007) Learning the value of information in an uncertain world. Nat Neurosci 10(9):1214–1221
52. Shenhav A, Straccia MA, Cohen JD, Botvinick MM (2014) Anterior cingulate engagement in a foraging context reflects choice difficulty, not foraging value. Nat Neurosci 17(9):1249–1254
53. Lin H, Saunders B, Hutcherson CA, Inzlicht M (2018) Midfrontal theta and pupil dilation parametrically track subjective conflict (but also surprise) during intertemporal choice. Neuroimage 172:838–852
54. Fouragnan E, Retzler C, Philiastides MG (2018) Separate neural representations of prediction error valence and surprise: evidence from an fMRI meta-analysis. Hum Brain Mapp 39(7):2887–2906

Michael McCormick received his Ph.D. in 2013 from the University of North Carolina at Greensboro and subsequently completed postdoctoral research positions at Cornell and Auburn Universities. His "Decision Neuroscience" research delves into the intersections of economics, health, and risky behaviors, while seeking to comprehend the intricacies of the neural network connectivity that underlies decision-making and advanced cognition. Dr. McCormick has published broadly in decision making and neuroscience journals and continues to collaborate with scholars at leading institutions. Current projects include an eye-tracking investigation of the risky-choice framing effect and an fMRI study on resting-state functional connectivity underlying the framing effect. His laboratory actively involves students interested in pursuing graduate studies in psychology, and several have successfully been accepted to competitive programs. Feel free to contact Dr. McCormick at mjmccormick@fgcu.edu.

Applications of Handedness Consistency: An Individual Difference Variable for Exploring the Neuroscience of Decision Making

5

John D. Jasper and Stephen D. Christman

Abstract

This chapter explores handedness consistency as a neuropsychological individual difference variable that predicts decision-making patterns. Consistent-handers (CH) and inconsistent-handers (ICH) differ in corpus callosum size and interhemispheric interaction, leading to predictable variations in cognitive processing. When tasks require integration of lateralized dual processes, ICH show superior performance and larger experimental effects due to increased interhemispheric communication. Conversely, when tasks demand hemispheric independence, CH demonstrate advantages. These differences manifest across various decision-making domains including risk perception, framing effects, loss aversion, and belief updating. The research reveals that ICH are more susceptible to belief updating, persuasion, and anchoring effects, while CH show greater loss aversion, discount rates, and hindsight bias under certain conditions. Practical applications include tailored public health messaging, enhanced eyewitness testimony protocols, and improved understanding of individual differences in decision-making. The findings suggest that handedness consistency serves as an accessible proxy for underlying neurological differences, offering both theoretical insights and practical applications for predicting and potentially modifying decision-making behaviors through targeted interventions.

Graphical Abstract

J. D. Jasper (✉) · S. D. Christman
University of Toledo, Toledo, USA
e-mail: john.jasper@utoledo.edu

© The Author(s), under exclusive license to Springer Nature Switzerland AG 2025

T. McElroy (ed.), *Decision-Making in Life and Work*, Integrated Science 36,
https://doi.org/10.1007/978-3-031-88415-3_5

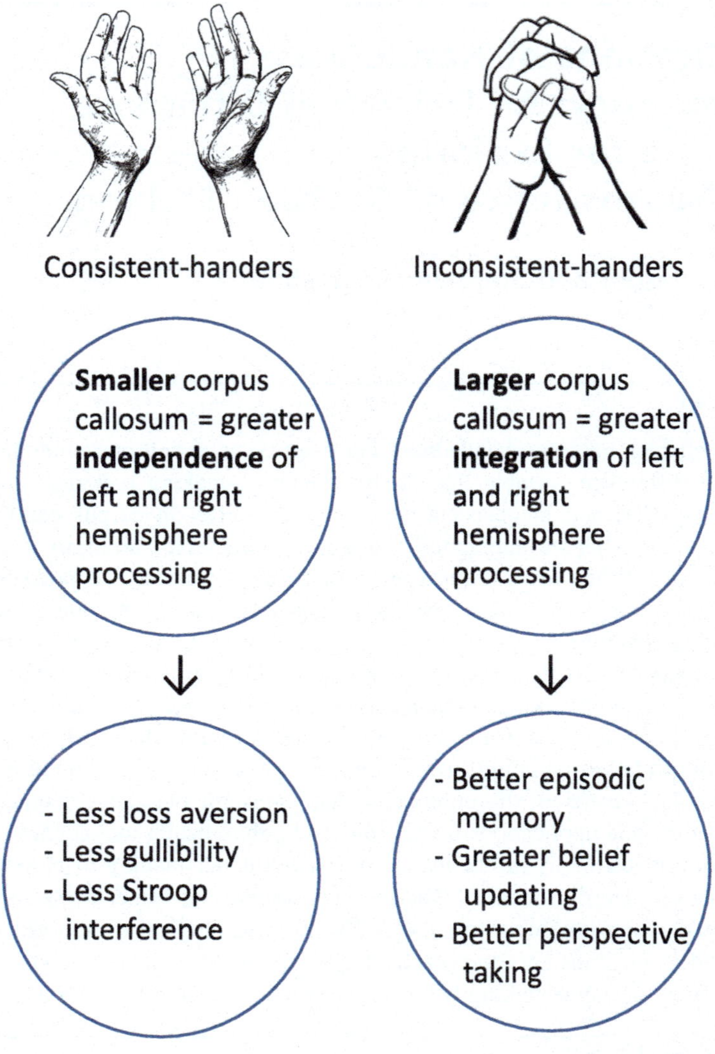

Keywords

Handedness consistency • Interhemispheric interaction • Decision making • Individual differences • Corpus callosum • Dual-process integration • Cognitive neuroscience • Framing effects • Loss aversion • Belief updating

Quote: "Hand preference has long been viewed as a troublesome variable in much research in psychology; left-handers in particular have been perceived as a noisy and unpredictable population and have often been excluded from studies."

– Scott Hardie and Lynn Wright [18]

Quote: "Handedness should join sex and age as one of the primary biological dimensions of individual differences. It hasn't yet because researchers have been comparing left- versus right-handers instead of inconsistent- versus consistent-handers."

– Stephen Christman

Traditional studies in cognitive neuroscience use methods like PET, fMRI, and ERP to explore the relationship between brain activity and various cognitive processes. While the findings from this research (some of it cited in this book) have been insightful, we take a different tact. One thing you quickly discover as a decision researcher is that not everyone exhibits various biases and effects to the same degree. For example, some are susceptible to information framing while others are not. The question is why? One way to explore this is through individual differences. In terms of framing, Smith and Levin [45], for example, were the first to identify need for cognition (NFC) as a generalizable and stable trait predictive of responses to two risky choice framing problems developed by Kahneman and Tversky [27]. Specifically, the usual framing effects were found in both problems, but only for those low on NFC, high need for cognition subjects appeared to be less susceptible to frame of reference manipulations. As for other examples, Levin and his colleagues [29, 30] and Stanovich and West [48] have demonstrated the value of using a variety of personality characteristics (e.g., neuroticism and openness to new experience) and cognitive ability, respectively, to identify the decision makers who are susceptible to these effects and those who are not. Even we have found many individual differences. For example, nonusers of sunscreen are more affected by message framing than users [26], blacks seem to show more own race bias than whites Voss et al. [54], and those lower in numeracy tend to exhibit more ratio bias than those higher in numeracy [16].

As with traditional methods of neuroscience, exploring individual differences is very useful, but it occurred to us that there might be something even more fundamental, in this case something more neurologically-based, to these differences, and that something might be how one's brain is wired. As most know, the brain hemispheres are specialized for different tasks or at least different ways of (or different components involved in) doing the same task. What fewer people probably appreciate is that sometimes cognitive tasks rely on one or the other hemisphere to accomplish them while at other times cognitive tasks require the integration of both hemispheres. For instance, brain imaging evidence indicates that episodic memories are preferentially encoded by left prefrontal regions, while subsequent retrieval is handled by right prefrontal areas (e.g., Babiloni et al. [1], Habib et al. [17], Tulving et al. [52]). Thus, episodic retrieval is dependent on the integration of left-lateralized encoding and right-lateralized retrieval processes. In contrast, to perform well in a traditional Stroop task requires separation (or independence) between the left hemisphere which processes the reading of the word (e.g., red) while the right hemisphere processes the naming of the color (e.g., green). And, of course, this creates conflict in an individual's mind, thus exhibiting what's known as the Stroop effect.

What one also might not know is that some individuals are predisposed to combining (or integrating) the hemispheres while others are predisposed to keeping them separate (what we call independence). Is this good or bad? It depends on the individual and on the task. Regarding individuals, stronger degrees of handedness are associated with smaller corpus callosum size [31], thus, strong- versus mixed—or what is now referred to as consistent—versus inconsistent-handedness (CH vs. ICH) is associated with decreased versus increased interhemispheric interaction (e.g., Prichard et al. [39]). And in terms of tasks, some require the two hemispheres to be integrated, while others require the two hemispheres to be kept independent. For example, when it comes to episodic memory, mixed- or inconsistent-handers (ICH, those who use their non-dominant hand for at least some tasks) have the upper hand, given their larger corpus callosum which encourages cross-talk (or process integration) between the two hemispheres. Specifically, research shows that inconsistent-handers have superior episodic retrieval of both lab-based [32] and real-world memories [40]. But those same individuals are at a disadvantage when it comes to the verbal Stroop task which requires separation (or process independence). Here, strong- or consistent-handers (CH, those who use their dominant hand for all tasks) outperform inconsistent-handers, given their smaller corpus callosum and reduced cross-talk (and subsequent conflict) between the RH and LH.

This is where our almost twenty years of research on handedness and decision making comes in. The first thing we can definitively conclude is that consistency (or strength) of handedness is a reliable proxy (or outside marker) of one's neurological predisposition to integrate the left and right brain hemispheres. That means we can, with pretty good accuracy, predict who will show (or drive) an effect and who will not. We have outlined this elsewhere in Jasper et al. [21], but in sum, when a task requires lateralized, dual processes to be integrated (like episodic memory) it is routinely the inconsistent-handers who are more affected by experimental manipulations, while for tasks involving mutually exclusive (independent) dual processes (like the Stroop task) it is the consistent-handers who are more affected by experimental manipulations. These effects typically manifest themselves in the form of interactions between handedness and the particular manipulation or sometimes even other individual difference variables (e.g., Jasper and Clarkson [22]). Listed below are a few of the things we have found, separated by the nature of the dual-process task.

1 Integrated Dual-Tasks Where ICH Do Better and/or Show Bigger Effects

A. **Integration of LH encoding and RH retrieval**
- ICH have better source memory [14]

- ICH have better episodic memory for lab-based and real-world events [36, 37, 40]
- ICH have an earlier offset of childhood amnesia [13]
- ICH demonstrate decreased semantic clustering during verbal fluency tasks [47]

B. **Integration of LH belief maintenance and RH belief updating**
- ICH exhibit equivalent priming for dominant and subordinate word meanings [46]
- ICH have a greater appreciation for/understanding of paradoxical concepts than CH [35]
- ICH are more open to persuasion than CH [10]
- ICH are more 'open-eared' [6]
- ICH are more likely than CH to update from creationist to evolutionary beliefs [34]
- ICH exhibit fewer eating disorder symptoms and have more accurate body image than CH [7]
- ICH show more behavioral change than CH in response to cognitive dissonance [24]
- ICH counterfactualize more than CH [19]
- ICH show larger anchoring effects if one uses "relevant" (informative) anchors; the findings reverse if the anchors are traditional, random anchors [20]
- ICH show greater Stroop interference effects than CH [5], Capizzi et al. 3
- ICH are more susceptible to the fake hand illusion [33]
- ICH show bigger attribute framing effects than CH [23, 49]
- ICH are more affected by placebos than CH [9]

C. **Integration of LH self-directed representations and RH other-directed representations**
- ICH show smaller above and below average effects in the Dunning-Kreuger paradigm [28, 42]
- CH weight social comparative risk (whether their risks are above or below average) less heavily, whereas ICH weight absolute risk (whether their risks are high or low) less heavily [43]
- Relative to descriptive norms (what others do), ICH are less sensitive to prescriptive norms (what others *should* do) [28]
- ICH are more likely to forgive and accept apologies [50]
- ICH show smaller mere ownership effects [25]

2 Independent Dual-Tasks Where CH Do Better and/or Show Larger Effects

A. **Independence of LH approach-behavior and RH withdrawal behavior**
- CH tend to be less loss averse than ICH [11]
- CH show smaller sunk cost and inaction inertia effects than ICH [56]

- CH show smaller endowment effects than ICH [25]
- CH show more status quo in negative frames; ICH show more status quo in positive frames [55]
- CH show larger discount rates than ICH [57]
- CH are more influenced by potential benefits, ICH are more influenced by potential risks [26]

B. **Independence of LH verbal processing and RH spatial/temporal processing**
- CH show lesser Stroop interference effects than ICH [5], Capizzi et al. 3
- CH perceive distant events (e.g., 25 years away) as further away than ICH [57]

C. **Independence of LH control of the right hand and RH control of the left hand**
- CH are more likely to play keyboard instruments, which have separate right- and left-hand parts [4]

So, you might ask, what does this have to do with decision making, particularly from a practical standpoint? The fact that many decision biases can be predicted by handedness suggests that including handedness in one's studies may prove fruitful in explaining otherwise unaccounted for variance. Even for researchers who are not interested in individual differences, including handedness as a factor in analyses will likely lead to variability moving from the error term to an effect term, thereby providing greater power to detect other effects of interest. At best, though, including strength of handedness as a variable could shed light on theoretical issues concerning how different components of a task that are lateralized to opposite hemispheres interact, since such components are more likely to operate independently in consistent handers.

To an individual, these findings also hold promise. Take, for instance, someone who has relatively poor episodic memory. While degree of handedness is an outward proxy of an inward trait relating to amount of connection between the LH and RH, the degree of interhemispheric connection (one's state) can be temporarily altered, thereby changing one's responses. For example, episodic memory performance can be enhanced by having participants move their eyes saccadically for as little as 30 s (which promotes increased interhemispheric activation) [8, 14, 32]. A more naturalistic and practical way of increasing eye movements would be to reduce the column width of print text, as seen in most newspapers [15, 38]. In decision making, this could affect one's susceptibility to something like hindsight bias. Our research (Bhattacharya and Jasper [2], see Fig. 1) shows that CH typically show more of this bias under the context of what's called the "memory design" (where one is asked to give an initial estimate—to be remembered later). We predict that increasing the connection between the LH and RH (using eye movements or reduced column width) will significantly reduce the bias (particularly in CH), while the same eye movement manipulation in the more popular "hypothetical design" context (where one is never asked for an initial estimate) will enlarge the bias (particularly in ICH, since it involves a form of belief updating, in this case, anchoring and adjustment).

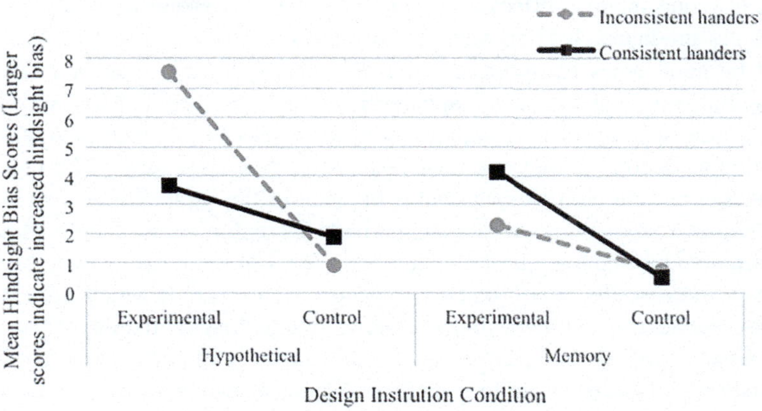

Fig. 1 ICH show more hindsight bias under hypothetical design instructions, while CH show more hindsight bias under memory design instructions [2]

Our research also holds tantalizing possibilities for public policy makers. For example, perhaps one wants to reduce the public's direct exposure to the sun, thereby reducing the rates of skin cancer, saving lives and perhaps government money in the process. One way to do that would be to encourage greater use of sunscreen. Our research [26], though, indicates that a one-size fits all message, likely, will not work, at least not for everyone. Instead, one needs to identify sub-groups of decision makers according to how they process information and then tailor different ads or messages to each of these separate groups, which shouldn't be as difficult today (given various computer technologies) as it was in years past. Specifically, inconsistent-handers appear to be more persuaded by the negative frame or the goal of avoiding a negative consequence or loss, consistent with reports of greater baseline right-hemisphere activation (associated with greater withdrawal tendencies and loss-aversion) in inconsistent-handers [11, 56]. And in the same study, consistent-handers appeared to be nominally more persuaded by the positive frame, consistent with reports of greater baseline left-hemisphere activation (associated with greater approach tendencies).

In a more recent example, with real world implications, our research group found that states in the U.S. with relatively large percentages of ICH were more sensitive to calls for mask wearing and vaccination during the COVID-19 pandemic than were states with relatively large percentages of CH [12]. Given the "if it bleeds, it leads" philosophy behind much mass media news coverage, it is possible that media coverage emphasizing the risks of *not* wearing masks and *not* getting vaccinated had more effect on ICH. By the same reasoning, if media coverage had also emphasized the benefits of wearing masks and getting vaccinated, perhaps there would have been better compliance in the states dominated by CH.

Finally, we have brand new (currently unpublished) data exploring handedness differences in something referred to as own race bias. This bias describes the tendency that people have to more accurately identify and remember the faces of racial in-group members (ie, one's own race) than out-group members (ie, other races). As predicted, ICH showed less of an own-race bias than CH. While this might be good news for those studying this bias (with the hope of reducing or even eliminating it altogether), it apparently comes with a price. More specifically, ICH showed a reduced bias and sometimes no bias at all, they also showed a reduction in accuracy for same race targets (ie, their own race). The take away here might be a bit controversial, but if the goal of the legal system is to increase the accuracy of eyewitness testimony, perhaps law enforcement should administer a one-minute handedness questionnaire and select CH same race witnesses. While this may increase the "bias" that some want to eliminate, it would also serve to increase the accuracy of eyewitness identification as long as the targets are of the same (one's own) race. Future research, of course, is required to determine the effectiveness of such an action within an actual eyewitness context before any firm recommendations can be made, as it could backfire. For example, it's easy to imagine the situation where a lawyer might argue that a person's testimony is inaccurate simply because of their race or handedness, which may not be true.

3 Limitations & Final Thoughts

As interesting as some of these findings are, we'll end by pointing out three things to keep in mind. First, some of these effects are temporary. For instance, while one may be able to enhance episodic memory in CH via saccadic eye movements (in fact, even alternating left and right hand clenches or ball squeezing works! [41], it's not likely to be a permanent enhancement (although research has never tested this). Thus, some of the things we mentioned might be viewed more appropriately as nudges [51]. Second, more of something (in this case, interhemispheric activation) can be a good thing, but only to a point. As with arousal [58], you don't want too little or too much, the sweet spot is somewhere in the middle, ie, with a moderate level of activation [41]. Third, these effects can be enhanced with other variables/manipulations as long as they align. For example, the relevant textual message in the sunscreen study cited earlier was presented in either blue/purple or red/orange ink in a second, related study [53]. Overall, a match between the representations of the target behavior, the color, and the frame resulted in the greatest amount of persuasion. The take-away message is that behavioral compliance can be increased, but the "planets must align" as we like to say, Seta et al. [44] come to a similar conclusion by arguing that many decision-making effects are grounded in priming but formally refer to "planets aligning" as unification. In this case, those who have greater sensitivity to positive information need to be given the critical information in a positive frame and in a positive color. To do otherwise, will weaken and perhaps even cancel the effects.

References

1. Babiloni C, Vecchio F, Cappa S, Pasqualetti P, Rossi S, Miniussi C, Rossini PM (2006) Functional frontoparietal connectivity during encoding and retrieval processes follows HERA model: a high-resolution study. Brain Res Bull 68:203–212
2. Bhattacharya C, Jasper JD (2018) Degree of handedness: a unique individual differences factor for predicting and understanding hindsight bias. Personality Individ Differ 125:97–101
3. Capizzi M, Ambrosini E, Vallesi A (2017) Individual differences in verbal and spatial Stroop tasks: interactive role of handedness and domain. Front Human Neurosci 11
4. Christman SD (1993) Handedness in musicians: bimanual constraints on performance. Brain Cogn 22:266–272
5. Christman SD (2001) Individual differences in Stroop and local-global processing: a possible role of interhemispheric interaction. Brain Cogn 45:97–118
6. Christman SD (2013) Handedness and 'earedness': strong right-handers are less likely to prefer obscure musical genres. Psychol Music 41:89–96
7. Christman SD, Bentle M, Niebauer CL (2007) Handedness differences in body image distortion and eating disorder symptomatology. Int J Eat Disord 40:247–256
8. Christman SD, Garvey KJ, Propper RE, Phaneuf KA (2003) Bilateral eye movements enhance the retrieval of episodic memories. Neuropsychology 17:221–229
9. Christman SD, Geers AL, Kosbab K, Weiland PE (2006) Placebos and belief updating: strong right-handers do not show placebo effects. In: Presented at the 35th annual meeting of the international neuropsychological society, Boston
10. Christman SD, Henning BR, Geers AL, Propper RE, Niebauer CL (2008) Mixed -handed persons are more easily persuaded and are more gullible: interhemispheric interaction and belief updating. Laterality 13(5):403–426
11. Christman SD, Jasper JD, Sontam V, Cooil B (2007) Individual differences in risk perception versus risk taking: handedness and interhemispheric interaction. Brain Cogn 63:51–58
12. Christman SD, Prichard EC (2023) Righty-tighty, lefty-loosey: relation between societal tightness scores, left-handedness rates, and COVID-19 outcomes in US states. Laterality 28:285–304
13. Christman SD, Propper RE, Brown TJ (2006) Increased interhemispheric interaction is associated with earlier offset of childhood amnesia. Neuropsychology 20:336–345
14. Christman SD, Propper RE, Dion A (2004) Increased interhemispheric interaction is associated with decreased false memories in a verbal converging semantic associates paradigm. Brain Cogn 56:313–319
15. Christman SD, Sabry D (2009) Eye movements and interhemispheric interaction: newspapers may be more persuasive than books. In: Presented at the 21st annual convention of the association for psychological science, San Francisco
16. Corser R, Voss Jr RP, Jasper JD (2024) Do errors on classic decision biases happen fast or slow? Numeracy and decision time predict probability matching, sample size neglect, and ratio bias. J Num Cogn 10, e12473
17. Habib R, Nyberg L, Tulving E (2003) Hemispheric asymmetries of memory: the HERA model revisited. Trends Cogn Sci 7:241–245
18. Hardie SM, Wright L (2014) Differences between left- and right-handers in approach/avoidance motivation: influence of consistency of handedness measures. Front Psychol 5, 134
19. Jasper JD, Barry K, Christman SD (2008) Individual differences in counterfactual production. Personality Individ Differ 45:488–492
20. Jasper JD, Christman SD (2005) A neuropsychological dimension for anchoring effects. J Behav Decis Mak 18:343–369
21. Jasper JD, Christman SD, Clarkson E (2021) Predicting interactions in handedness research: the role of integrated versus independent dual processes. Laterality 26(6):607–623

22. Jasper JD, Clarkson E (2024) Degree of handedness (a proxy for interhemispheric interaction) interacts with religiosity to predict responses to moral dilemmas. Manuscript submitted for publication
23. Jasper JD, Fournier C, Christman SD (2014) Handedness differences in information framing. Brain Cogn 84:85–89
24. Jasper JD, Prothero M, Christman SD (2009) I'm not sexist!!! Cognitive dissonance and the differing cries of mixed- and strong-handers. Personality Individ Differ 47:268–272
25. Jasper JD, Sanchez K, Corser R, Christman SD (2024) A reversed endowment effect for objects of little value. Unpublished manuscript
26. Jasper JD, Woolf, Christman SD (2014) Responding to framed health messages: different strokes for different (handedness) folks. Psychol Health 29:671–686
27. Kahneman D, Tversky A (1984) Choices, values, and frames. Am Psychol 39:341–350
28. Lanning MD, Christman SD, Fugget A (2023) Social comparison and handedness: mixed handers are less susceptible to egocentric biases in judgments about others' performance. Personality Individ Differ 205:1–4
29. Lauriola M, Levin JP (2001) Personality traits and risky decision-making in a controlled experimental task; an exploratory study. Personality Individ Differ 31:215–226
30. Levin LP, Gaeth GJ, Schreiber J, Lauriola M (2002) A new look at framing effects: distribution of effect sizes, individual differences, and independence of types of effects. Organ Behav Hum Decis Process 88:411–429
31. Luders E, Cherbuin N, Thompson PM, Gutman B, Anstey KJ, Sachdev P, Toga AW (2010) When more is less: associations between corpus callosum size and handedness lateralization. Neuroimage 52:43–49
32. Lyle KB, Hanaver-Torrez SD, Hacklander RP, Edlin JM (2012) Consistency of handedness, regardless of direction, predicts baseline memory accuracy, and potential for memory enhancement. J Exp Psychol Learn Mem Cogn 38(1):187–193
33. Niebauer CL, Aselage J, Schutte C (2002) Interhemispheric interaction and consciousness: degree of handedness predicts the intensity of a sensory illusion. Laterality 7:85–96
34. Niebauer CL, Christman SD, Reid SA, Garvey K (2004) Interhemispheric interaction and beliefs on our origin: degree of handedness predicts beliefs in creationism versus evolution. Laterality 9:433–447
35. Niebauer CL, Garvey K (2004) Gödel, Escher, and degree of handedness: differences in interhemispheric interaction predict differences in understanding self-reference. Laterality: Asymmetries Body, Brain Cogn 9(1):19–34
36. Parker A, Dagnall N (2010) Effects of handedness and saccadic bilateral eye movements on components of autobiographical recollection. Brain Cogn 73(2):93–101
37. Prichard EC, Christman SD (2017) Inconsistent-handed advantage in episodic memory extends to paragraph-level materials. Memory 25:1063–1071
38. Prichard EC, Christman SD (2021) Memory effects of manipulating text column width: eye-movement induced attentional processes interfere with prose encoding among consistent handers. Percept Mot Skills 128:560–577
39. Prichard E, Propper RE, Christman SD (2013) Degree of handedness, but not direction, is a systematic predictor of cognitive performance. Front Psychol 4:1–6
40. Propper ER, Christman SD, Phaneuf KA (2005) A mixed-handed advantage in episodic memory: a possible role of interhemispheric interaction. Mem Cognit 33(4):751–757
41. Prunier S, Christman SD, Jasper JD (2018) The effects of varying levels of hemispheric activation on episodic memory. Laterality 23(4):409–421
42. Rose JP, Jasper JD, Corser R (2012) Interhemispheric interaction and egocentrism: the role of handedness in social comparative judgement. Br J Soc Psychol 51(1):111–129
43. Rose JP, Nagel B (2013) Relation between comparative risk, absolute risk, and worry: the role of handedness strength. J Health Psychol 18:866–874
44. Seta JJ, Seta CE, McCormick M (2017) Commonalities and differences among frames: a unification model. J Behav Decis Mak 30:1113–1130

45. Smith SM, Levin IP (1996) Need for cognition and choice framing effects. J Behav Decis Mak 9:283–290
46. Sontam V, Christman SD (2012) Semantic organization and handedness: mixed-handedness is associated with more diffuse activation of ambiguous word associates. Laterality 17:38–50
47. Sontam V, Christman SD, Jasper JD (2009) Individual differences in semantic switching flexibility: effects of handedness. J Int Neuropsychol Soc 15:1023–1027
48. Stanovich KE, West RF (1998) Individual differences in rational thought. J Exp Psychol Gen 127:161–188
49. Stein JM (2012) Framing effects: the influence of handedness and access to right hemisphere processing. Laterality 17(1):98–110
50. Stein JM, Miller NS (2019) Individual differences in apology acceptance and forgiveness: the influence of handedness. Laterality 24:559–581
51. Thaler RH, Sunstein CR (2008) Nudge: improving decisions about health, wealth, and happiness. Yale University Press, New Haven, CT
52. Tulving E, Kapur S, Craik FI, Moscovitch M, Houle S (1994) Hemispheric encoding/retrieval encoding asymmetry in episodic memory: positron emission tomography findings. Proc Nat Acad Sci USA 91:2016–2020
53. Voss R, Corser R, McCormick M, Jasper JD (2018) Influencing health decision making: a study of color and message framing. Psychol Health 33(7):941–954
54. Voss RP, Prunier S, Corser RJ, Jasper JD (2024) The effects of handedness on the identification of own- and cross-race faces. Manuscript submitted for publication
55. Westfall JE, Corser R, Jasper JD (2014) Handedness predicts motivation to switch from the status quo. Laterality 19(6):745–764
56. Westfall J, Jasper JD, Christman SD (2012) Inaction inertia, the sunk cost effect, and handedness: avoiding the losses of past decisions. Brain Cogn 80:192–200
57. Westfall JE, Jasper JD, Zelmanova Y (2010) Differences in time perception as a function of strength of handedness. Personality Individ Differ 49(6):629–633
58. Yerkes RM, Dodson JD (1908) The relationship of strength of stimulus to rapidity of habit-formation. J Comp Neurol Psychol 18:459–482

Half-Decided: Bilateral or Unilateral Stimulation and Susceptibility to Persuasion

6

Ruth E. Propper, Magen Daniela Teasley, and Emily Lynch

Social scientists have determined that we accept inner responsibility for a behavior when we think we have chosen to perform it in the absence of strong outside pressures.

Robert B. Cialdini, *Influence: The Psychology of Persuasion*

I'm not in love, but I'm open to persuasion.

Boy George

Abstract

The current chapter examines two interrelated questions: (1) Can physical manipulation of the body directly or via sensory input, alter susceptibility to persuasion? (2) Can these manipulations be done without the awareness of the persuaded? These issues are considered in the context of known and proposed neurophysiological substrates involved in persuasion, and evidence from several different motor and sensory domains is provided. Specifically, research is reviewed supporting the hypothesis that increased access to right hemispheric processing, either via callosally-mediated interhemispheric interaction or via hemispheric activation, results in increased susceptibility to persuasive

R. E. Propper (✉) · M. D. Teasley · E. Lynch
Department of Psychology, Montclair State University, Montclair, NJ, USA
e-mail: propperr@montclair.edu

M. D. Teasley
e-mail: propperr@mail.montclair.edu

E. Lynch
e-mail: lynche2@montclair.edu

© The Author(s), under exclusive license to Springer Nature Switzerland AG 2025

T. McElroy (ed.), *Decision-Making in Life and Work*, Integrated Science 36,
https://doi.org/10.1007/978-3-031-88415-3_6

messages. A range of body movements and sensory experiences, including unilateral nostril breathing, unilateral gaze, unilateral hand clenching, bilateral eye movements, unilateral or bilateral auditory presentation, and body position are examined for their potential to alter hemispheric interaction and/or activity and to change performance on tasks across various domains such as memory, creativity, the placebo effect, and other cognitive processes. Direct examinations of the impact of these manipulations on susceptibility to persuasion are few, but of those, auditory stimulation in the form of selective frequency amplification may be most promising. Practical relevance for consumer behavior and potential clinical applications in stroke victims are discussed, as are directions for future research. It is concluded that the answers to both questions posed is 'Yes' and that future research would be valuable to further delineate the conditions under which these manipulations are effective, as well as for determining the best methods of presentation.

Graphical Abstract/Art Performance

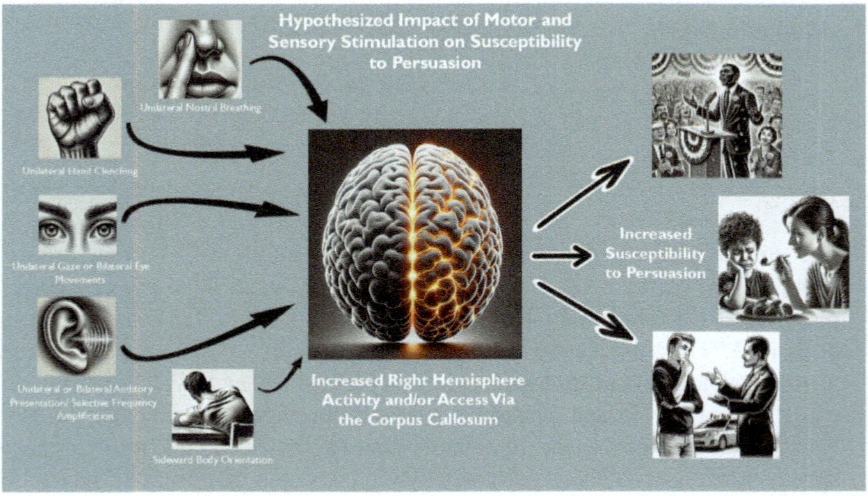

Impact of stimulation on persuasion

Keywords

Hemisphere • Handedness • Lateralization • Persuasion • Bilateral eye movements • Corpus callosum

1 Introduction

Robert Cialdini is arguably one of the most *influential* scientists in the study of persuasion. His book, *Influence: The Psychology of Persuasion*, captivated the public, selling millions of copies, and his ideas stimulated thousands of studies by social scientists. In *Influence* Cialdini outlines seven principals of persuasion: reciprocation, liking, social proof, authority, scarcity, commitment/consistency, and unity. Undoubtedly, these methods impact recipients' decision making via biasing information processing without conscious awareness [1].

Certainly there have been many other books and scientific articles written as well that address how to persuade- that is, how to get someone else to do what you want them to do, or to believe what you want them to believe. Indeed, *How to Win Friends and Influence People* [2], one of the earliest self-help books including ways to persuade, has been re-issued multiple times and has been listed among the *Time's* 100 most influential books. The popularity of the subject is a testament to how much we, as humans, want to influence others, and to the importance of persuasion in everyday life. From parents who try to convince their children to eat vegetables, to politicians who want you to vote for them, the need to persuade others to act or think in a particular way is arguably woven into the fabric of society itself- so much so in fact that there has been controversy over the persuasive influence internet advertisers wield, having honed the ability to market products so well, their persuasion has been deemed by some to be unethical and intrusive.

As the quote by Cialdini above suggests, knowing that someone is trying to persuade us reduces our sense of 'ownership' of the subsequent behavior. Ownership of our choices is important for both internal and external reasons; internally, choice ownership can increase sense of self-efficacy and positive affect about a decision. Externally, ownership can help with behavioral change maintenance (e.g. eating health foods, avoiding nicotine) over time. Processes that provide the- even illusory- experience that a choice is 'voluntary' are likely to be more successful over the long and short term, than are 'forced' choices.

While Cialdini discusses persuasion in decision making from a cognitive perspective- that is, how to change someone's mind by impacting the various cognitive and perceptual biases to which humans are likely to fall victim- the focus of this current chapter is on manipulations of the motor or sensory systems to impact decision making. The question is: Can physical manipulation, of the body directly or via sensory input, alter susceptibility to persuasion? Secondarily, can this be done without the awareness of the persuaded? Unfortunately for Boy George we won't be able to directly determine if one can be persuaded to be in love, though to the extent how we determine how we feel about a person is similar to how we think about arguments for/against various topics, then love may indeed be persuadable. Hopefully we will have some suggestions one way or the other by the end of this chapter.

Ultimately, the orientation here stems from at least two main ideas: (i) persuasion has a neuroanatomical/physiological basis and (ii) manipulating motor or sensory systems can effect those neurophysiological processes, thereby impacting the ability to be persuaded. The sections below outline the logic and support for these points.

2 Brain Bases of Persuasion: Hemispheric Asymmetry in Persuadibility

With regard to the brain bases of susceptibility to persuasion, the attention here is on general hemispheric asymmetries. There is evidence that the two halves of the brain differ in the ease with which they can be persuaded. Specifically, there is strong evidence that the right hemisphere is more persuadable than the left. Ramachandran [3, 4], in work with stroke patients suffering from right hemisphere damage resulting in unilateral neglect and anosognosia for left-sided paralysis, suggested that the right hemisphere serves as an 'anomaly detector' while the left hemisphere tends to favor 'the status quo'. In the case of paralysis anosognosia, the previous 'status quo'—that movement is possible—is maintained due to right hemisphere damage, resulting in a failure of belief updating. Accordingly, the ability to update one's beliefs in the face of new evidence is thought to result from an interaction between the cerebral hemispheres such that the right hemisphere's processing is favored. Ramachandran reported that increasing the activity of the right hemisphere temporarily reduced or eliminated the anosognosia, supporting this interpretation. More recent work supports this distinction between the hemispheres in individuals with anosognosia [5].

Other work further bolsters this analyses of the right hemisphere being more likely to update its beliefs in the face of new evidence- of being more persuadable- relative to the more conservative, stable, left hemisphere. For example, research with left versus right temporal lobectomy patients indicates that left lobectomy vs right lobectomy patients switch hypotheses frequently even when told they are correct, vs maintaining their hypotheses despite being told they incorrect [6]. Additional work with individuals with brain damage indicates that a damaged right hemisphere leads to increased perseveration on the Wisconsin Card Sort and on fluency tasks, and decreased mental set switching. Together these examinations indicate that belief updating vs belief maintenance may be biased toward right vs left hemisphere processing, respectively.

Furthermore, increased access to these right hemisphere processes is predicted to increase the likelihood of 'changing one's mind'. That is, in the face of new evidence that may be contradictory to beliefs currently held, right hemisphere processes, via either increased hemispheric interaction or a bias to process information via the right hemisphere, should be associated with increased susceptibility to persuasion, given the right hemisphere's penchant for belief updating.

Several lines of investigation support this interpretation. For example, individual differences in handedness can be used to indicate individual differences in interhemispheric interaction and access to right hemisphere processes. First, a note about handedness is in order, before additional discussion. Although traditionally handedness has been examined via comparing left- versus right-handers (a comparison which is between *directions* of hand-use preference), more contemporary research suggests that individual differences as a function of handedness may be more related to individual differences in the *consistency, or degree* of hand use preference. Individuals who are consistent in their hand use preference are generally defined as those who report *always* using their preferred hand across approximately 80% of items on various checklists, with the Edinburgh Handedness Inventory [7], being frequently used [8]. People who report that they *sometimes* use their preferred hand, or who report frequently preferring either, or neither hand, are considered inconsistently-handed. It is important to note that consistently-left handed people are exceedingly rare, comprising only about 1–2% of the population [9]. Therefore, in many studies these individuals are either excluded from analyses or placed with those who are consistently-right-handed, thereby collapsing across handedness direction (left versus right) altogether, and resulting in comparisons between consistent and inconsistent handedness groups.

Multiple lines of research have supported the existence of neuroanatomical and related functional differences between the consistently- and inconsistently-handed. Of relevance here, the inconsistently-handed have a larger corpus callosum, the major tract of fibers that connects the hemispheric cortices [10], increased right, compared with left, hemisphere activation, and a right hemispheric bias in interhemispheric asymmetry [11]. These physical differences have been proposed to underlie the behavioral and cognitive disparities between these handedness groups across such wide-ranging domains as memory, sleep, musical preferences, and various personality factors [8]. Ultimately, greater interhemispheric communication between the left and right halves of the brain via the corpus callosum, either in tandem with, or resulting in, increased right hemisphere access and presumably right hemisphere task dominance, in inconsistently- versus consistently-handed individuals, has been proposed to result in the behavioral differences between groups.

Supporting the hypotheses that increased access to right hemisphere processes is associated with increased susceptibility to persuasion, inconsistent-handers are more persuadable than are consistent-handers. For example, Christman et al. [12] reported that inconsistent handers were more likely to change their beliefs to be more positive about comprehensive college final exams following *either* weak or strong arguments, indicating increased belief updating, or persuadibility, in these individuals. Similar findings have been reported across other paradigms [12, 13]. For example, Christman et al. reported increased 'gullibility' in inconsistent-handers, such that they were more likely to agree that the output of a (false) personality questionnaire was accurate, than were consistent-handers. Inconsistent-handers also demonstrate an increased susceptibility to the placebo effect [14]. Though these latter two lines of research do not use persuasive messaging per se it is reasonable to suggest that susceptibility to the Barnum Effect (likelihood of agreeing with false statements about oneself) and the placebo effect, both require 'agreement' with a previously un-held belief. Overall, inconsistent-handers, potentially via their increased access to right hemisphere processing, demonstrate increased belief updating and susceptibility to persuasion.

Other work has altered hemispheric activity using electrical or magnetic stimulation in order to more directly examine hemisphere-belief updating relationships. For example, Schulreich and Schwabe [15], using transcranial direct current stimulation (tCDS), reported that increased right hemisphere dorsolateral prefrontal cortex activity was associated with increased belief updating. Mengotti, Dombert, Fink, and Vossel [16], demonstrated that transcranial magnetic stimulation (TMS), used to disrupt activity within the right temporoparietal junction, prevented the updating of beliefs about the location of a cue following false information. In a somewhat more socially-oriented paradigm, Sharot et al. [17] used TMS to increase acceptance of 'bad news'. Sharot et al. exploited previous findings that individuals are more likely to 'update their beliefs' and alter their behavior following good, but not bad, news. Participants were asked to estimate how likely negative/positive outcomes were, for example, their likelihood of getting cancer or of being robbed. Following TMS to either the left or right inferior frontal gyrus, participants were presented with the actual likelihood of these events and asked if they would like to revise their previous estimate. People were more likely to re-estimate their likelihood of 'bad' events when the left hemisphere was deactivated, demonstrating a right-hemisphere modulation of belief updating. No such impact on revision of estimates following 'good news' was demonstrated. It may be that estimates of 'good' information are more consistent with a person's typical belief system, representing a maintenance of the status quo, and therefore do not require updating beliefs substantially.

Taken together, these results indicate that the right hemisphere is involved in belief updating, and that increased reliance on, or access to, right hemisphere processes increases changes in beliefs. Given that persuasion, definitionally, requires belief change, these findings suggest that increased right hemisphere access or activity is associated with increased susceptibility to persuasion.

3 Motor/Sensory Manipulation to Alter Hemispheric Activity

With the above in mind, it makes sense then that manipulations designed to alter hemispheric activity might ultimately alter the experience of lateralized cerebral functions, including susceptibility to persuasion. In fact, several studies have demonstrated that simple unilateral or bilateral motor movements, or presentation of stimuli, result in alterations in cognition, emotion, and task performance. Importantly, these subsequent alterations of performance following motor/sensory stimulation are in line with known differences is asymmetrically lateralized processes, and presumably result in changes in brain activity.

For example, Jella and Shannahoff-Khalsa [18] examined the impact of unilateral nostril breathing on verbal versus spatial task performance. Left nostril breathing increased mental rotation performance, while right nostril breathing increased performance of a verbal analogy task. These results were interpreted as being consistent with known lateralization of left hemisphere language, and right hemisphere spatial, hemispheric lateralization, suggesting changes in hemispheric activity. Schiffer, Anderson, and Teicher [19] used specially altered glasses to restrict vision to one versus the other visual field. Glasses limiting vision to the left visual field increased anxiety, and altered hemispheric activity as measured via electroencephalograph (EEG) and tympanic membrane temperature. Given a right hemisphere bias in negative affect, these results are supportive of the effectiveness of unilateral stimuli to alter hemispheric activity and emotion. Using similar glasses, Propper, Brunyé, Christman, and Januszewski [20] reported superior memory for semantic information, the location and names of the 50 United States, as a function of rightward versus leftward gaze, a result interpreted as indicative of left-hemisphere verbal/semantic memory bias.

Unilateral hand clenching has been investigated relatively frequently as a method by which experience and task performance can be altered as a result of changing hemispheric activity. In a review of 24 articles examining the impacts of hand clenching, Schierenbeck and Propper [21] report that hand clenching changes cognition across a wide range of domains, including memory, local–global processing, social decision making, creativity, and emotion. For example, Goldstein et al. [22] examined creativity, measured via performance on the remote associates task (RAT) following left versus right unilateral hand clenching. Results demonstrated superior RAT scores following left hand clenching, in line with right hemisphere bias toward divergent and wider associative connections. Research using EEG has supported that unilateral hand clenching increases activity in the contralateral cortex [23], however not all physiological measures are in agreement with the physiological mechanisms responsible [24].

Bilateral eye movements have also been used to alter cognition and emotion, and studies using physiological measures support that these eye movements alter hemispheric activity [25], ultimately resulting in the changes in experience and task performance consistent with typical lateralization of hemispheric function. Eye movements have been shown to result in increased hemispheric activity in the contralateral hemisphere [26]; presumably leftward-rightward repeated eye movements, such as those used in many of the research paradigms here, result in sustained contralateral hemispheric activity. The idea then, is that this increased and ostensibly equalized activity across the cerebral hemispheres allows for greater accessibility of right hemisphere processes and greater interhemispheric communication [27].

It is notable that Propper et al. [25] reported a change in frontal interhemispheric gamma activity pursuant to bilateral eye movements, indicating a change in hemispheric communication. Interestingly, the findings were in the opposite direction of that predicted, with bilateral eye movements reducing interhemispheric coherence in this EEG band. One suggestion is that the neurophysiological correlates of behavior are more complex than typically assumed; for example, functionally lateralized behaviors might not necessarily demonstrate physiological or structural lateralization. Bilateral eye movements decreasing interhemispheric coherence could result in decreased functional lateralization via a reduction in excitatory inhibition. Conversely, increased interhemispheric interaction might increase callosally-mediated inhibition, thereby decreasing functional lateralization. Generally, bilateral eye movements produce the same, or similar, changes in behavior and cognition as those found in inconsistently-handed individuals (discussed above), further supporting the proposed neurophysiological mechanisms (though other work does not [28–30].

Christman et al. [27] had participants engage in leftward-rightward eye movements prior to episodic memory tasks (memory for a word list and for personal real-world events). These bilateral eye movements increased episodic, but not semantic or implicit memory, findings interpreted as indicative of a reliance of episodic memory on interhemispheric, and of semantic/implicit memory on unilateral hemispheric processes. Bilateral eye movements have also been found to increase memory across a wide range of stimuli and tasks, including eye-witness testimony type information [31], age of earliest memory [32], and spatial information [33]. Using the RAT, Shobe et al. [34] reported some measures of increased creativity in consistently-right-handed people following bilateral eye movements, a result that, when integrated with that of unilateral hand clenching impacts, suggests a greater right-hemisphere accessibility following such eye movements. Xu et al. [35] report decreased emotional response to trauma films following bilateral, but not static, eye movements.

The take-home here is that unilateral or bilateral movements, or sensory information presentation, can result in changes in cognition, emotion, and behavior, with these changes consistent with known hemispheric differences in functional lateralization (note that this differs from structural or neurophysiological lateralization, which may not necessarily be neatly mapped directly onto function). Given the working hypothesis in this chapter that the right hemisphere is more easily persuaded than the left, findings suggest that similar situations or stimuli might be able to alter susceptibility to persuasion. And so, the remainder of this chapter asks, and hopefully answers: Can unilateral or bilateral movements/sensory stimulation alter susceptibility to persuasion? The next several sections discuss this possibility.

4 Motor/Sensory Manipulation to Alter Susceptibility to Persuasion

We examined literature for studies using motor or sensory stimulation, including that which was identical or similar to that outlined above, to examine persuasion, and we also discuss other similar- yet under-investigated- methods as well. We have found a limited number of representative research articles, and organize the remainder of this section by type of stimuli used. Note that the research included here is not comprehensive, given the many methods used to persuade, paradigms to measure persuasion, and the myriad underlying neural structures that have been the focus of research.

4.1 Unilateral Nostril Breathing

To our knowledge, there have been no studies examining the impacts of unilateral nostril breathing on susceptibility to persuasion. However, given that previous research supports that such breathing can alter cognition consistent with typical hemispheric lateralization of cortical functions [18], it seems reasonable that such a technique might be able to alter persuadibility. Additionally, it is interesting to note that across the day, typically one nostril is in fact more 'open' or 'closed' at any given moment in time. Furthermore, which nostril is open/closed cycles rhythmically throughout the day in tandem with hemispheric activity. Of interest then might be the question: Does natural cyclical variation in uni-nostril breathing result in circadian variation in susceptibility to persuasion? Other research might address the impact of various nasal sprays on susceptibility to persuasion. For example, nasal sprays that unclog one nostril may selectively increase hemispheric activity in the ipsilateral hemisphere; could it be that nasal spray of the right nostril would then increase persuadibility, while spraying the left nostril might make someone less persuadable?

4.2 Unilateral Gaze

As noted above, Schiffer's group [19, 36, 37] has been active examining the impact of unilateral gaze on cognition, emotion, and brain activity. While their work supports hemispheric and cognitive changes in response to such stimuli, no work has examined susceptibility to persuasion per se. It is interesting to note however that such unilateral gaze has been used to alter clinical conditions, including depression [38], and that later work suggests the possibility that unilateral stimulation (directly with 'photobiomodulation') may hold promise as a method for decreasing addictive drug use [39], arguably a behavior for which change may require persuasion. Given that some aspects of addictive behavior may be susceptible to modification using unilateral gaze, determination if these impacts may be due to changes in persuasion susceptibility might be of interest.

4.3 Unilateral Hand Clenching

A comprehensive literature review [21] of the impacts of unilateral hand clenching on behavior, cognition, and emotion revealed that such movements can under some circumstances in fact alter experiences and performance in these domains. The most frequently used method involved 45 s of unilateral hand clenching, followed by a 15 s rest, followed by another 45 s of clenching, with these movements resulting in changes to hemispheric activity. However, no work explicitly examining persuasion was found. Another similar line of research has examined unilateral arm flexion (considered an approach movement) and extension (considered a withdrawal movement) to examine whether these actions impact hemispheric activity and cognition [40–42]. They report impacts of motor movements on judgements of attractiveness, likelihood of cancer, and other characteristics, interpreted in part as being consistent with hemispheric lateralization of withdrawal versus approach emotional states. Unfortunately, they have not, to our knowledge, used their technique to investigate persuasion (perhaps a member of their laboratory will read this chapter and decide to do so!).

4.4 Bilateral Eye Movements

A group led by Dijkstra [43–45] has examined the impact of eye movements on persuasive messaging for healthy lifestyle choices, particularly involving fruit and vegetable intake. Although their primary hypothesis- that voluntary eye movements disrupt working memory and prevent counter-arguments to persuasive messages- does not provide neurobiological/neurophysiological explanation, the methodology used closely (though not exactly) replicates the bilateral eye movements used in the above studies of cognitive/emotional change and so likely involves the same or similar neural substrates. Additionally, the stimuli they use- leftward-rightward eye movements at the speed of 4 movements per second- was explicitly chosen to be similar to that used in the therapy of Eye Movement Desensitization and Reprocessing (EMDR) [46, 47], which is also similar to the designs used in those other studies of bilateral eye movements, described within this chapter earlier (which was at the rate of 2 eye movements per second). Across studies, viewing a moving dot while listening to a negatively framed persuasive message about the harms that can come from not eating enough fruits and vegetables resulted in intent to change behavior [43] or in increased actual self-reported ingestion [44, 45].

Several caveats are in order here regarding these results and their implications for the use of bilateral eye movements in persuasion. First, although Dijkstra et al. report impacts of bilateral eye movements on persuasion, these effects were moderated by several variables, including gender, health status, 'cognitive self-affirmation inclination', and actual self-affirmation. Such results would not necessarily be expected if the neurobiological/neurophysiological model of persuasion proposed previously (e.g., that increased right hemisphere activity or access increases susceptibility to persuasion) was applied to these results. That is, there is currently no framework that would account for differences in the impact of bilateral eye movements on persuasibility as a function of 'cognitive self-affirmation inclination', for example.

However, it is interesting to note that Dijkstra and Elbert [45] found that men were more susceptible to the effects of bilateral eye movements on persuasion than were women. Men tend to be more hemispherically functionally lateralized than women [48], and have a smaller corpus callosum [49, 50], which is presumably associated with decreased interhemispheric interaction and accessibility to right hemisphere processes [51]. For example, Ardekani et al. [50] examined brain Magnetic Resonance Images (MRI), of men and women and reported, when controlling for intercranial capacity, smaller corpus callosum in men relative to women, across ages. Similarly, Ingalhalikar et al. [51] using Diffusion Tensor Imaging (DTI), reported significantly lower interhemispheric connectivity and higher intrahemispheric connections in men compared to women. Thus, it may be that if there is an optimal level of hemispheric communication/right hemisphere access for persuasion, then those individuals who are below that threshold would benefit (be more likely to be susceptible to persuasive messaging) from eye movements. Therefore, bilateral eye movements would be expected to be more impactful on men compared to women [52].

This argument is similar to that discussed in research examining the effects of bilateral eye movements on performance of right- and non-right-handers [28, 53]. For example, with regard to memory, bilateral eye movements increase memory performance in right-, but not in non-right-handed individuals [31], with the logic being that bilateral eye movements in non-right-handers cause a move away from optimal hemispheric functioning due to their typically high trait levels of interhemispheric interaction/right hemisphere access [52]. Parker et al. [31] had participants watch a video of bank robberies, and reported that bilateral eye movements enhanced recognition task performance in right, but not non-right-handed participants. Lyle et al. [53] examined bilateral eye movements prior to a word memory assessment of a list of words, finding that bilateral eye movements negatively impacted recall and recognition in non-right handers, in comparison to their right-handed counterparts, in which retrieval was enhanced. Thus, it could be that women, with higher state interhemispheric interaction, are already in the 'optimal' zone for persuasive messaging and therefore are not impacted positively by bilateral eye movements.

Interestingly, Dijkstra et al. examined only negatively framed persuasive messages. There is evidence that negatively framed health messages are more likely to be persuasive when the right hemisphere is particularly active [54]. Again, perhaps it is an optimal level of activation that is required to demonstrate an effect of eye movements on persuasion, resulting in the mediating/moderating effects found by Dijkstra and company. These suggestions are purely speculative, but lend themselves to empirical examination via, for example physiological measures recorded concurrently with persuasive messaging and bilateral eye movements.

On the other hand, it is possible that rather than producing too much hemispheric activity or interaction, the opposite took place, with the stimuli not resulting in much change. This is an alternative, second possibility: Specifically, the eye movements used by Dijkstra's lab involved following a stimulus continuously, thereby producing smooth pursuit eye movements. Smooth pursuit eye movements differ substantially from those used in other studies described above, which require saccadic eye movements, in hemispheric activation [55]. Importantly, activation of the frontal eye fields was greater, as measured via positron emission tomography (PET) during saccadic, compared to smooth pursuit, eye movements. These findings suggest that Dijkstra and colleagues [43–45] may have underestimated the potential impact that eye movements can have on persuasive messaging by having participants employ eye movements that provide lower neurophysiological influence; it may be that the other conditions wherein no effect or a moderating effect of variables was found was due to limited, relatively speaking, change in brain activity in response to the moving stimulus.

In sum, Dijkstra et al. report an impact of bilateral eye movements on persuadibility. Future research could use saccadic eye movements, rather than smooth pursuit movements, to investigate whether or not these effects might be more robust than reported in these studies.

4.5 Unilateral or Bilateral Auditory Presentation and Body Position

At least two different groups have used unilateral or bilateral presentation of auditory information to alter susceptibility to persuasion with or without body position alteration. The first is from Drake [56–58], who presented arguments for and against student-relevant topics (discounted ski passes for students, the lengthening or shortening of classes to enable a week between the end of classes and final exams) to either the left or the right ear of college student participants. Importantly, participants were asked to orient their head/body toward the side of stimuli presentation, in addition to the unilateral auditory presentation. The goal was to alter hemispheric activity via attentional allocation in order to alter susceptibility to persuasion, rather than to use unilateral auditory information as the method of causing asymmetrical activation. Drake and Bingham [58] auditorily and unilaterally presented college student participants with arguments against the discounting of ski tickets for students, while having their heads turned toward the side to which the information was being presented. Participants laterally oriented toward the left (increasing right hemisphere activity) reported agreeing more with the message than did participants oriented toward the right (increasing left hemisphere activity). Drake [56] presented participants with arguments for and against the inclusion of a study week prior to final exams. Arguments were presented to the left or right ear, or bilaterally, while body position was oriented away from the ear of presentation. Participants recalled more arguments against their position when they were presented to the left ear/right hemisphere and more arguments with which they agreed when they were presented to the right ear/left hemisphere. The same pattern of results was also found for recognition of arguments, though those results were not significant. Drake suggests that these findings reflect the lateralized nature of emotional experience along with hemispheric specialization for cognitive processes involved in persuasion. Other work [57] using the same stimuli presentation techniques resulted in increased belief in the relevance of information with which one agrees, and increased discounting of information with which one disagrees, when the right vs the left hemisphere was selectively targeted for increased activation via auditory and physical orientation.

Second, the use of sound itself to alter hemispheric activity, rather than relying on attentional direction, to examine susceptibility to persuasion, also demonstrates promise. A novel, centrally presented, auditory manipulation, developed by McCormick and Seta [59], Selective Frequency Amplification (SFA), takes advantage of hemispheric differences in processing relatively low versus high frequencies. McCormick and Seta amplified the relatively high (2080–4000 Hz) versus low frequencies (80–2080 Hz) of a voice in an audio message filtered to between 80 and 4000 Hz in order to enhance left versus right hemispheric processing, respectively. SFA altered the influence of a message's context on decision making, with right hemisphere processing (due to a low frequency message amplification) being associated with framing effects [59]. It is important to make clear that participants were unaware that the audio had been manipulated, and that the audio manipulation altered decision making without participants' awareness. In their study, they had participants rate the appeal of meat labeled either 85% lean or 15% fat. An impact of framing based on anchor was found when the right hemisphere was (assumedly) more active than the left. These results suggest that the right hemisphere is more likely to be influenced by messages, and by message context, than the left.

Consistent with this research, work from the first author's lab indicates that centrally presented information that has been frequency altered can impact cognition. We presented participants with an altered classical music score that had been high, low, or no frequency amplified. Prior to and during stimulus presentation, individuals rated their mood using the Brief Mood Introspection Scale. Results revealed greater change in mood pre- to concurrent-stimulus in individuals exposed to relatively lower frequency amplification (increased right hemisphere activity), compared to those exposed to non-amplified frequencies. These results again suggest the right hemisphere is more malleable than is the left (see Fig. 1, and Implications/Application section). Similarly, Gallagher and Dagenbach [60] and Seta et al. [61] altered the frequency with which information was presented about the "Asian Disease Problem" and reported a framing effect with low frequency presentation, which presumably biased participants for right hemisphere information processing, but not with high frequency, left hemisphere biasing.

Although SFA and similar technologies have not been used to directly examine the impact of a persuasive message, the fact that the right hemisphere is susceptible to context effects, such as anchor, while the left hemisphere may not be, suggests that factors influencing decision making- such as persuasive arguments- may be more effective when presented to the right hemisphere via SFA. Additionally, presumably this change in decision making is the result of changes in hemispheric activity, and can occur without explicit awareness of the individual being manipulated.

At a practical level, the implications for SFA in real-world situations are extensive. For example, SFA in background music played at political venues, or SFA altering narrator voiceovers in commercials, might directly impact likelihood of agreement with the messaging.

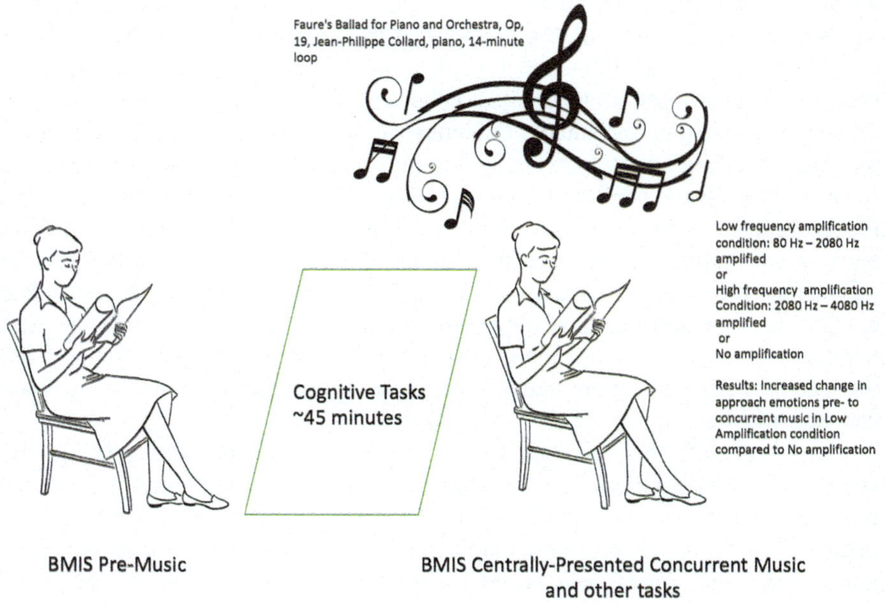

Fig. 1 Schematic of protocol using high, low, and no frequency amplification to examine mood

5 Implications/Applications

This survey of the research examining the impact of unilateral or bilateral motor movements/sensory stimulation on susceptibility to persuasion indicates that such methods can alter decision making. In particular, bilateral eye movements, unilateral presentation of auditory information with body position/attentional allocation consistent with proposed hemispheric activation, and central presentation of auditory, digitally altered information designed to take advantage of left hemisphere versus right hemisphere preference for high versus low frequencies may be able to increase or decrease persuasion.

Regarding the quotes at the beginning of this chapter, is it possible to persuade—without awareness—using any of these methods? Would such attempts at persuasion work, and if so, would the persuaded attribute any change of mind to internal evaluation or to external pressure? We can imagine several scenarios in fact where the methods here could be used to surreptitiously change someone's mind, though clearly more research is necessary.

First, in the domain of advertising, we can think of at least two different formats that could be used to persuade. We can imagine advertising that presents information with enhanced low frequencies. Such presentation could, without the listener's awareness, increase the likelihood of positive evaluation of the advertised item. As discussed briefly above, in a recent (unpublished) study from this chapter's first author's lab [62], participants were provided with centrally presented classical music, Faure's Ballad for Piano and Orchestra, Op, 19, Jean-Philippe Collard, piano (auditory stimulus downloaded from Youtube, https://www.youtube.com/watch?v=tZcKii9Sd-o) that had been low, high, or not frequency amplified, following the protocol of McCormick and Seta [59]. Participants completed the Brief Mood Introspection Scale (BMIS) [63] twice; once prior to, and once during presentation of, and listened to the music, while engaging in other tasks. Although mood did not change as a function of amplification (it was hypothesized that high frequency/left hemisphere and low frequency/right hemisphere amplification would result in increased approach vs withdrawal mood, respectively), there was greater change in approach mood from pre- to concurrent music when the music was low frequency amplified (right hemisphere processing) compared with non-amplification (see Fig. 1).

Though not examining persuasive messages *per* se, these results again indicate that the right hemisphere is more likely to change its perspective in response to digitally altered auditory, central stimuli than is the left hemisphere. Notable was that participants experienced the impacts of the music even while not explicitly focusing on it. In theory then, the ubiquitous background music present in advertising could be modified, potentially resulting in more favorable opinions of the information presented. Certainly research in this domain is warranted (we have no comments about the ethics involved in such manipulations, though we would like to point out that many methods of persuasion are *already* being used in advertising to encourage people to behave in particular ways. It is not clear how or if such sound manipulations are more or less ethical than these other methods. We leave that question for others to consider). Whether or not such auditory manipulations in the home setting could impact the persuasiveness of a message enough to result in actual behavioral change requires additional study, and practical issues, such as the level of background noise and sound quality of various home electronics, would likely need to be considered as well.

Another potential way to subtly manipulate might be via presenting information to the left, and then then to the right, side of an electronic screen, following the typical protocols (e.g. a left–right eye movement every second for 30 s). We can imagine a scenario where such movements would arise 'naturally' from the visual information presented, and with it integrated with the item- or perspective- being 'sold', for example. Again though, practical issues would impact the utility here, including distance from the screen (which impacts visual angle and the 'length' of the eye movements), which in turn would impact the neurophysiological mechanisms involved (for better or worse, depending on the situation). Other issues involve the necessary sustaining of attention, that may or may not occur in the home or other non-laboratory locations, and individual differences in handedness or sex that could impact hemispheric activity in response to bilateral eye movements. Again, future research using these methods under ecological conditions would be useful.

In our day to day lives, we do find ourselves in situations where our body positions, and the sound to which we are paying attention, are lateralized. Is it possible under real-world conditions that we are more or less likely to be persuaded? One such situation occurs when driving with a passenger. The driver is (hopefully only slightly) attending to the right (left hemisphere) of a speaking passenger, thereby potentially increasing left hemisphere activity, and resulting in less susceptibility to persuasive messaging. The passenger, on the other hand, attends toward the left side of space (right hemisphere) toward a speaking driver, potentially resulting in increased likelihood of being persuaded. Thus, in the 'real world' it may be beneficial to be 'in the driver's seat' in order to increase the chance that a passenger will agree with a particular point of view/argument. Conversely, if you want to convince someone of something, presenting an argument to a driver may be less helpful.

As a side note, there are many other basic avenues for future research. For example, as noted above, although unilateral nostril breathing and unilateral gaze have been shown to alter cognition and emotion they have not, to our knowledge, been investigated in the context of persuasion. Extension of unilateral hand clenching to the persuasion domain as well could be worthwhile. See Table 1 for general overview of stimuli and potential directions for future research.

Additionally, given that one impetus for the hypotheses here stems from work by Ramachandran [3, 4] on individuals with anosognosia, one practical application could potentially be for the treatment of this condition. In fact, Ramachandran [4] alludes to the use of bilateral eye movements for the updating of the belief of paralysis in individuals who are injured by stroke. In his description, he uses a variation of the caloric reflex test, whereby cold water is slowly injected into one or the other ear. Doing so results in a form of nystagmus, causing repeated rapid eye deflection to the side away from the infused ear. The nystagmus was postulated by Ramachandran to potentially 'wake up' the non-accessible, damaged, right hemisphere, enabling belief updating, and allowing anosognosics to recognize their own illness temporarily. We would add that this nystagmus is similar to the bilateral eye movements reviewed within this chapter, and therefore increased interhemispheric communication via the bilateral, leftward-rightward eye movements. Future work could determine if it is nystagmus per se or the resultant eye movements that impact anosognosia in these patients. Unfortunately, as per Ramachandran [4], the effects lasted only about an hour. Its not known if repeated use of this technique could be beneficial longer term, though certainly additional research is warranted.

Although unilateral movements and body position would not be appropriate for this population (due to the paralysis), digital amplification of low frequencies in music, or even in messaging, could potentially be used to 'persuade' anosognosics of the existence of their own illness. Whether or not such manipulations could be beneficial in other hemispherically lateralized damage is not clear either. For example, individual suffering from unilateral neglect ignore the left side of the world, dressing one half their bodies, eating half (the right) of the food on their plates, or brushing only the right half of their hair. Thought to be caused by right hemisphere damage to parietal areas, could frequency manipulated music reduce or even eliminate this hemispheric neglect? Would it be possible to combine methods (such as music and body position, or hand clenching) to increase the potential therapeutic effects? Additional work could test these impacts in clinical populations.

Table 1 General overview of stimuli and directions for future research

Stimuli	Investigated persuasion?	General Findings	Suggestions for Future Work	Can impact without awareness? Other factors?	Representative References
Unilateral Nostril Breathing	Not Investigated	Increased contralateral hemisphere-specific task performance- verbal vs spatial tasks	Nostril open/closed cycles throughout the day in tandem with hemispheric activity; does natural cyclical variation result in circadian variation in susceptibility to persuasion?	Unlikely. Other factors, such as allergies, illness, septum deviations, etc., might impact influence on effectiveness	Jella, S. A., & Shannahoff-Khalsa, D. S. (1993). The effects of unilateral forced nostril breathing on cognitive performance. *International Journal of Neuroscience*, *73* (1–2), 61–68. https://doi.org/10.3109/00207459308987211
Unilateral Gaze	Not Investigated	Alteration in emotion, brain activity, and memory consistent with theories of hemispheric processing	Some aspects of addictive behavior may be susceptible to modification using unilateral gaze; determination if these impacts may be due to changes in persuasion susceptibility	No	Propper, R. E., Brunyé, T. T., Christman, D., & Januszewski, A. (2012). Look over there! Sustained unilateral gaze increases geographical memory for the 50 United States. *Brain and Cognition*, *78*, 59–62. Schiffer, F., Anderson, C. M., & Teicher, M. H. (1999). Electroencephalogram, bilateral ear temperature, and affect changes induced by lateral visual field stimulation. *Comprehensive Psychiatry*, *40* (3), 221–225. https://doi.org/10.1016/s0010-440x(99)90007-x Schiffer, F., Mottaghy, F. M., Pandey Vimal, R. L., Renshaw, P. F., Cowan, R., Pascual-Leone, A., Teicher, M., Valente, E., & Rohan, M. (2004). Lateral visual field stimulation reveals extrastriate cortical activation in the contralateral hemisphere: An fMRI study. *Psychiatry Research*, *131* (1), 1–9. https://doi.org/10.1016/j.pscychresns.2004.01.003 Schiffer, F., & Teicher, M. H. (1997). Affect changes observed with right versus left lateral visual field stimulation in psychotherapy patients: Possible physiological, psychological, and therapeutic implications. *Comprehensive Psychiatry*, *38* (5), 289–295. https://doi.org/10.1016/s0010-440x(97)90062-6

(continued)

Table 1 (continued)

Stimuli	Investigated persuasion?	General Findings	Suggestions for Future Work	Can impact without awareness? Other factors?	Representative References
Unilateral Hand Clenching	Not Investigated	Physiological activity, memory, emotion, creativity consistent with theories of hemispheric lateralization; flexion and extension associated with changes in judgements	Extend paradigms to persuasive messaging protocols	Unlikely without decption	Cretenet, J., & Dru, V. (2005). Influence of bilateral motor actions on judgments: Additional evidence for the motor congruence hypothesis. *Laterality*, *10* (5), 389–398. https://doi.org/10.1080/13576500500143314 Cretenet, J., Mullet, E., & Dru, V. (2015). Motor and cognitive integration: effect of bilateral behaviors on judgment. *Acta psychologica, 161,* 64–72. Goldstein, A., Revivo, K., Kreitler, M., & Metuki, N. (2010). Unilateral muscle contractions enhance creative thinking. *Psychonomic Bulletin & Review*, *17* (6), 895–899. https://doi.org/10.3758/PBR.17.6.895 Mullet, E., Cretenet, J., & Dru, V. (2014). Motor influences on judgment: Motor and cognitive integration. *British Journal of Psychology, 105* (1), 69–91. https://doi.org/10.1111/bjop.12022 Schierenbeck, S., & Propper, R. E. (2022). Integrative research review of the impacts of unilateral hand clenching on behavior: Clinical implications. In C. Pracana & M. Wang (Eds.), Impact: Psychological Applications and Trends (pp. 230-234). Lisbon, Portugal: World Institute for Advanced Research and Science
Bilateral Eye Movement	Yes	Increased persuasiveness of health-related messages	Extend research to other persuasive messages, delineate effectiveness of variations in protocol	Potentially. Distance from stimuli, attentional factors impact effectiveness	Dijkstra, A., & Elbert, S. (2019). Eye movement inductions influence health behaviour: the working memory account of persuasion. *Psychology & Health*, *34* (11), 1378–1394. https://doi.org/10.1080/08870446.2019.1609678 Dijkstra, A., & Elbert, S. P. (2021). Detecting and preventing defensive reactions toward persuasive information on fruit and vegetable consumption using induced eye movements. *Frontiers in psychology*, *11*, 578287. https://doi.org/10.3389/fpsyg.2020.578287 Dijkstra, A., & van Asten, R. (2014). The eye movement desensitization and reprocessing procedure prevents defensive processing in health persuasion. *Health Communication*, *29* (6), 542–551. https://doi.org/10.1080/10410236.2013.779558 Lyle, K. B., Logan, J. M., & Roediger, H. L., 3rd (2008). Eye movements enhance memory for individuals who are strongly right-handed and harm it for individuals who are not. *Psychonomic Bulletin & Review*, *15* (3), 515–520 https://doi.org/10.3758/pbr.15.3.515 Parker, A., & Dagnall, N. (2012). Effects of saccadic bilateral eye movements on memory in children and adults: an exploratory study. *Brain and Cognition*, *78* (3), 238–247. https://doi.org/10.1016/j.bandc.2012.01.007 Shobe, E. R., Ross, N. M., & Fleck, J. I. (2009). Influence of handedness and bilateral eye movements on creativity. Brain and Cognition, 71(3), 204–214.

(continued)

Table 1 (continued)

Stimuli	Investigated persuasion?	General Findings	Suggestions for Future Work	Can impact without awareness? Other factors?	Representative References
Frequency Altered Auditory Presentation	Yes, decision making, emotion	Right hemisphere especially likely to change in response to stimuli	Explicitly examine persuasive arguments with auditory frequencies augmented, use neurophysiological measures	Yes	Gallagher, P., & Dagenbach, D. (2007). Manipulating noise frequencies alters hemispheric contributions to decision making. *Brain and Cognition*, *64* (1), 42–49. https://doi.org/10.1016/j.bandc.2006.10.004 McCormick, M., & Seta, J. J. (2012). A new method for selectively enhancing hemisphere processing: voice frequency amplification influences the strength of attribute framing. *Laterality*, *17* (6), 727–735. https://doi.org/10.1080/1357650X.2011.626559 Seta, J. J., McCormick, M., Gallagher, P., McElroy, T., & Seta, C. E. (2010). Voice frequency impacts hemispheric processing of attribute frames. *Journal of Experimental Social Psychology*, *46* (6), 1089–1092. https://doi.org/10.1016/j.jesp.2010.06.009
Unilateral Auditory Presentation With Body Orientation	Yes	Increased agreement and recall of/with persuasive messages	Examine other domains, such as health-related messaging, timing of any actual changes, use neurophysiological measures	Yes	Drake, R. A. (1991). Processing persuasive arguments: Recall and recognition as a function of agreement and manipulated activation asymmetry. *Brain and Cognition*, *15* (1), 83–94. https://doi.org/10.1016/0278-2626(91)90017-3 Drake, R. A. (1993). Processing Persuasive Arguments: 2. Discounting of Truth and Relevance as a Function of Agreement and Manipulated Activation Asymmetry. *Journal of Research in Personality*. *27* (2), 184–196. doi:10.1006/jrpe.1993.1013 Drake, R. A., & Bingham, B. R. (1985). Induced lateral orientation and persuasibility. *Brain and Cognition*, *4* (2), 156–164. https://doi.org/10.1016/0278-2626(85)90067-3

Lastly, while the main focus here has been on *increasing* susceptibility to persuasion, we want to point out that the methods discussed (with the possible exception of bilateral eye movements) lend themselves also to being used to *decrease* persuadability. There are likely many instances where a person would like to prevent themselves from being persuaded, such as when making a large purchase that involves negotiations (e.g.; buying a car). Such a use could also possibly benefit those in military situations wherein adversaries attempt to 'recruit' someone to their cause (though of course there are many other factors that influence susceptibility to persuasion in those instances in particular).

6 Conclusion

To summarize, it does appear that external stimuli or body movements can differentially impact hemispheric processing and therefore can potentially alter a recipient's susceptibility to persuasion, possibly without awareness. Additional research is necessary in order to further delineate the conditions under which such enhanced persuasiveness might occur, as well as those situations that might be more likely to be useful in an applied setting. To paraphrase Boy George as quoted at the beginning of this chapter, you may be surprisingly open to persuasion, and more so than you might think.

References

1. Cialdini RB (2007) Influence: the psychology of persuasion. Harper Collins, New York
2. Carnegie D (1936) How to win friends and influence people (1st edn). Simon & Schuster, New York
3. Ramachandran VS (1995) Anosognosia in parietal lobe syndrome. Conscious Cogn 4(1):22–51. https://doi.org/10.1006/ccog.1995.1002
4. Ramachandran VS, Blakeslee S (1999) Phantoms in the brain: probing the mysteries of the human mind. William Morrow, New York
5. Kirsch LP, Mathys C, Papadaki C, Talelli P, Friston K, Moro V, Fotopoulou A (2021) Updating beliefs beyond the here-and-now: the counter-factual self in anosognosia for hemiplegia. Brain Commun 3(2), fcab098. https://doi.org/10.1093/braincomms/fcab098
6. Rausch R (1977) Cognitive strategies in patients with unilateral temporal lobe excisions.Neuropsychologia 15:385–395
7. Oldfield R (1971) The assessment and analysis of handedness: the Edinburgh inventory. Neuropsychologia 9:97–113
8. Prichard E, Propper RE, Christman SD (2013) Degree of handedness, but not direction, is a systematic predictor of cognitive performance. Front Cognit 4(9). https://doi.org/10.3389/fpsyg.2013.00009
9. Lansky LM, Feinstein H, Peterson JM (1988) Demography of handedness in two samples of randomly selected adults (N = 2083). Neuropsychologia 26:465–477. https://doi.org/10.1016/0028-3932(88)90099-1
10. Luders E, Cherbuin N, Thompson PM, Gutman B, Anstey KJ, Sachdev P, Toga AW (2010) When more is less: associations between corpus callosum size and handedness lateralization. Neuroimage 52(1):43–49. https://doi.org/10.1016/j.neuroimage.2010.04.016

11. Propper RE, Pierce J, Geisler MW, Christman SD, Bellorado N (2012) Asymmetry in resting alpha activity: effects of handedness. Open J Med Psychol 1(4):86–90. https://doi.org/10.4236/ojmp.2012.14014
12. Christman SD, Henning BR, Geers AL, Propper RE, Niebauer CL (2008) Mixed-handed persons are more easily persuaded and are more gullible: Interhemispheric interaction and belief updating. Laterality 13(5):403–426. https://doi.org/10.1080/13576500802079646
13. Christman SD, Propper RE, Dion A (2004) Increased interhemispheric interaction is associated with decreased false memories in a verbal converging semantic associates paradigm. Brain Cogn 56(3):313–319. https://doi.org/10.1016/j.bandc.2004.08.005
14. Christman SD, Geers AL, Kosbab K, Weiland PE (2006) Placebos and belief updating: strong right-handers do not show placebo effects. In: Presented at the 35th annual meeting of the international neuropsychologial society, Boston
15. Schulreich S, Schwabe L (2021) Causal role of the dorsolateral prefrontal cortex in belief updating under uncertainty. Cereb Cortex 31(1):184–200. https://doi.org/10.1093/cercor/bhaa219
16. Mengotti P, Dombert PL, Fink GR, Vossel S (2017) Disruption of the right temporoparietal junction impairs probabilistic belief updating. J Neurosci 37(22):5419–5428. https://doi.org/10.1523/JNEUROSCI.3683-16.2017
17. Sharot T, Kanai R, Marston D, Korn CW, Rees G, Dolan RJ (2012) Selectively altering belief formation in the human brain. Proc Natl Acad Sci 109(42):17058–17062. https://doi.org/10.1073/pnas.1205828109
18. Jella SA, Shannahoff-Khalsa DS (1993) The effects of unilateral forced nostril breathing on cognitive performance. Int J Neurosci 73(1–2):61–68. https://doi.org/10.3109/00207459308987211
19. Schiffer F, Anderson CM, Teicher MH (1999) Electroencephalogram, bilateral ear temperature, and affect changes induced by lateral visual field stimulation. Compr Psychiatry 40(3):221–225. https://doi.org/10.1016/s0010-440x(99)90007-x
20. Propper RE, Brunyé TT, Christman D, Januszewski A (2012) Look over there! Sustained unilateral gaze increases geographical memory for the 50 United States. Brain Cogn 78:59–62
21. Schierenbeck S, Propper RE (2022) Integrative research review of the impacts of unilateral hand clenching on behavior: clinical implications. In: Pracana C, Wang M (eds) Inpact: psychological applications and trends. World Institute for Advanced Research and Science, Lisbon, Portugal, pp 230–234
22. Goldstein A, Revivo K, Kreitler M, Metuki N (2010) Unilateral muscle contractions enhance creative thinking. Psychon Bull Rev 17(6):895–899. https://doi.org/10.3758/PBR.17.6.895
23. Harmon-Jones E (2006) Unilateral right-hand contractions cause contralateral alpha power suppression and approach motivational affective experience. Psychophysiology 43(6):598–603. https://doi.org/10.1111/j.1469-8986.2006.00465.x
24. Propper RE, Dodd K, Christman SD, Brunye TT (2017) Relationship between sustained unilateral hand clench, emotional state, line bisection performance, and prefrontal cortical activity: a functional near-infrared spectroscopy study. Laterality 22(6):671–689. https://doi.org/10.1080/1357650X.2016.1268148
25. Propper RE, Pierce J, Geisler MW, Christman SD, Bellorado N (2007) Effect of bilateral eye movements on frontal interhemispheric gamma EEG coherence: implications for EMDR therapy. J Nervous Mental Disease 195(9):785–788. https://doi.org/10.1097/NMD.0b013e318142cf73
26. Bakan P, Svorad D (1969) Resting EEG alpha asymmetry of reflective lateral eye movements. Nature 223:975–976
27. Christman S, Garvey JK, Propper RE, Phaneuf K (2003) Bilateral eye movements enhance the retrieval of episodic memories. Neuropsychology 17(2):221–229. https://doi.org/10.1037/0894-4105.17.2.221
28. Lyle KB, Orsborn AE (2011) Inconsistent handedness and saccade execution benefit face memory without affecting interhemispheric interaction. Memory (Hove, England) 19(6):613–624. https://doi.org/10.1080/09658211.2011.595418

29. Polden M, Crawford TJ (2022) On the effect of bilateral eye movements on memory retrieval in ageing and dementia. Brain Sci 12(10):1299
30. Samara Z, Elzinga BM, Slagter HA, Nieuwenhuis S (2011) Do horizontal saccadic eye movements increase interhemispheric coherence? Investigation of a hypothesized neural mechanism underlying EMDR. Front Psychiatry 2(4). https://doi.org/10.3389/fpsyt.2011.00004
31. Parker A, Parkin A, Dagnall N (2023) Saccade induced retrieval enhancement, handedness & the retrieval of central & peripheral details in eyewitness memory. J Cogn Enhancement. Advance online publication. https://doi.org/10.1007/s41465-023-00282-x
32. Christman SD, Propper RE, Brown TJ (2006) Increased interhemispheric interaction is associated with earlier offset of childhood amnesia. Neuropsychology 20(336–345)
33. Brunye TT, Mahoney CR, Augustyn JS, Taylor HA (2009) Horizontal saccadic eye movements enhance the retrieval of landmark shape and location information. Brain Cogn 70:279–288
34. Shobe ER, Ross NM, Fleck JI (2009) Influence of handedness and bilateral eye movements on creativity. Brain Cogn 71(3):204–214. https://doi.org/10.1016/j.bandc.2009.08.017
35. Xu Z, Hu J, Wang Y (2023) Bilateral eye movements disrupt the involuntary perceptual representation of trauma-related memories. Behav Res Ther 165:104311. https://doi.org/10.1016/j.brat.2023.104311
36. Schiffer F, Teicher MH (1997) Affect changes observed with right versus left lateral visual field stimulation in psychotherapy patients: possible physiological, psychological, and therapeutic implications. Comprehensive Psychiatry 38(5):289–295. https://doi.org/10.1016/s0010-440x(97)90062-6
37. Schiffer F, Mottaghy FM, Pandey Vimal RL, Renshaw PF, Cowan R, Pascual- Leone A, Teicher M, Valente E, Rohan M (2004) Lateral visual field stimulation reveals extrastriate cortical activation in the contralateral hemisphere: An fMRI study. Psychiatry Res 131(1):1–9. https://doi.org/10.1016/j.pscychresns.2004.01.002
38. Schiffer F, Glass I, Lord J, Teicher MH (2008) Prediction of clinical outcomes from rTMS in depressed patients with lateral visual field stimulation: a replication. J Neuropsychiatry Clin Neurosci 20(2):194–200. https://doi.org/10.1176/jnp.2008.20.2.194
39. Schiffer F, Khan A, Bolger E, Flynn E, Seltzer WP, Teicher MH (2021) An effective and safe novel treatment of opioid use disorder: unilateral transcranial photobiomodulation. Front Psychiatry 12(713686). https://doi.org/10.3389/fpsyt.2021.713686
40. Cretenet J, Dru V (2005) Influence of bilateral motor actions on judgments: additional evidence for the motor congruence hypothesis. Laterality 10(5):389–398. https://doi.org/10.1080/13576500500143314
41. Cretenet J, Mullet E, Dru V (2015) Motor and cognitive integration: effect of bilateral behaviors on judgment. Acta Physiol (Oxf) 161:64–72
42. Mullet E, Cretenet J, Dru V (2014) Motor influences on judgment: motor and cognitive integration. Br J Psychol 105(1):69–91. https://doi.org/10.1111/bjop.12022
43. Dijkstra A, van Asten R (2014) The eye movement desensitization and reprocessing procedure prevents defensive processing in health persuasion. Health Commun 29(6):542–551. https://doi.org/10.1080/10410236.2013.779558
44. Dijkstra A, Elbert S (2019) Eye movement inductions influence health behaviour: the working memory account of persuasion. Psychol Health 34(11):1378–1394. https://doi.org/10.1080/08870446.2019.1609678
45. Dijkstra A, Elbert SP (2021) Detecting and preventing defensive reactions toward persuasive information on fruit and vegetable consumption using induced eye movements. Front Psychol 11, 578287. https://doi.org/10.3389/fpsyg.2020.578287
46. Shapiro F (1989) Efficacy of the eye movement desensitization procedure in the treatment of traumatic memories. J Trauma Stress 2(2):199–223. https://doi.org/10.1002/jts.2490020207
47. Shapiro F (1989) Eye movement desensitization: a new treatment for post-traumatic stress disorder. J Behav Ther Exp Psychiatry 20:211–217. https://doi.org/10.1016/0005-7916(89)90025-6

48. Hirnstein M, Hugdahl K, Hausmann M (2019) Cognitive sex differences and hemispheric asymmetry: a critical review of 40 years of research. Laterality 24(2):204–252. https://doi.org/10.1080/1357650X.2018.1497044
49. Allen LS, Richey MF, Chai YM, Gorski RA (1991) Sex differences in the corpus callosum of the living human being. J Neurosci 11(4):933–942. https://doi.org/10.1523/JNEUROSCI.11-04-00933.1991
50. Ardekani BA, Figarsky K, Sidtis JJ (2012) Sexual dimorphism in the human corpus callosum: an mri study using the OASIS brain database. Cereb Cortex 23(10):2514–2520. https://doi.org/10.1093/cercor/bhs253
51. Ingalhalikar M, Smith A, Parker D, Satterthwaite TD, Elliott MA, Ruparel K, Hakonarson H, Gur RE, Gur RC, Verma R (2014) Sex differences in the structural connectome of the human brain. Proc Nat Acad Sci 111(2):823–828. https://doi.org/10.1073/pnas.1316909110
52. Prunier S, Christman S, Jasper J (2018) The effects of varying levels of hemispheric activation on episodic memory. Laterality: Asymmetries of Body, Brain and Cognition 23(4):409–421
53. Lyle KB, Logan JM, Roediger HL 3rd (2008) Eye movements enhance memory for individuals who are strongly right-handed and harm it for individuals who are not. Psychon Bull Rev 15(3):515–520. https://doi.org/10.3758/PBR.15.3.515
54. McCormick M, Seta JJ (2016) Lateralized goal framing: How health messages are influenced by valence and contextual/analytic processing. Psychol Health 31(5):535–548. https://doi.org/10.1080/08870446.2015.1117082
55. O'Driscoll GA, Strakowski SM, Alpert NM, Matthysse SW, Rauch SL, Levy DL, Holzman PS (1998) Differences in cerebral activation during smooth pursuit and saccadic eye movements using positron-emission tomography. Biol Psychiat 44(8):685–689. https://doi.org/10.1016/s0006-3223(98)00047-x
56. Drake RA (1991) Processing persuasive arguments: recall and recognition as a function of agreement and manipulated activation asymmetry. Brain Cogn 15(1):83–94. https://doi.org/10.1016/0278-2626(91)90017-3
57. Drake RA (1993) Processing persuasive arguments: 2. Discounting of truth and relevance as a function of agreement and manipulated activation asymmetry. J Res Person 27(2):184–196. https://doi.org/10.1006/jrpe.1993.1013
58. Drake RA, Bingham BR (1985) Induced lateral orientation and persuasibility. Brain Cogn 4(2):156–164. https://doi.org/10.1016/0278-2626(85)90067-3
59. McCormick M, Seta JJ (2012) A new method for selectively enhancing hemisphere processing: voice frequency amplification influences the strength of attribute framing. Laterality 17(6):727–735. https://doi.org/10.1080/1357650X.2011.626559
60. Gallagher P, Dagenbach D (2007) Manipulating noise frequencies alters hemispheric contributions to decision making. Brain Cogn 64(1):42–49. https://doi.org/10.1016/j.bandc.2006.10.004
61. Seta JJ, McCormick M, Gallagher P, McElroy T, Seta CE (2010) Voice frequency impacts hemispheric processing of attribute frames. J Exp Soc Psychol 46(6):1089–1092. https://doi.org/10.1016/j.jesp.2010.06.009
62. DeSimone N, Lall S, Vallafana M, McCormick M, Propper RE (2020) Selective auditory frequency amplification may increase right hemisphere activity. (Poster presentation). Presentation at the annual meeting of the New England Psychological Association
63. Mayer JD, Gaschke YN (1988) The experience and meta-experience of mood. J Pers Soc Psychol 55(1):102

Ruth E. Propper obtained her Ph.D. in Cognitive Psychology with specialization in hemispheric lateralization at the University of Toledo, followed by a Postdoctoral Research Fellowship in the Laboratory of NeurophysiologyMass Mental Health, Harvard Medical School. She is a full Professor at Montclair State University. She has published more than 60 peer-reviewed manuscripts, book chapters, abstracts, and encyclopedia entries, has presented research at more than 60 national, regional, and international conferences, has mentored more than 100 graduate and undergraduate students in her lab, and has received ~ $500,000 in Department of Defense funding for her research. In alignment with her views that scientific information should be accessible to non-scientists, she has also published three children's books; two that teach the scientific method in the context of light mysteries, and one that encourages STEM fields for girls. She has been a consultant for both private industry and the U.S. government. Her research typically examines state and trait differences in functional and structural hemispheric lateralization.

Magen Daniela Malka-Teasley is a Master's of Arts in Applied Linguistics candidate at Montclair State University. She holds a Bachelor's of Arts degree in Psychology and another in Spanish, having graduated *Summa Cum Laude* from the same institution. Her research engages the intersection of cognitive science and pragmatic linguistics: language learning, multilingualism, language atrophy, natural language processing, and literacy. Additionally, Ms. Malka-Teasley is editor of the academic journal *Inquiry: Critical Thinking Across The Disciplines*. She lives, works, sings, and does gymnastics in Montclair, NJ. This is her first publication.

Emily Lynch is a master's student in the Psychological Sciences program at Montclair State University. She earned her bachelor's degree in Psychology with a minor in Child Advocacy and Policy from Montclair State University. Emily has presented four first-authored posters at local and national conferences and co-authored 10 additional posters and three manuscripts. Emily is actively involved in research through the Cerebral Lateralization Lab and the Autism and Neurodevelopment Lab at Montclair State University and plans to pursue a PhD in Clinical Psychology. Her clinical training focuses on the assessment and identification of emotion dysregulation, anxiety, and psychopathology in youth, including individuals within the neurodiverse population. Her research interests lie in exploring cognitive and behavioral differences in the psychopathology of neurodiverse populations to inform interventions and treatment strategies that improve their quality of life.

Decision Making and Working Memory: Bridging the Gap

7

Nikolay R. Rachev and Nelson Cowan

Thinking cannot occur in a vacuum; it relies on the long-term knowledge base and a temporary workspace. [1]

Abstract

Decision making and working memory are conceptually related: inasmuch as the former involves integration of beliefs and preferences for the purpose of actions, it must rely on the latter. Both fields have generally acknowledged this relationship but have largely refrained from delving into the details. Research studies on working memory expressed interest in reasoning much more than in judgment and decision making. Research studies on judgment and decision making, dominated by the heuristics-and-biases tradition, have long tackled issues related to working memory only implicitly, by assuming that cognitive limitations are a main reason for using heuristics (mental shortcuts). More recently, the dual-process view of thinking has explicitly stated that access to working memory is a defining feature of the putative Type 2 processes. However, the statement was not followed by extensive research testing specific hypotheses related to working memory. We propose ways in which this gap could be bridged from both sides and consider the embedded-processes approach to working memory (Cowan in Psychol Bull 104:163–191, 1988 [2];

N. R. Rachev (✉)
Department of General, Experimental, Developmental, and Health Psychology, Sofia University St. Kliment Ohridski, Sofia, Bulgaria
e-mail: nrrachev@phls.uni-sofia.bg

N. Cowan
Department of Psychological Sciences, University of Missouri, Columbia, MO, USA
e-mail: CowanN@missouri.edu

© The Author(s), under exclusive license to Springer Nature Switzerland AG 2025

T. McElroy (ed.), *Decision-Making in Life and Work*, Integrated Science 36,
https://doi.org/10.1007/978-3-031-88415-3_7

Cowan in Models of working memory: Mechanisms of active maintenance and executive control. Cambridge University Press, New York, NY, 1999 [3]; Cowan et al. in Annu Rev Psychol 75:183–214, 2024 [4]) as a framework for merging decision making and working memory.

Keywords

Working memory • Decision making • Heuristics and biases • Dual-process theories • Dual-task paradigm • Cognitive load • Strategies • Mental representations

1 Introduction

Research on decision making and working memory share similar developmental trajectories: both have been a subject of decade-long research, both had their first peak in the 1970s [5, 6]. Both witnessed further refinement in the decades to follow, resulting in competing accounts of the existing evidence [vs. 2, 3, in the field of working memory, e.g., 7, 8, vs. 9, in the field of judgment and decision making]. Both are surrounded by continuing debates; debates that are, however, unlikely to lead to a unanimously accepted theory anytime soon. Despite these commonalities, the two research traditions unfolded largely independently of one another for a long period of time.

This situation is unnatural because decision making and working memory are conceptually interleaved: engagement of working memory is indispensable for any decision that requires active maintenance of information. Yet, researchers on both sides avoided tackling issues concerning the other side. Researchers on working memory have long pointed out, at a general level, the role of working memory in higher order cognition, such as reasoning, judgment, and decision making. In fact, what we believe to be the first use of the term working memory in the field of psychology occurred within a book on making plans [10]. However, research on working memory has focused much less on decision making than one would expect, and particularly less so compared to other aspects of higher cognition, including language comprehension, reasoning, and problem solving [5, 11–16]. Similarly, research on judgment and decision making, inspired by the heuristics-and-biases tradition, has largely sidestepped issues related to working memory. The state of affairs seemed to change when proponents of the dual-process view of higher cognition, an influential account of thinking stemming from heuristics-and-biases research, pointed to access to working memory as a key distinguishing feature between the two types of thinking this view posits [8]. Yet, this strong theoretical commitment was not coupled with a substantial body of empirical research linking judgment and decision making to working memory.

In this chapter, we show how each field has largely overlooked the other over the years and suggest ways to bridge the gap between them. In Sect. 2, we define working memory and decision making, and show how they are conceptually related. In

Sect. 3, we explore the state of affairs from the perspective of working memory research, and argue that decision making has received much less attention than deserved, particularly compared to the long tradition of research on working memory and reasoning. In Sect. 4, we take the perspective of the judgment and decision-making tradition, with a focus on the dual-process framework. We suggest that, although essential correlational and some experimental evidence exists linking decision making to working memory, it is insufficient because it fails to selectively support the dual-process account to the exclusion of alternative accounts. We also argue that the theoretical move of linking Type 2 processes to working memory was underspecified and thus insufficient. In Sect. 5, we suggest ways of bridging the gap from both sides and reflect on the potential benefits of such a bridge for both theory and everyday life. We propose the embedded-processes view of working memory [2–4] as an overarching framework that could help build the bridge.

2 Decision Making and Working Memory Are Conceptually Related

2.1 Working Memory

The term *working memory* has been around since the dawn of the cognitive revolution [5, 10, 17, 18], and precursors of it could be traced back to the work on consciousness and attention by the founder of experimental psychology, Wilhelm Wundt [19]. A current, broadly agreed upon and theory-neutral definition of working memory is "a system, or a set of processes, holding mental representations temporarily available for use in thought and action" [20, cf. 21, Definition 6]. Beyond this broad definition, theories of working memory diverge from one another in various respects, such as whether the function of working memory is mere maintenance of information [21], or, rather, also includes processing [e.g., positing a central executive subsystem within working memory, 22, or emphasizing on executive control as a crucial aspect of working memory, 23]; whether it is distinct from long-term memory [22] or, rather, an activated part of it [3]; whether it includes distinct modules for processing and storing different types of information [22] or, rather, a single domain-general store [3]. Recently, working memory researchers acknowledged that an integrated theory of working memory is nowhere near and proposed a number of robust phenomena, or benchmarks, which working memory models should aim to account for, with differing priority [20].

An important feature of working memory is its limited capacity, which has been established to be around four coherent items or chunks [20, 24] though others suggest that capacity is a fluid resource rather than a fixed number of chunks [25]. However, people differ in their working memory capacity. These differences go together with differences in various domains of thinking. Indeed, one of the benchmarks established by Oberauer et al. [20] concerns individual differences, a benchmark acknowledging the robust evidence that working memory correlates with indicators of attention and executive control, and reasoning abilities, most often those used to measure fluid intelligence. This finding is not surprising given thinking refers to the mental process of manipulating and "changing the currently active mental representations", that is, "representations held in working memory" [26].

2.2 Decision Making, Heuristics, and Biases

Decision making is a type of thinking that involves choosing one out of multiple available courses of action, based on one's expectations about how probable and attractive the consequences of each course of action are. In other words, decisions arise in situations with (1) many possible options, varying in (2) uncertainty about the outcome and (3) a positive or negative value attached to each option [27]. Some examples of decisions in everyday life include choosing what car to buy, whether to get married, and whether to undergo a surgery (see also Table 1). Decision-making researchers have invented a host of imaginary scenarios requesting choices to be made, in the hope that those hypothetical choices evoke processes similar to real-life choices and actions.

Decision scientists have shown that if economic agents followed a small number of axioms of rational choice, they would be able to arrange the choice options on a continuum from the least desirable to the most desirable in a non-controversial and coherent way. Put more technically, by following the axioms economic agents would maximize expected utility. Unlike hypothetical economic agents', however, real humans' choices often deviate from rational norms. The influential heuristics-and-biases tradition launched by the pioneering work of Tversky and Kahneman [6, 28] posits that, when reasoning, judging, or deciding, humans are designed to use mental shortcuts, or heuristics, which are often useful (for example, requiring less computation than other solutions), but also imprecise. To be sure, humans have also at their disposal the so-called normative models, formal systems that have been devised to guide and assess thinking according to standard norms. Examples of normative models are deductive logic in reasoning, probability theory in probability judgments, and expected utility theory in decision making. However, using normative models is cognitively demanding and requires specialized knowledge that some might not have. Therefore, the heuristics-and-biases view predicts that humans will use heuristics whenever possible, especially when heuristic processing provides a fast and intuitively compelling response. The basic research strategy of this view has been to present participants with cleverly invented problems pitting against each other a compelling but wrong response, supposedly produced by heuristic processing, and a correct response derived from a normative model. Results have shown that people's responses often systematically deviate from normative responses. These systematic deviations are called biases. In short, proneness to biases, including decision biases, is built in the human cognitive system [29]. Since the 1970s, various decision biases of reasoning, judgment, and decision making have been established (see Table 2). Whether they represent deviations from

Table 1 An example of a single decision problem, whether to bring an umbrella to work, carried through all stages

Knowledge	Working memory	Reasoning	Judgment	Decision making
• What it's like to get wet on the way to work • What is the formal and informal attitude to arriving wet at work • What the inconvenience of carrying an umbrella entails • How to balance these factors based on probability of rain, etc.	• What does the weather report say • How accurate have the reports been • The factors activated from the knowledge base A scenario is formed by linking these facts together using the focus of attention and then this new knowledge is held in activated long-term memory as a newly-learned scenario while the focus of attention is shifted to the reasoning, judgment, and decision-making portions of the process	Use relevant activated knowledge to draw conclusions • **Deductive**: Wearing sloppy attire in the office is considered inappropriate, wet clothing seems sloppy, therefore I will look inappropriately dressed • **Inductive**: The boss expressed discomfort when people came into work with wet clothing several times before, and this is a likely outcome again if I get wet on the way to work	• The likelihood of rain is about 40% • The consequences of getting wet at work seem relatively large • The hassle of carrying an umbrella seems relatively mild	Decide to carry the umbrella and act on that decision

Table 2 Example effects from reasoning, judgment, and decision making, and the proposed normative principles they violate

Thinking function	Definition	Example normative principle	Example problem	Typical effect[a]
Reasoning	Arrive to a conclusion based on premises	**Deductive logic: categorical syllogisms** Premise 1: *All C are B* Premise 2: *All B are A* Conclusion: *All A are B* If the above premises are true, then the conclusion is true as well, regardless of the specific content of A, B, and C	**Problem 1** Premise 1: All birds can fly Premise 2: Penguins are birds Conclusion: Penguins can fly	**Belief bias** Judging the validity of a syllogism based on the believability of the conclusion [67]. Rejecting the valid but unbelievable conclusion in Problem 1 is an example of belief bias
		Deductive logic: conditional statements Premise 1: *If p then q*, and: (1) Premise 2: *p* Conclusion: *q* (Modus Ponens) (2) Premise 2: *not p* *No valid conclusion* (3) Premise 2: *q* *No valid conclusion* (4) Premise 2: *not q* Conclusion: *not p* (Modus Tollens)	**Problem 2 (Wason selection task)** Imagine you are presented with four cards with a letter on one side and a number on the other: A C 4 7 Which cards do you have to turn over to assess if the following rule is followed: "If a card has A on one side, it has 4 on the other side."	**Matching bias** Choosing the cards that are mentioned in the problem (i.e., A and 4 in Problem 2) is an example of matching bias [66]. Following Modus Ponens and Modus Tollens, the two cards one has to turn are A (equivalent to p) and 7 (equivalent to not q)
Judgment	Estimate an unknown quantity (e.g., probability)	**Bayes rule** The posterior probability of an event is directly related to its prior probability, or its base rate [32]	**Problem 3** A study of 1000 people contains 5 clowns and 995 accountants. Person *L* is funny. Is Person *L* more likely to be: A. A clown B. An accountant	**Base-rate neglect** Judging probability by stereotypical associations (e.g., "Clowns are funny, accountants are not"), thereby ignoring important prevalence information [74]. Choosing A in Problem 3 is an example of base-rate neglect

(continued)

Table 2 (continued)

Thinking function	Definition	Example normative principle	Example problem	Typical effect[a]
		Conjunction rule The probability of a conjunction of events, P(A&B), cannot be higher than the probabilities of each of its constituents, P(A) or P(B)	**Problem 4 (The Linda problem)** Linda is 31 years old, single, outspoken, and very bright. She majored in philosophy. As a student, she was deeply concerned with issues of discrimination and social justice, and also participated in antinuclear demonstrations. Which of the following is more likely? A. Linda is a bank teller B. Linda is a bank teller and is active in the feminist movement	**Conjunction fallacy** Judging probability by stereotypical associations (e.g., "This personality profile reminds me of a feminist"), thereby ignoring the conjunction rule [66]. Choosing B. in Problem 4 is an example of conjunction fallacy

(continued)

Table 2 (continued)

Thinking function	Definition	Example normative principle	Example problem	Typical effect[a]
Decision making	Choose among alternative courses of action	**Description invariance** Different representations of the same choice options should yield the same preferences. That is, the same thing is not evaluated differently just because it is described differently [29]	**Problem 5** **The Disease problem** Imagine that your country is preparing for the outbreak of an unusual disease, which is expected to kill 600 people. Two alternative programs to combat the disease have been proposed. Assume that the exact scientific estimate of the consequences of the programs are as follows: **5.1 (Gain frame)** 5.1A If Program A is adopted, 200 people will be saved 5.1B If Program B is adopted, there is 1/3 probability that 600 people will be saved, and 2/3 probability that no people will be saved **5.2 (Loss frame)** 5.2A If Program C is adopted 400 people will die 5.2B If Program D is adopted there is 1/3 probability that nobody will die, and 2/3 probability that 600 people will die Which of the two programs would you favor?	**Framing effect** The phenomenon whereby different representations of the same options lead to preference shifts [50, 77, 80, 104, 105]. Preferring 5.1A to 5.1B but 5.2B to 5.2A is an example of a risky-choice framing effect

(continued)

Table 2 (continued)

Thinking function	Definition	Example normative principle	Example problem	Typical effect[a]
		Constant discounting Rewards lose their utility at constant rate over time. Consequently, preferences for a sooner or a later reward should be time-consistent, or independent of the time the decision is made [53, 119]	**Problem 6.1** What do you prefer? 6.1A $100 now 6.1B $110 in one month? **Problem 6.2** What do you prefer? 6.2A $100 in 12 months 6.2B $ 110 in 13 months	**Hyperbolic discounting** Discount rates are higher for shorter than for longer delays **Dynamic inconsistency** One consequence of hyperbolic discounting, whereby intertemporal choices are inconsistent over time. Choosing both 6.1A and 6.2B is an instance of dynamic inconsistency [75, 79]

[a]Referenced studies in these columns are studies mentioned in the text

rationality is a subject to heated debates both regarding specific effects [e.g., 30, vs. 31, regarding framing effects, 32, vs. 33, regarding the base rate neglect in judgment] and in general [34, 35].

2.3 How Decision Making and Working Memory Are Related

From the above definitions, it is easy to see that decision making critically depends on working memory. The integration of mental representations, be it in the form of beliefs, evaluations or something else, is working memory's job. Making good decisions is thus possible only to the extent that working memory does its job successfully. The conceptual relationship between working memory and decision making within a larger context of the interdependence of knowledge and thinking is illustrated in Fig. 1 and an example is provided in Table 1. Working memory involves holding in active state knowledge that is relevant for the ongoing thinking processes. In other words, knowledge can be used for thinking only to the extent to which it is held in working memory, even though the amount of readily available relevant knowledge can be dramatically enhanced with practice [36–38]. Often, the purpose of keeping in mind several pieces of knowledge simultaneously is to combine them to come up with a new conclusion, a process that is typically referred to as reasoning. Sometimes the conclusion that follows from combining the pieces of knowledge, or the premises, is a hundred per cent certain, and other times it is only most probable. The former case is called deductive reasoning and the latter is called inductive reasoning [e.g., 39]. A particular type of inductive

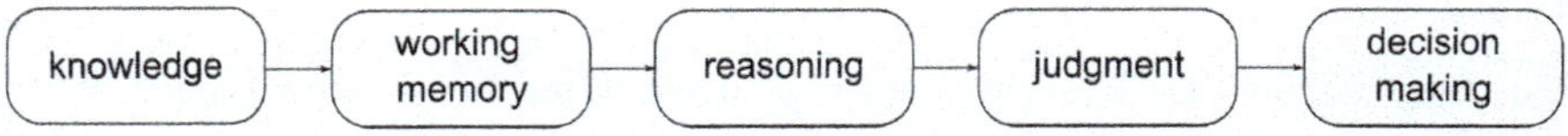

Fig. 1 Conceptual relationship between working memory and decision making in the process of using knowledge for thinking purposes. *Note* Arrows denote dependence: the concept to the left of the arrow is dependent on the concept to the left

reasoning involves judgment, including judgment about unknown aspects of reality, both qualitative (e.g., *what is this?*) and quantitative (*how likely is this?*) ones. Sometimes judgments are not enough; one has to also decide on one's course of action. Yet, judgments are an essential part of this decision-making process. In order to choose one's best option, one has to make both qualitative judgments (what the state of affairs is, what my options are) and quantitative judgments (how probable each option is, how desirable each one is).

The diagram clearly depicts that decision making is largely dependent on working memory. Working memory is critically involved whenever one deals with maintaining information in mind and combining it to choose one's best option. Some routine or well-rehearsed decisions may not require working memory, such as decisions in familiar situations with known frequencies of events. However, the vast majority if everyday situations involve novelty, hence considerable uncertainty. For instance, a decision may depend on the depth of one's understanding of a text passage, or on keeping in mind another person's point of view, or combining in mind probabilities and values. In more specialized settings, a simultaneous interpreter has to retrieve relevant terminology, keep in mind the phrases that are being translated, comprehend and store new phrases for upcoming translation, and discard already translated phrases from memory while deciding on where to start or stop speaking, whether to translate or skip a word or phrase, and whether to convey the message literally or paraphrase. A recruiter conducting a semi-structured interview has to decide what part, if any, of the current response provided by the interviewee has to be written down into the report while also planning her next question based on both her knowledge of the mandatory questions to be asked and her real-time decisions whether the current response requires further investigation, however tricky delving into that may be. A dispatcher receiving an emergency call has to record the information and quickly decide on a course of action based on constantly monitoring the route, traffic, and status of the vehicles. A large amount of decision making in everyday life and laboratory more generally rely critically on working memory.

3 Decision Making in Working Memory Research

3.1 A Long Tradition of Relating Working Memory to Reasoning

The role of working memory in higher-order information processing, including reasoning, problem-solving, and decision making, has been recognized from the beginning of research on working memory [5, 10, e.g., 40]. Reasoning has received much attention from working memory theorists and, as a result, considerable empirical evidence exists for the relationship between working memory and reasoning. The role of working memory has been investigated in various aspects of reasoning, such as verbal reasoning [5, 14, 15], deductive reasoning such as syllogistic reasoning [12, 41] and propositional reasoning [42, 43], matrix reasoning [13], analogical reasoning [44, 45], and related areas such as spatial abilities [15] and algebraic performance [16]. Evidence from experimental research has been corroborated by a body of converging evidence from the individual-difference perspective [11, 39, 46, 47] showing strong correlations between working memory capacity and different aspects of reasoning abilities.

Not only did such studies expand the empirical knowledge base with a range of robust phenomena but they also advanced the theory behind working memory by testing predictions of competing working memory accounts. For instance, in their seminal work on working memory, Baddeley and Hitch [5] used verbal tasks involving sentence verification to show that a limited-capacity central executive aspect of working memory, rather than just an articulatory system, is involved in verbal reasoning. More recently, Ünal et al. [16] established that both verbal and spatial algebraic problems engaged a domain-general attentional system in working memory but a verbal maintenance mechanism was also selectively available for verbal tasks. Drawing on individual-differences research, Oberauer et al. [39] argued that working memory is distinct from the so-called executive functions, despite the often cited association between the two constructs. Thus, the considerable body of research that has been conducted to link working memory to reasoning had direct implications for advancing theories of working memory.

3.2 Comparison to Research on Working Memory and Decision Making

Although the role of working memory in decision making has also been long recognized in general, it has received much less attention from both empirical researchers and theorists. In their seminal book, Miller et al. [10] argued that behavioral "plans" are stored in a "quick-access, working memory" (p. 65), and that deciding on which plan to execute is a crucial aspect of turning intentions into behavior. However, they treated both working memory and decision making in a rather general way. The authors of the multistore memory model, Atkinson and Shiffrin [17, 40] explicitly mentioned decision making as a process occurring in a short-term memory store. According to them, "the short-term store is considered a working memory: a system in which decisions are made, problems are solved and information flow is directed." [17]. However, the range of decisions they discussed was limited to decision about retrieval of information from the long-term store. None of ten major models of working memory, assembled in an influential volume [48], dealt with decision making in detail, despite an explicit question from the editors about the role of working memory in complex cognitive activities. In a collection of writings dedicated to the relationship between memory and thinking [49], various authors listed decision making as a crucial thinking function among others. Yet, in the entire collection of 165 pages the word *decision* was used only 31 times while *reasoning* was used 248 times. A more recent overview on thinking in working memory [98] mentioned decision making only once. These numbers appear to reflect the disproportionately small attention that has been dedicated on decision making relative to reasoning on the part of working memory researchers. Unlike reasoning, decision making has not become an integral part of testing hypotheses related to working memory.

4 Working Memory in Decision-Making Research

4.1 The Heuristics-and-Biases Tradition and the Overlooking of Working Memory

According to the heuristics-and-biases tradition, humans use heuristics as mental tools for two main reasons. First, proper solutions to many problems require computations or integration of knowledge exceeding the limited human cognitive capacity. Second, even when calculating the proper solution is within human cognitive limits, humans often opt for sacrificing accuracy to save mental effort. For instance, in their paper reviewing framing effects, Tversky and Kahneman [50] argued that "people generally evaluate acts in terms of a minimal account, which includes only the direct consequences of the act" (p. 456). A major reason for using a minimal account is that it "simplifies evaluation and reduces cognitive strain" (p. 457). Similarly, Thaler [51] coined the term *mental accounting* to refer to various mental operations that people use to organize and keep track of their mental activities, arguing that those "have evolved to economize on time and thinking costs and also to deal with self-control problems" (p. 202).

Despite this posited central role of cognitive limitations as a main driving source of heuristics, the accounts of working memory in the heuristics-and-biases literature have remained rather sketchy. For instance, the best-selling *Thinking, Fast and Slow* [52] contains a chapter dedicated to attention and effort that eloquently demonstrates how limited mental resources are. Yet, working memory is not discussed beyond this chapter. A seminal book on judgment and decision making largely adopting the heuristics-and-biases tradition [27] depicts Baddeley's model of working memory in the opening chapter but refers to the term only three times in subsequent 13 chapters, most often to remind about its limited capacity. Another often-cited book on thinking and deciding [53] cites the term working memory four times. In sum, the heuristics-and-biases tradition, and research in judgment and decision making more generally, have largely overlooked working memory.

4.2 The Dual-Process Solution: Access to Working Memory as a Defining Feature of Type 2 Processes

More recently, the heuristics-and-biases approach has been incorporated into the dual-process view of thinking [34, 52, 54]. According to this view, humans have two types of thinking processes at their disposal. One, called System 1 or Type 1, is typically fast, automatic, and effortless. The other, called System 2 or Type 2, is typically slow, controlled, and effortful. The two lists of attributes are typically longer [e.g., 54]. Different subtypes of this view exist, the most popular being the default-interventionist subtype. It posits that System 1 proposes a fast default response to a question, which is often a good approximation, but also prone to errors. This is when System 2 could, and should intervene to double-check the System 1 response and correct it if needed. A major account of many decision biases, prospect theory [6], has also been integrated into the dual-process account, as Kahneman [52] stated that its three main principles, reference point, diminishing sensitivity, and loss aversion, should be treated as features of the intuitive System 1.

Defining Type 1 and 2 processes in terms of lists of attributes implies a strong claim that those attributes always go together, a claim that is hardly defendable [55]. To address this criticism, Evans and Stanovich [8] proposed using new defining features of Type 1 and Type 2 processes. In particular, unlike Type 1 processes, Type 2 processes require access to WM.[1]

4.3 Evidence Supporting the Dual-Process Solution

To support their statement, Evans and Stanovich presented three types of empirical evidence, pertaining to individual differences, experimental manipulations, and neuroscience. Research on individual differences has typically revealed that there are substantial correlations between performance on judgment and decision-making tasks and measures of cognitive ability, including intelligence, executive control, and working memory capacity [56–58], although the association might be absent when the two alternative versions of a decision problem are administered between subjects [59, 60]. In the decade since the publication of Evans and Stanovich's article, this line of evidence has been ever growing in the same direction [e.g., 61, 62], with some researchers arguing that general intelligence and

[1] The other defining features distinguishing the two systems are autonomy (System 1) and mental simulation or cognitive decoupling (System 2).

decision-making competence are empirically indistinguishable [63]. It is important to note, however, that although the general pattern of correlations between performance on decision-making tasks and measures of cognitive capacity supports the dual-process view, it does not rule out alternative explanations [64, 65, see Sect. 4.4].

The second line of evidence cited by Evans and Stanovich concerns experimental studies showing that the so-called heuristic responses increase under memory load [66, 67] or time pressure [68], manipulations that ostensibly disrupt the corrective operations of System 2 [69]. Compared to the evidence from individual differences studies, experimental evidence for impaired decision making under memory load is much scarcer, although such evidence exists regarding syllogistic reasoning [67, 68, see Table 66, see Table 66, see Table 2], conditional reasoning and hypothesis testing [e.g., the Wason selection task, 2], probability judgment [e.g., problems eliciting the conjunction fallacy, 2], judgments evoking counterfactual thinking [70] or math problems with lures [71]. Although evidence has generally supported the dual-process assumptions, there are also notable exceptions [65, 72]. Furthermore, recent studies using both time pressure and cognitive load during the same task show that a considerable proportion of participants are nevertheless able to provide a correct solution in two-alternative versions of those problems [73, 74]. This evidence challenges the traditional dual-process assumption that corrective operations from System 2 are necessary to override the default, incorrect, heuristic response.

Moving on to decision-making tasks, Benjamin et al. [75, Study 3] found that cognitive load (memorizing seven numbers) generally reduced normative choices (defined as preference for the option with a larger expected value in small-stakes risky monetary choices or as preference for the delayed option in intertemporal choices between an immediate and a delayed reward), although this reduction was statistically significant in only one out of four types of choices. Contrary to predictions of the dual-process view, cognitive ability did not moderate the effect of cognitive load. However, this study with a sample of 37 Chilean high school students did not have enough power to reliably detect interaction effects [76] and maybe also the main effect of load. In a more powered study using 60 participants and a computerized set of 64 framing tasks (32 gain framed and 32 loss framed), Whitney et al. [77] found that cognitive load (memorizing a letter sting) made participants less willing to gamble but did not increase the framing effect. This finding challenges a dual-process assumption that an automatic affective response to the frames can sometimes be overridden by the deliberate system [78], although it is compatible with a weaker assumption that the framed response is guided by a strategy to minimize cognitive effort [77].

The third line of evidence concerns neuroimaging studies showing that different brain areas are activated when the heuristic (associated with Type 1 processing) or normative (associated with Type 2 processing) response is provided.[2] For instance, using fMRI, McClure et al. [79] found that choices involving immediate monetary awards are selectively associated with activation in parts of the limbic system. In contrast, activation regions of the lateral prefrontal cortex and posterior parietal cortex is related to choices favoring both immediate and delayed rewards, and this activation is larger when participants choose the delayed reward. De Martino et al. [80] found that the framing effect in monetary choices is associated with amygdala activity while a reduced framing effect is associated with orbital and medial prefrontal cortex activity.

Despite these promising findings, the available neuroscientific evidence seems even scarcer than the available experimental evidence supporting the dual-process account, as it encompasses fewer studies using fewer participants each. The small sample sizes also means that fMRI studies are often underpowered, leading to inflated correlations [81]. Furthermore, interpretations separating the brain into different systems seem to overlook the dynamic interaction between brain regions, as in the case of cognition and emotion [64, 82]. Finally, inferring cognitive processes from neuroscientific evidence is often dependent on the preexisting cognitive theory [83]. For instance, one cannot simply rule out the possibility that different brain regions are activated by the same cognitive processes. In other words, similar to evidence presented from individual differences, neuroscientific evidence is also compatible with competing approaches to decision making [55, 84].

4.4 Alternative Explanations and Objections to the Dual-Process Solution

Defining the putative Type 2 processes by their unique access to working memory was intended to not only deal with the criticisms of competing approaches to higher order cognition [9, 55, 85] but also show that the evidence supporting dual-process theories outweighs the evidence for those competing approaches. However, most of the evidence presented to support the dual-process approach is also compatible with more parsimonious single-process accounts [64, 65] or memory-driven approaches to judgment and decision making such as the fuzzy-trace theory [86, 87]. In other words, the existing evidence does not constrain theories enough to the point of critically pitting them against each other. Although abundant, correlational evidence does not speak directly to the posited underlying processes. Experimental evidence, which could potentially do the job, is much scarcer and

[2] Evans and Stanovich [8] warn against the so-called normative fallacy, or assuming that Type 1 processing always leads to incorrect responses while Type 2 processing always leads to normative (i.e., correct) responses. Yet, the dual-process view they defend predicts that these pairings are strongly correlated, thereby justifying the use of accuracy as the primary outcome variable in research validating this view.

has rarely been used to disentangle competing predictions. Neuroscientific evidence is also far from being conclusive. Similar to much areas in psychology, theories are underdetermined by present evidence [88].

Besides the empirical gap, there is also a theoretical gap impeding a genuine integration between the dual-process approach and working memory. Every thinking task depends to some degree on working memory [89], so it is hard to argue that some tasks only trigger Type 1 processing. The question is, therefore, what specific role working memory plays during different stages of the reasoning process, a question the dual-process approach has failed to answer yet. For instance, the default-interventionist dual-process approach posits that, when solving framing problems, people make decisions based on Type 1 processes, which are occasionally overridden by Type 2 processes. This purported mechanism suggests that Type 2 processing might not be switched on at all during the decision. However, understanding and mentally representing the problem requires working memory, that is, Type 2 processing. It turns out that Type 2 processing has already been switched on before arriving to the decision stage. Does the dual-process account maintain that Type 2, that is, working-memory related, processing switches off the moment a representation has been formed and does not participate in the actual decision-making phase? Why would the information-processing system rather switch off its most powerful tool *before* rather than *after* the decision is made? Is it that the cost of switching off is typically lower than the added value of keeping it switched on? By solving one problem, the posited access to working memory raised new questions that remain unresolved.

Part of the problem lies in the fact that the dual-process account has referred to the term 'working memory' at a rather general level, largely ignoring the fact that different views of working memory exist [21, 90]. Thus, the dual-process account failed to posit specific processes pertaining to working memory that are involved during reasoning, judgment, and decision making, processes that might at least partially depend on the definition of working memory that one adopts. Alternative approaches to decision-making, such as the evolutionary approach promoting the adaptive nature of heuristics, or their ecological rationality [35], have also used the term working memory sparingly, and typically at a general level. However, this fact is less problematic for these approaches because they never relied on working memory as a crucial concept to help make a central theoretical distinction. In sum, the proposal that access to working memory is a defining feature of Type 2 processes fell short of its promise, partly because it refrained from dealing with working memory in detail, and was thus not followed by research benefiting from the advances in the study of working memory.

5 Building Bridges from Both Sides

Our short review showed that despite some good efforts, neither field has done enough to incorporate the accomplishments of the other to their advantage. In this section, we suggest ways in which this gap can be addressed from both sides. The list of suggestions is not exhaustive; we point out several possibilities that are most obvious to us, hoping that once researchers appreciate the value of such bridges, they will come up with ideas that are most suitable to their research.

5.1 Incorporating Decision Making into Working Memory Research

As mentioned earlier, working memory theorists were much less interested in testing their assumptions using decision-making tasks compared to reasoning tasks. There might be different reasons for this situation. First, there is a long theoretical and empirical tradition linking reasoning to short-term memory, most notably within the psychometric tradition of intelligence tests, which often include measures of both [e.g., 91]. Therefore, to many working memory researchers, reasoning tasks might come to mind faster and be easier to justify than decision-making tasks. Second, unlike reasoning tasks, many classical decision-making tasks lack a correct answer in a single administration. For instance, frame-consistent responses (i.e., choosing the safe option when the options are framed as potential gains or the risky options when the options are framed as losses) are not incorrect per se. The inconsistency is only visible when one compares the proportion of safe choices across the two frame conditions. In other words, reasoning tasks are the most typical choice for studying reasoning rationality, or the ability to solve individual problems when the information is provided, while decision-making tasks often test coherence rationality, or the coherence of preferences and beliefs [92]. However, even tasks that were originally designed to be presented between-subjects and thus measure coherence rationality can be adapted to measure reasoning rationality. For instance, the same framing problem could be presented twice, once in a gain frame and later in a loss frame, to measure individual susceptibility to framing. While presenting exactly the same problem twice might be controversial for both theoretical and methodological reasons [93], varying the cover story while using the same numerical estimates might be a successful research strategy to reliably measure susceptibility to framing [94]. Alternatively, one can administer a longer series of brief tasks in both frames [77, 80]. As Whitney et al.'s [77] study demonstrates, the latter strategy was successfully used to test hypotheses concerning the role of working memory in decision-making.

In short, there are no insurmountable methodological obstacles to using decision-making tasks to test theoretical predictions about working memory. Most of the questions researchers ask could be reasonably answered by using decision-making tasks. Decision-making tasks can be used to test whether findings from reasoning problems can be generalized to other aspects thinking. For instance, one can ask whether the typical finding that processing impedes retention of spatial information more than verbal information [16] will be replicated using decision-making tasks that are predominantly verbal, spatial, or both [see 95, 96, for examples of spatially presented information in decision-making tasks]. Decision-making tasks can also contain unique properties that other thinking problems may lack. In a way, choosing between available options, which is the essence of decision making, resembles the hypothetical job of the central executive, which is at the core of major working memory models [2, 7]. Whether this is a superficial analogy or a promising way for testing hypotheses related to the central executive is yet to be answered. The point is that although there might be objective reasons why decision-making tasks have been overlooked in working memory research, their relative unfamiliarity could be turned to their advantage by opening fresh avenues for new research.

5.2 Incorporating Working Memory Into Decision Making Research

Incorporating working memory in decision making research is even more important than the other way around because, even though various aspects of working memory can be investigated with no explicit regard to decision making, investigating decision making with no regard to working memory seems rather incomplete. Here, we suggest five ways to integrate the two in decision-making research. Most of them were already used in previous research, but we believe all need to be approached more systematically to help fill the gap. Although all concern methodological considerations, all are also theoretically important and some bear practical implications for decision making in everyday life and in the workplace.

5.2.1 Justifying Methods of Imposing Cognitive Load

Dual-task paradigms have proved very useful in both working memory [14, 16, 97] and decision-making research [77]. However, they have been used sparingly in research on judgment and decision making. Furthermore, most studies have used various ways to impose memory load. Although choices of a particular type of load were methodologically justified by the relative difficulty of the task, other aspects, such as the modality of the task, have been often overlooked. For instance, in different studies using reasoning and judgment problems, De Neys and his colleagues used a complex finger-tapping pattern [72], a spatial pattern to hold in memory [67], or both, in different experiments [66]. In light of findings that different types of secondary tasks can differentially affect processing performance [98], this heterogeneity of measures makes it difficult to generalize across findings.

Importantly, De Neys [66] found that, in two separate experiments, two different secondary tasks affected differently performance on two versions of the Wason selection task. These findings corroborate findings from research on reasoning that different secondary tasks, aiming to disrupt maintenance of verbal or spatial information more specifically, or the processing of the central executive more generally, may selectively interfere or not with different aspects of reasoning [e.g., deductive and spatial reasoning vs. analogical reasoning, 98]. The need for a systematic choice of secondary tasks seems especially relevant to studies aiming to test dual-process theoretical predictions [89]. Researchers might benefit from explicitly basing their choices of secondary task on specific hypotheses of why and how a specific memory load would affect a specific aspect of decision making. One benefit of such an additional effort is that it will clarify how the choice was motivated by a particular view of working memory rather than relying on a received, unified view of working memory that does not reflect the state of the art in working memory research [20, 21]. Knowing how different types of memory load affect decision-making processes may also have practical benefits. Decisions under predominantly verbal load, such as those made by an interpreter or a recruiter, may be differently affected than decisions under substantial visuospatial load such as that made by a dispatcher.

5.2.2 Measuring Dual Costs

Another important aspect of dual-task paradigms that has rarely been investigated in judgment and decision-making research is that the dual-task interference can go in both directions. While most studies in reasoning, judgment, and decision making have focused on the effect of memory load on performance on the problems, hardly any to our knowledge has addressed the effect on processing the problems on maintaining the information in memory. To do so, researchers need to include trials including single memory tasks in the procedure, in addition to single processing and dual memory-and-processing tasks. This additional burden seems to pay off given the extended number of possibilities one can test [e.g., 16]. In addition, including single tasks of both memory and processing allows researchers to check if participants have strategically prioritized one at the expense of the other in the dual-task phase, trade-offs that most of the extant studies on reasoning and working memory have missed to measure [26]. Such a check might be important both methodologically and theoretically because these so-called dual task costs may come from different sources. For instance, it might be useful to distinguish between prioritization costs, or the costs resulting from participants intentionally emphasizing one task rather than the other, and concurrence costs, or the costs of performing dual versus single tasks, irrespective of intentional emphasis [97]. Researchers might want to be explicit about which type of dual cost, if any, they expect to find, and design their manipulations accordingly. The nature of trade-offs might be also important in everyday life and in the workplace, where it is important that concurrent decision making does not lead to loss of critical information from working memory (e.g., an interpreter who misses a phrase to translate, a recruiter who omits to record crucial details about a candidate, or a dispatcher who fails to update his mental model of the locations of the vehicles).

5.2.3 Imposing Cognitive Load and also Measuring Working Memory Capacity

Another underexplored but promising way to incorporate working memory in decision-making research is to impose cognitive load in a task while also measuring working memory capacity as a covariate. Such a manipulation would allow researchers to combine experimental and individual-differences perspectives by investigating whether the effect of memory load is moderated by the individual's capacity to actively maintain relevant information. The one decision-making study we are aware of that incorporates all these aspects [75] had a sample size too small to reliably detect any interaction. Apart from ensuring a decent sample size, it is also important to specify which interaction, out of many possibilities, would support, or even better, disconfirm the theories that are tested. Consider again predictions of dual-process theories. One possibility, investigated using reasoning and probability judgment tasks [67] is that individuals with lower working memory capacity are less affected by cognitive load because their reasoning is usually dominated by System 1, which does not require access to working memory [see 99, for a similar finding with respect to retrieval from long-term memory]. Yet, it is also possible that all individuals use both types of processes but Type 2 processing is more easily disrupted among individuals of lower capacity, as they find it harder to deal with the additional load; therefore individuals of lower rather than higher working memory capacity would be more affected by cognitive load [as in 70]. These two predictions entail interactions going in opposite directions, yet both are compatible with the dual-process account in general. To complicate things further, even a lack of interaction, which is what De Neys found, was interpreted as compatible with a so-called "weak" dual-process version, according to which people of higher capacity are more efficient analytic thinkers regardless of cognitive load [67]. To help advance our understanding of phenomena, theories need to be better specified and make clearer and testable predictions, for instance regarding the boundary conditions for a directional effect. Studies testing those theoretical predictions also need to address the intricacies related to studying interactions [100], including but not limited to, ensuring a sufficient power to detect the interaction they imply [76, 101]. Some relevant practical questions are, to what extent working memory capacity matters in real-life settings involving cognitive load, and to what extent limited cognitive capacity can be aided by external tools such as specialized software.

5.2.4 Working Memory, Decision-Making Strategies, and Mental Representations

Despite the claim that the limited cognitive capacity is a main reason for the mental shortcuts people use, relatively little research has directly investigated the causal link between cognitive load and strategy use. Yet, strategies are an important aspect of decision making. A manager under high memory load and time pressure may prefer hiring the first available candidate meeting the minimal requirements despite the fact that the same manager is generally highly aspiring and, under less stressing circumstances, would rather choose the best from a larger pool of candidates. These two different strategies (satisficing vs. choose the best) can have long-lasting effects for the company.

Studies sometimes infer strategies from response patterns. For instance, as already mentioned, Whitney et al. [77] interpreted the lack of effect of cognitive load on the framing effect as evidence against affective processing but in line with a heuristic of minimizing cognitive effort. However, researchers have also devised various ways to go beyond the end result (i.e., the choice made) and also measure the underlying strategy more directly, using techniques such as verbal protocols, patterns of information search by participants, or regression-based classification [102, 103]. One such study [102] did not find any effect of memory load on strategy selection but it seems unlikely that this single null finding can be generalized to various situations, strategies, and types of load.

To be clear, defining and measuring strategies is difficult. Gilhooly [26] has demonstrated, in the case of syllogistic reasoning, that the same strategy, defined at a general level, could depend more or less on working memory, depending on its specific implementation. Conversely, different strategies may make similar demands on working memory. However, the question of how limits on working memory, including working memory load by a secondary task, impose constrains on possible strategies seems too important to be sidestepped. Many decision-making researchers have demonstrated how content can dramatically affect choices on problems that are mathematically equivalent [104–107]. These studies added up to earlier studies showing the limited capacity of prospect theory to predict choices based on its assumed mental operations [108]. Such theoretical and empirical demonstrations challenged a basic assumption in prospect theory that people make decisions by automatically converting problems into expected utilities. Instead, they emphasized the importance of studying mental representations during decision making, advocating for a top-down rather than bottom-up processing during decision making, in line with a wide range of research in cognitive psychology [105, 106]. Emphasis on mental representations implicitly places a larger role of working memory in decision making than assumed by expected utility approaches such as prospect theory. These divergent assumptions lead to competing predictions regarding the effect of additional cognitive load on decision making, and vice versa, that could be empirically tested using the dual task paradigm. More generally, emphasis on strategies and mental representations during decision making can help go beyond quantitative questions, such as whether and how much cognitive load affects performance, and also address qualitative questions, such as how cognitive processing changes in relation to various task demands.

5.2.5 Task Components and Information Processing

Answering such questions may require a careful analysis of the tasks used to tap decision-making phenomena. Often, decision-making biases have been measured using single tasks, under the tacit assumption that the cover story is of little importance. The content effects on decision making show that this is not the case. Not surprisingly, therefore, using different problems intended to tap the same bias has led to different factor structures of measures of decision biases, and differing correlations with criteria such as the cognitive reflection test [109]. Even sets of tasks intended to measure the same bias, such as the Resistance to Framing scale [56] sometimes have low internal consistency and questionable factor structure [93], a finding corroborating the evidence that participants' response strategies vary depending on aspects of the problems that have been so far overlooked. Researchers might want to hypothesize what these aspects are, and how they affect decision-makers' mental representations, strategies, and, correspondingly, the work of their working memory.

Researchers might want to focus not only on task aspects; they might also want to be more explicit in positing the cognitive processes involved in decision making, and their relationship with working memory. For instance, Maule and Villejoubert [105] proposed a process model of framing effects that involved a serial involvement of mental processes including editing operations, mental representation, and evaluation operations. Although prospect theory posits that editing and evaluation operations occur largely automatically, Maule and Villejoubert presented evidence that people have a larger set of available editing strategies than considered by prospect theory. Different editing strategies can lead to different mental representations, affecting evaluation and, ultimately, action. A vacant work position can be represented, in the manager's mind, as a potential threat to productivity that needs urgent solution or as a chance to improve the company requiring careful choice. Two emergent questions are, therefore, what determines the relative involvement of working memory during the processes underlying decision making, and how this relative involvement affects decisions. It is already clear that different aspects of decision making, different decision-making phenomena, and decision-making skills are differently related to working memory [58]. What is yet to be understood is the variety of ways the same decision situation can be processed, and how this processing is related to working memory. Such an understanding would have not only theoretical but also practical implications for decisions in everyday life.

5.3 Beyond Heuristics and Biases: A Framework for Merging Decision Making and Working Memory

Most of the research strategies suggested in Sects. 5.1 and 5.2 concerned the question of *whether* and *when* working memory is engaged in a decision-making task. Yet, to better understand *how* working memory is involved in the decision-making process, one also needs a better specification of what role working memory is playing in different stages of the process. To this aim, one needs a theoretical framework of working memory. We use the embedded-processes approach to working memory [2–4] to illustrate a way to integrate decision making within working memory theorizing. The embedded-processes approach proposes that working memory consists of a capacity-limited focus of attention embedded in activated long-term memory (Fig. 2). The focus of attention is the place of conscious concurrent processing of 3–5 chunks of information, often leading to formation of new chunks. Outside the focus of attention, a larger subset of concepts and representations is activated for a limited time period to aid ongoing information processing, including concepts primed by external cues above or below conscious awareness. Given the capacity limits of the focus of attention, which is constantly needed for both storage and processing purposes, new representations that are formed there are often off-loaded into the activated long-term memory. What information will enter the focus of attention is jointly controlled by stimuli in the environment and voluntary central executive processes.

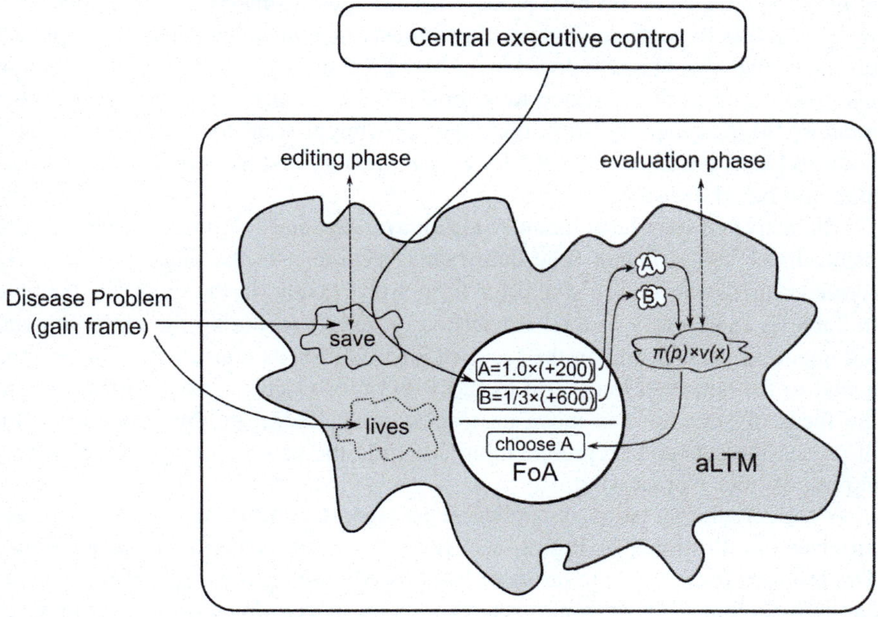

Fig. 2 *Schematic representation of the processing of the disease problem (gain frame) posited by prospect theory, incorporated into the embedded-processes view.* According to prospect theory, during the editing phase, the problem is coded as potential gains. Then, during the evaluation phase, the edited options are evaluated using decision weights $\pi(p)$ and the subjective value of the outcome $v(x)$. Since the subjective value function is concave for gains, option A will be chosen. This sequence of steps can be mapped into the embedded-processes approach by positing that both the editing and evaluation phases occur largely in the activated subset of long-term memory (aLTM, represented as a grey irregular shape). What enters in the focus of attention (FoA, white circle) is the end result of each phase, as an interim output or a final decision. In the process, the newly formed mental representations A and B resulting from the editing phase are off-loaded into aLTM to free space in the capacity-limited FoA. While the prospect theory posits that the problem formulation predetermines the end result of the editing phase, alternative, top-down approaches [105] propose a more active processing during the editing phase, which could be modeled as a greater involvement of FoA. An important issue for further research, therefore, involves the interplay between aLTM and FoA, akin to the relative role of System 1 and System 2, during decision making. Small irregular shapes represent activated representations in aLTM. Small rectangles represent representations in FoA. Solid arrows represent the flow of information processing. Dashed arrows link the hypothetical processing to the operations posited by prospect theory

For example, imagine a soccer player trying to decide whether to evade the opponent by dribbling the ball or passing it to a teammate. Working memory is involved in terms of keeping track of who is at what position in the field, what the goal is, what phase of the game one is in, etc. However, the final decision must be made in such an event-dependent manner, so quickly, that there may be aspects of the decision that cannot be mulled over in conscious working memory, but go rather from a standing plan to a final decision in a split-second based on cues the player is barely aware of. Many might agree that the player's decision is made intuitively, or, in dual-process terminology, using System 1, supposedly without access to working memory. In contrast, the embedded-processes approach proposes that the player's decision is based on the most highly primed action plan coming from the activated long-term memory, conceived as part of working memory. More generally, from this view, one might want to associate System 1 with decisions based on activated long-term memory and System 2 with focus-of-attention-based decisions.

The activated long-term memory also plays a crucial role in decisions that are made under less stringent time constraints because it is the place where newly formed representations are stored for further processing. In the umbrella example in Table 1, each newly formed conclusion or deduction during the reasoning and judgment stages depends on the focus of attention for its construction and then is added to the activated long-term memory and off-loaded out of the focus so that the focus of attention can move on to new work. Therefore, by definition, a lot of processing acts upon representations stored in the activated long-term memory, akin to System 1 processing.

A similar interpretation is applicable to laboratory tasks such as the Disease problem (as illustrated in Fig. 2, see Table 2 for the contents of the problem). The problem formulation in terms of lives saved might prime encoding of the situation as potential gains from a reference point of zero lives saved (as predicted by prospect theory). Some of those coding processes in the editing phase of the choice process [6] are not made consciously but are a part of the activated long-term memory. What enters the focus of attention is the output of the editing phase, in terms of mental representations [105] of a certain smaller gain and an uncertain larger gain with the same expected value.[3] To overcome the capacity limits of the focus of attention, the newly formed representations are off-loaded into the activated long-term memory. There, they are subjected to further processing, corresponding to the evaluative phase in prospect theory, ultimately leading to the conscious choice as an output in the focus of attention.

Such an interplay between decision making and working memory is also compatible with a recent development of dual-process theorizing [110] proposing concurrent activation and competition between several Type 1 intuitions and an uncertainty monitoring mechanism that switches on deliberative processing when the conflict between intuitions reaches a certain threshold. In terms of the embedded-processes view, the competing intuitions are part of the activated long-term memory, while the conflict monitoring mechanism is carried out by the central executive processes that also direct the focus of attention.

The proposal that Type 1 processing occurs in the activated long-term memory while Type 2 processing occurs in the focus of attention is also in keeping with the view that automatic learned processes are stored in long-term memory while

[3] For the sake of simplicity, the internal representations in Fig. 2 coincide with those predicted by prospect theory but they might be different depending on the particular editing processes, as argued by Maule and Villejoubert [105].

controlled processing is carried out in short-term memory [37, 38]. The latter view also emphasizes the role of expertise in decision making, among other aspects of thinking. Looking back to the soccer player example, a professional player would presumably take less time and make a better decision than a lay player because the former has stored considerable declarative and procedural knowledge in long-term memory that is easily activated. Slow thinking relying on the focus of attention is replaced with fast thinking where valid cues call for attention automatically [37]. Similarly, less experienced jazz improvisers show increased activity in frontoparietal brain regions associated with executive control while more experienced improvisers showed increased activity in central and posterior brain regions associated with "unconscious, associative, Type 1 processes" [111]. Deciding on the "right" note when improvising imposes technical and memory demands that easily overload the focus of attention, but those demands may be met, to a creative result, by keeping relevant representations in long-term memory in an activated and highly-accessible state.

The distinction between a conscious focus of attention and activated long-term memory beyond conscious awareness is also relevant to other issues pertaining to decision-making, such as the debate between intuitionism, the stance that heuristic processing usually outsmarts deliberation [35], and reflectionism, the view that decisions are improved by double-checking one's intuitions; claims that complex decisions are better made under unconscious rather than conscious thought [112, 113, for large-scale 114, and 115unsuccessful replications of the unconscious thought theory, see]; or findings showing brain changes predictive of the decisions seconds ahead of conscious awareness of that decision, thereby bearing on the notion of free will [116]. All these are intriguing avenues for future research integrating neuroscientific and behavioral evidence of laboratory and everyday decisions. Such evidence could not only benefit from but also advance evolving conceptions of working memory.

6 Conclusion

Despite the advancement of research in both working memory and decision making, effective communication between the two fields has been rare. While the relative lack of communication so far might be justified by the focus researchers needed to craft their theories, it is a good time for integration between the two. The need for integration is more palpable in the field of decision making where it is hard to ignore the role of working memory, and even recognizing it at a general level is not enough. Explicit consideration of the role of working memory in different aspects of decision making might be one way of critically pitting opposing theoretical approaches against each other. So far, researchers were primarily preoccupied to confirm their theories, and sometimes to disconfirm others' theories. Often, disconfirming evidence is easily dismissed by proponents of a theory [e.g., 117]. Maybe it is time that researchers on opposing camps unite and agree in advance on the predictions of each approach, as in an adversarial collaboration. Working memory is a crucial aspect supposed to engender diverging predictions by different approaches to decision making. Research in working memory has also shown that adversarial collaboration is possible and can successfully advance knowledge [97, 118]. Research in working memory can also benefit from studying decision making. Such benefits range from checking the generality of working memory effects across different types of thinking tasks, to establishing phenomena that might be unique to aspects of decision making. Approaching the issue from both sides can help fill out the theoretical and empirical vacuum, towards a better understanding of how decision making is "bounded" by human information-processing constraints.

References

1. Gilhooly KJ, Logie RH (1998) Thinking in working memory. In: Logie RH, Gilhooly KJ (eds) Working memory and thinking. Psychology Press, pp 1–4
2. Cowan N (1988) Evolving conceptions of memory storage, selective attention, and their mutual constraints within the human information-processing system. Psychol Bull 104:163–191
3. Cowan N (1999) An embedded-processes model of working memory. In: Miyake A, Shah P (eds) Models of working memory: mechanisms of active maintenance and executive control. Cambridge University Press, New York, NY, pp 62–101
4. Cowan N, Bao C, Bishop-Chrzanowski BM et al (2024) The relation between attention and memory. Annu Rev Psychol 75:183–214. https://doi.org/10.1146/annurev-psych-040723-012736
5. Baddeley AD, Hitch G (1974) Working memory. In: Bower GH (ed) Psychology of learning and motivation. Academic Press, New York, NY, pp 47–89
6. Kahneman D, Tversky A (1979) Prospect theory: an analysis of decision under risk. Econom J Econom Soc 263–291
7. Baddeley AD (1986) Working memory. Clarendon Press/Oxford University Press, Oxford, NY
8. Evans JStBT, Stanovich KE (2013) Dual-process theories of higher cognition: advancing the debate. Perspect Psychol Sci 8:223–241.https://doi.org/10.1177/1745691612460685
9. Kruglanski AW, Gigerenzer G (2011) Intuitive and deliberate judgments are based on common principles. Psychol Rev 118:97
10. Miller GA, Galanter E, Pribram KH (1960) Plans and the structure of behavior. Holt, Rinehart and Winston Inc, New York, NY
11. Capon A, Handley S, Dennis I (2003) Working memory and reasoning: an individual differences perspective. Think Reason 9:203–244. https://doi.org/10.1080/13546781343000222
12. Gilhooly KJ, Logie RH, Wynn V (1999) Syllogistic reasoning tasks, working memory, and skill. Eur J Cogn Psychol 11:473–498. https://doi.org/10.1080/095414499382264
13. Harrison TL, Shipstead Z, Engle RW (2015) Why is working memory capacity related to matrix reasoning tasks? Mem Cognit 43:389–396. https://doi.org/10.3758/s13421-014-0473-3
14. Hitch GJ, Baddeley AD (1976) Verbal reasoning and working memory. Q J Exp Psychol 28:603–621. https://doi.org/10.1080/14640747608400587
15. Salthouse TA, Mitchell DR, Skovronek E, Babcock RL (1989) Effects of adult age and working memory on reasoning and spatial abilities. J Exp Psychol Learn Mem Cogn 15:507–516. https://doi.org/10.1037/0278-7393.15.3.507

16. Ünal ZE, Forsberg A, Geary DC, Cowan N (2022) The role of domain-general attention and domain-specific processing in working memory in algebraic performance: an experimental approach. J Exp Psychol Learn Mem Cogn 48:348–374. https://doi.org/10.1037/xlm0001117
17. Atkinson RC, Shiffrin RM (1971) The control of short-term memory. Sci Am 225:82–91. https://doi.org/10.1038/scientificamerican0871-82
18. Newell A, Simon H (1956) The logic theory machine: a complex information processing system. IRE Trans Inf Theory 2:61–79
19. Cowan N, Rachev NR (2018) Merging with the path not taken: Wilhelm Wundt's work as a precursor to the embedded-processes approach to memory, attention, and consciousness. Conscious Cogn 63:228–238. https://doi.org/10.1016/j.concog.2018.06.001
20. Oberauer K, Lewandowsky S, Awh E et al (2018) Benchmarks for models of short-term and working memory. Psychol Bull 144:885–958. https://doi.org/10.1037/bul0000153
21. Cowan N (2017) The many faces of working memory and short-term storage. Psychon Bull Rev 24:1158–1170. https://doi.org/10.3758/s13423-016-1191-6
22. Baddeley AD, Logie RH (1999) Working memory: the multiple-component model. In: Miyake A, Shah P (eds) Models of working memory: mechanisms of active maintenance and executive control. Cambridge University Press, New York, NY, US, pp 28–61
23. Engle RW, Kane MJ (2003) Executive attention, working memory capacity, and a two-factor theory of cognitive control. In: Ross BH (ed) Psychology of learning and motivation. Academic Press, pp 145–199
24. Cowan N (2001) The magical number 4 in short-term memory: a reconsideration of mental storage capacity. Behav Brain Sci 24:87–114
25. Ma WJ, Husain M, Bays PM (2014) Changing concepts of working memory. Nat Neurosci 17:347–356. https://doi.org/10.1038/nn.3655
26. Gilhooly KJ (1998) Working memory, strategies, and reasoning tasks. In: Logie RH, Gilhooly KJ (eds) Working memory and thinking. Psychology Press, pp 5–20
27. Hastie R, Dawes RM (2010) Rational choice in an uncertain world: the psychology of judgment and decision making. SAGE, Thousand Oaks, CA
28. Tversky A, Kahneman D (1974) Judgment under uncertainty: heuristics and biases. Science 185:1124–1131. https://doi.org/10.1126/science.185.4157.1124
29. Shafir E, Tversky A (1995) Decision making. In: Smith EE, Osherson DN (eds) An invitation to cognitive science. Thinking, vol 3, 2nd ed, pp 77–100. The MIT Press
30. Tversky A, Kahneman D (1986) Rational choice and the framing of decisions. J Bus 59:S251–S278. https://doi.org/10.1086/296365
31. Mandel DR (2014) Do framing effects reveal irrational choice? J Exp Psychol Gen 143:1185–1198. https://doi.org/10.1037/a0034207
32. Kahneman D, Tversky A (1973) On the psychology of prediction. Psychol Rev 80:237–251. https://doi.org/10.1037/h0034747
33. Mueser PR, Cowan N, Mueser KT (1999) A generalized signal detection model to predict rational variation in base rate use. Cognition 69:267–312. https://doi.org/10.1016/S0010-0277(98)00072-9
34. Stanovich KE, West RF (2000) Individual differences in reasoning: implications for the rationality debate? Behav Brain Sci 23:645–665. https://doi.org/10.1017/S0140525X00003435
35. Todd PM, Gigerenzer G (2012) Ecological rationality: intelligence in the world. Oxford University Press, Oxford
36. Ericsson KA, Kintsch W (1995) Long-term working memory. Psychol Rev 102:211–245. https://doi.org/10.1037/0033-295X.102.2.211
37. Shiffrin RM, Schneider W, Logan GD (2023) The distinction between long-term knowledge and short-term control processes is valid and useful: behavioral & brain sciences. Behav Brain Sci 46:1–71. https://doi.org/10.1017/S0140525X2200317X
38. Shiffrin RM, Schneider W (1977) Controlled and automatic human information processing: II. Perceptual learning, automatic attending and a general theory. Psychol Rev 84:127

39. Oberauer K, Süß H-M, Wilhelm O, Sander N (2008) Individual differences in working memory capacity and reasoning ability. In: Conway ARA, Jarrold C, Kane MJ, Towse J (eds) Variation in working memory, pp 49–75
40. Atkinson RC, Shiffrin RM (1968) Human memory: a proposed system and its control processes. Psychol Learn Motiv 2:89–195
41. Gilhooly KJ, Logie RH, Wetherick NE, Wynn V (1993) Working memory and strategies in syllogistic-reasoning tasks. Mem Cognit 21:115–124. https://doi.org/10.3758/BF03211170
42. Evans JStBT, Brooks PG, (1981) Competing with reasoning: a test of the working memory hypothesis. Curr Psychol 1:139–147. https://doi.org/10.1007/BF02684486
43. Klauer KC, Stegmaier R, Meiser T (1997) Working memory involvement in propositional and spatial reasoning. Think Reason 3:9–47. https://doi.org/10.1080/135467897394419
44. Morrison RG, Holyoak KJ, Troung B (2001) Working-memory modularity in analogical reasoning. In: Proceedings of the annual meeting of the cognitive science society
45. Waltz JA, Lau A, Grewal SK, Holyoak KJ (2000) The role of working memory in analogical mapping. Mem Cognit 28:1205–1212. https://doi.org/10.3758/BF03211821
46. Nettelbeck T, Burns NR (2010) Processing speed, working memory and reasoning ability from childhood to old age. Personal Individ Differ 48:379–384. https://doi.org/10.1016/j.paid.2009.10.032
47. Süβ H-M, Oberauer K, Wittmann WW et al (2002) Working-memory capacity explains reasoning ability—and a little bit more. Intelligence 30:261–288
48. Miyake A, Shah P (1999) Models of working memory: mechanisms of active maintenance and executive control. Cambridge University Press, Cambridge, UK
49. Logie RH, Gilhooly KJ (1998) Working memory and thinking. Psychology Press
50. Tversky A, Kahneman D (1981) The framing of decisions and the psychology of choice. Science 211:453–458. https://doi.org/10.1126/science.7455683
51. Thaler RH (1999) Mental accounting matters. J Behav Decis Mak 12:183–206. https://doi.org/10.1002/(SICI)1099-0771(199909)12:3%3c183::AID-BDM318%3e3.0.CO;2-F
52. Kahneman D (2011) Thinking, fast and slow. Farrar, Straus and Giroux, New York, NY
53. Baron J (2008) Thinking and deciding, 4th edn. Cambridge University Press
54. Evans JStBT (2008) Dual-processing accounts of reasoning, judgment, and social cognition. Annu Rev Psychol 59:255–278.https://doi.org/10.1146/annurev.psych.59.103006.093629
55. Keren G, Schul Y (2009) Two is not always better than one: a critical evaluation of two-system theories. Perspect Psychol Sci 4:533–550. https://doi.org/10.1111/j.1745-6924.2009.01164.x
56. Bruine de Bruin W, Parker AM, Fischhoff B (2007) Individual differences in adult decision-making competence. J Pers Soc Psychol 92:938–956. https://doi.org/10.1037/0022-3514.92.5.938
57. Del Missier F, Mäntylä T, Bruine de Bruin W (2012) Decision-making competence, executive functioning, and general cognitive abilities. J Behav Decis Mak 25:331–351. https://doi.org/10.1002/bdm.731
58. Del Missier F, Mäntylä T, Hansson P et al (2013) The multifold relationship between memory and decision making: an individual-differences study. J Exp Psychol Learn Mem Cogn 39:1344–1364. https://doi.org/10.1037/a0032379
59. Bogdanov B, Corbin JC, Dobreva S et al (2023) Working memory capacity and the risky-choice framing effect: a preregistered replication and extension of Corbin et al. (2010). Judgm Decis Mak 18:e39. https://doi.org/10.1017/jdm.2023.40
60. Stanovich KE, West RF (2008) On the relative independence of thinking biases and cognitive ability. J Pers Soc Psychol 94:672–695. https://doi.org/10.1037/0022-3514.94.4.672
61. Burgoyne AP, Mashburn CA, Tsukahara JS et al (2023) Understanding the relationship between rationality and intelligence: a latent-variable approach. Think Reason 29:1–42. https://doi.org/10.1080/13546783.2021.2008003
62. Erceg N, Galić Z, Bubić A (2022) Normative responding on cognitive bias tasks: some evidence for a weak rationality factor that is mostly explained by numeracy and actively open-minded thinking. Intelligence 90:101619. https://doi.org/10.1016/j.intell.2021.101619

63. Blacksmith N, Behrend TS, Dalal RS, Hayes TL (2019) General mental ability and decision-making competence: theoretically distinct but empirically redundant. Personal Individ Differ 138:305–311. https://doi.org/10.1016/j.paid.2018.10.024
64. Kruglanski AW (2013) Only one? The default interventionist perspective as a unimodel—commentary on Evans & Stanovich (2013). Perspect Psychol Sci 8:242–247. https://doi.org/10.1177/1745691613483477
65. Osman M (2013) A case study: dual-process theories of higher cognition—commentary on Evans & Stanovich (2013). Perspect Psychol Sci 8:248–252. https://doi.org/10.1177/1745691613483475
66. De Neys W (2006) Automatic–heuristic and executive–analytic processing during reasoning: chronometric and dual-task considerations. Q J Exp Psychol 59:1070–1100. https://doi.org/10.1080/02724980543000123
67. De Neys W (2006) Dual processing in reasoning: two systems but one reasoner. Psychol Sci 17:428–433. https://doi.org/10.1111/j.1467-9280.2006.01723.x
68. Evans JStBT, Curtis-Holmes J, (2005) Rapid responding increases belief bias: evidence for the dual-process theory of reasoning. Think Reason 11:382–389
69. Kahneman D, Frederick S (2005) A model of heuristic judgment. In: Holyoak KJ, Morrison RG (eds) The Cambridge handbook of thinking and reasoning. Cambridge University Press, New York, NY, pp 267–293
70. Goldinger SD, Kleider HM, Azuma T, Beike DR (2003) "Blaming the victim" under memory load. Psychol Sci 14:81–85. https://doi.org/10.1111/1467-9280.01423
71. Johnson ED, Tubau E, De Neys W (2014) The unbearable burden of executive load on cognitive reflection: a validation of dual process theory. In: Proceedings of the annual meeting of the cognitive science society, pp 2441–2446
72. De Neys W, Schaeken W, d'Ydewalle G (2005) Working memory and everyday conditional reasoning: retrieval and inhibition of stored counterexamples. Think Reason 11:349–381
73. Bago B, De Neys W (2019) The smart System 1: evidence for the intuitive nature of correct responding on the bat-and-ball problem. Think Reason 25:257–299. https://doi.org/10.1080/13546783.2018.1507949
74. Bago B, De Neys W (2020) Advancing the specification of dual process models of higher cognition: a critical test of the hybrid model view. Think Reason 26:1–30. https://doi.org/10.1080/13546783.2018.1552194
75. Benjamin DJ, Brown SA, Shapiro JM (2013) Who is 'behavioral'? Cognitive ability and anomalous preferences. J Eur Econ Assoc 11:1231–1255
76. Simonsohn U (2014) [17] No-way interactions. In: Data Colada. http://datacolada.org/17. Accessed 13 Apr 2022
77. Whitney P, Rinehart CA, Hinson JM (2008) Framing effects under cognitive load: the role of working memory in risky decisions. Psychon Bull Rev 15:1179–1184. https://doi.org/10.3758/PBR.15.6.1179
78. Kahneman D, Frederick S (2007) Frames and brains: elicitation and control of response tendencies. Trends Cogn Sci 11:45–46. https://doi.org/10.1016/j.tics.2006.11.007
79. McClure SM, Laibson DI, Loewenstein G, Cohen JD (2004) Separate neural systems value immediate and delayed monetary rewards. Science 306:503–507. https://doi.org/10.1126/science.1100907
80. De Martino B, Kumaran D, Seymour B, Dolan RJ (2006) Frames, biases, and rational decision-making in the human brain. Science 313:684–687. https://doi.org/10.1126/science.1128356
81. Yarkoni T (2009) Big correlations in little studies: inflated fMRI correlations reflect low statistical power: commentary on Vul et al. (2009). Perspect Psychol Sci 4:294–298. https://doi.org/10.1111/j.1745-6924.2009.01127.x
82. Pessoa L (2008) On the relationship between emotion and cognition. Nat Rev Neurosci 9:148–158. https://doi.org/10.1038/nrn2317
83. Henson R (2006) Forward inference using functional neuroimaging: dissociations versus associations. Trends Cogn Sci 10:64–69. https://doi.org/10.1016/j.tics.2005.12.005

84. Keren G (2013) A tale of two systems: a scientific advance or a theoretical stone soup? Commentary on Evans & Stanovich (2013). Perspect Psychol Sci 8:257–262. https://doi.org/10.1177/1745691613483474
85. Osman M (2004) An evaluation of dual-process theories of reasoning. Psychon Bull Rev 11:988–1010. https://doi.org/10.3758/BF03196730
86. Corbin JC, Liberali JM, Reyna VF, Brust-Renck PG (2015) Intuition, interference, inhibition, and individual differences in fuzzy-trace theory. In: Wilhelms EA, Reyna VF (eds) Neuroeconomics, judgment, and decision making. Psychology Press, pp 77–90
87. Reyna VF, Rahimi-Golkhandan S, Garavito DMN, Rebecca K. Helm (2018) The fuzzy-trace dual process model. In: De Neys W (ed) Dual process theory 2.0. Routledge, pp 82–99
88. Eronen MI, Bringmann LF (2021) The theory crisis in psychology: how to move forward. Perspect Psychol Sci 16:779–788. https://doi.org/10.1177/1745691620970586
89. Thompson VA, Newman IR (2020) Working memory, autonomy, and dual process theories: a roadmap. In: Logic and uncertainty in the human mind. Routledge, pp 227–242
90. De Neys W (2021) On dual- and single-process models of thinking. Perspect Psychol Sci 16:1412–1427. https://doi.org/10.1177/1745691620964172
91. Gibbons A, Warne RT (2019) First publication of subtests in the Stanford-Binet 5, WAIS-IV, WISC-V, and WPPSI-IV. Intelligence 75:9–18. https://doi.org/10.1016/j.intell.2019.02.005
92. Kahneman D (2000) A psychological point of view: violations of rational rules as a diagnostic of mental processes. Behav Brain Sci 23:681–683
93. Geiger SJ, Vintr J, Rachev NR (2022) A reassessment of the resistance to framing scale. Behav Res Methods. https://doi.org/10.3758/s13428-022-01876-7
94. Berthet V (2021) The measurement of individual differences in cognitive biases: a review and improvement. Front Psychol 12:419. https://doi.org/10.3389/fpsyg.2021.630177
95. Finucane ML, Gullion CM (2010) Developing a tool for measuring the decision-making competence of older adults. Psychol Aging 25:271–288. https://doi.org/10.1037/a0019106
96. Parker AM, Fischhoff B (2005) Decision-making competence: external validation through an individual-differences approach. J Behav Decis Mak 18:1–27
97. Rhodes S, Jaroslawska AJ, Doherty JM et al (2019) Storage and processing in working memory: assessing dual-task performance and task prioritization across the adult lifespan. J Exp Psychol Gen 148:1204–1227. https://doi.org/10.1037/xge0000539
98. Morrison RG (2005) Thinking in working memory. In: Holyoak KJ, Morrison RG (eds) Cambridge handbook of thinking and reasoning. Cambridge University Press, pp 457–473
99. Rosen VM, Engle RW (1997) The role of working memory capacity in retrieval. J Exp Psychol Gen 126:211–227. https://doi.org/10.1037/0096-3445.126.3.211
100. Rohrer JM, Arslan RC (2021) Precise answers to vague questions: issues with interactions. Adv Methods Pract Psychol Sci 4:25152459211007370. https://doi.org/10.1177/25152459211007368
101. Lakens D, Caldwell AR (2021) Simulation-based power analysis for factorial analysis of variance designs. Adv Methods Pract Psychol Sci 4:2515245920951503. https://doi.org/10.1177/2515245920951503
102. Bröder A (2003) Decision making with the "adaptive toolbox": influence of environmental structure, intelligence, and working memory load. J Exp Psychol Learn Mem Cogn 29:611–625. https://doi.org/10.1037/0278-7393.29.4.611
103. Payne JW (1976) Task complexity and contingent processing in decision making: an information search and protocol analysis. Organ Behav Hum Perform 16:366–387. https://doi.org/10.1016/0030-5073(76)90022-2
104. Fagley NS, Miller PM (1997) Framing effects and arenas of choice: your money or your life? Organ Behav Hum Decis Process 71:355–373. https://doi.org/10.1006/obhd.1997.2725
105. Maule J, Villejoubert G (2007) What lies beneath: reframing framing effects. Think Reason 13:25–44. https://doi.org/10.1080/13546780600872585
106. Rettinger DA, Hastie R (2001) Content effects on decision making. Organ Behav Hum Decis Process 85:336–359. https://doi.org/10.1006/obhd.2000.2948

107. Wagenaar WA, Keren G, Lichtenstein S (1988) Islanders and hostages: deep and surface structures of decision problems. Acta Psychol (Amst) 67:175–189. https://doi.org/10.1016/0001-6918(88)90012-1
108. Fischhoff B (1983) Predicting frames. J Exp Psychol Learn Mem Cogn 9:103–116. https://doi.org/10.1037/0278-7393.9.1.103
109. Aczel B, Bago B, Szollosi A et al (2015) Measuring individual differences in decision biases: methodological considerations. Front Psychol 6:1770. https://doi.org/10.3389/fpsyg.2015.01770
110. De Neys W (2023) Advancing theorizing about fast-and-slow thinking. Behav Brain Sci 46:e111. https://doi.org/10.1017/S0140525X2200142X
111. Rosen DS, Oh Y, Erickson B et al (2020) Dual-process contributions to creativity in jazz improvisations: an SPM-EEG study. Neuroimage 213:116632. https://doi.org/10.1016/j.neuroimage.2020.116632
112. Dijksterhuis A, Nordgren LF (2006) A theory of unconscious thought. Perspect Psychol Sci 1:95–109. https://doi.org/10.1111/j.1745-6916.2006.00007.x
113. Wilson TD, Schooler JW (1991) Thinking too much: introspection can reduce the quality of preferences and decisions. J Pers Soc Psychol 60:181–192. https://doi.org/10.1037/0022-3514.60.2.181
114. Huizenga HM, Wetzels R, van Ravenzwaaij D, Wagenmakers E-J (2012) Four empirical tests of unconscious thought theory. Organ Behav Hum Decis Process 117:332–340. https://doi.org/10.1016/j.obhdp.2011.11.010
115. Nieuwenstein MR, Wierenga T, Morey RD et al (2015) On making the right choice: a meta-analysis and large-scale replication attempt of the unconscious thought advantage. Judgm Decis Mak 10:1–17. https://doi.org/10.1017/S1930297500003144
116. Soon CS, Brass M, Heinze H-J, Haynes J-D (2008) Unconscious determinants of free decisions in the human brain. Nat Neurosci 11:543–545. https://doi.org/10.1038/nn.2112
117. Evans JStBT, Stanovich KE, (2013) Theory and metatheory in the study of dual processing: reply to comments. Perspect Psychol Sci 8:263–271
118. Cowan N, Belletier C, Doherty JM et al (2020) How do scientific views change? Notes from an extended adversarial collaboration. Perspect Psychol Sci 15:1011–1025. https://doi.org/10.1177/1745691620906415
119. Frederick S, Loewenstein G, O'donoghue T, (2002) Time discounting and time preference: a critical review. J Econ Lit 40:351–401. https://doi.org/10.1257/002205102320161311

Nikolay R. Rachev is an associate professor in cognitive psychology at Sofia University St. Kliment Ohridski, Sofia, Bulgaria. His research interests include individual and cultural aspects of judgment and decision making, and their relationship to working memory. He is also passionate about open and reproducible science and has participated in local and global replications of psychological effects pertaining to judgment and decision making.

Nelson Cowan is the Curators' Distinguished Professor of Psychological Sciences at the University of Missouri. He specializes in working memory, the small amount of information held in mind and used for language processing and various kinds of problem solving. To overcome conceptual difficulties that arise for models of information processing in which different functions occur in separate boxes, Cowan proposed a more organically organized "embedded processes" model. Within it, representations held in working memory comprise an activated subset of the representations held in long-term memory, with a smaller subset held in a more integrated form in the current focus of attention. Other work has been on the developmental growth of working memory capacity and the scientific method. His work has been cited over 57,000 times according to Google Scholar. The work has resulted in over 300 peer-reviewed articles, over 60 book chapters, 2 sole-authored books, and 5 edited volumes.

Attitudes to Uncertainty and Real-World Outcomes

8

Jozef Bavolar

> *"The oldest and strongest emotion of mankind is fear, and the oldest and strongest kind of fear is fear of the unknown."*
>
> Joshi and Schultz (2001)

Abstract

Attitudes to uncertainty describe perceptions of uncertainty and affect an individual's reaction to it, taking similar concepts such as ambiguity and risk into account. This chapter looks at the associations between these, and intolerance of uncertainty in particular, with a variety of decision outcomes. It first presents the main approaches to uncertainty and how it is perceived. The central part of the chapter is divided into four sections, which explore the decisional outcomes of attitudes to uncertainty in information processing, psychological disorders, well-being, and real life. The closing section presents the main conclusions and challenges for future research.

Graphical Abstract

J. Bavolar (✉)
Faculty of Arts, Department of Psychology, Pavol Jozef Safarik University in Košice, Košice, Slovakia
e-mail: jozef.bavolar@upjs.sk

© The Author(s), under exclusive license to Springer Nature Switzerland AG 2025

T. McElroy (ed.), *Decision-Making in Life and Work*, Integrated Science 36,
https://doi.org/10.1007/978-3-031-88415-3_8

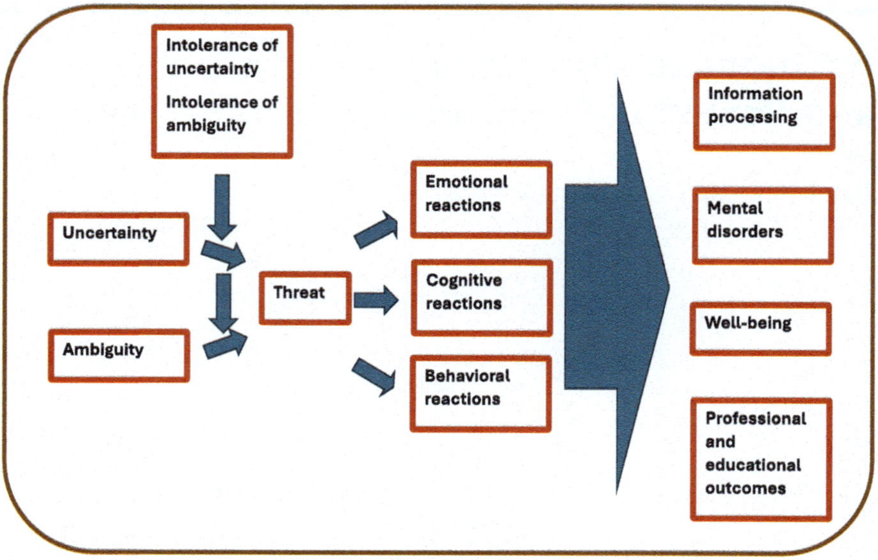

Keywords

Uncertainty • Intolerance of uncertainty • Decision outcomes • Subjective assessment • Mental health

1 Introduction

Every decision includes thinking about the future regarding what it will bring and the probabilities of the options available. These probabilities can only be estimated for very specific situations, such as an insurance incident or traffic accident. This is often done at a very vague level or even unconsciously [1]. This uncertainty varies across decisions as well as attitudes towards it. While some people view uncertainty related to their future as an inherent feature of their lives, others perceive it as something to be avoided. This is mainly due to their lack of control over the consequences of their choices. There are several personality characteristics related to attitudes (approaches) to uncertainty [2] that have been subject to research in various areas of psychology (e.g., judgment and decision-making, clinical and personality psychology). These have frequently focused on their role in anxiety-related disorders [3] or in general information processing [4]. However, attitudes to uncertainty play a role in everyone's lives and multiple studies have shown their associations with subjectively as well as objectively assessed decision outcomes [4, 5].

We generally speak of uncertainty when faced with the inability to predict the future. It is an incompatibility between different cognitions, between cognitions and experiences, or between cognitions and behaviors [6]. Uncertainty occurs when the outcome of an event is unknown due to insufficient or ambiguous information and can also be characterized as a state of doubt about whether or not a particular phenomenon will occur [7]. It is often equated in the literature with concepts such as ambiguity, unpredictability, vagueness, variability, ignorance, and unreliability [8, 9].

Uncertainty can be either operationalized a) objectively as the actual state of (un)knowledge about the likelihood of the outcome of an event, or b) subjectively as a cognitive or affective response to an unclear or ambiguous situation [10]. Its sources can be probability, ambiguity, and complexity. The first is a situation where the probability of an event occurring in the future is known. The probability is most often expressed numerically (e.g., there is a 35% probability of winning). It is also referred to as known uncertainty or risk [11]. Secondly, ambiguity occurs when there is more than one possible explanation of an event, with each possibility having varying degrees of uncertainty [1]. This can result from conflicting, unreliable, inaccurate, insufficient, or incomplete information. An example might be a situation with a 10–30% probability of treatment success, but where the scientific evidence is inadequate and experts have not yet reached a consensus [11]. The third source or type of uncertainty is complexity, where the available information is too complex to understand. In this case, uncertainty arises not from the phenomenon (probability) or the lack of reliable and trustworthy information (ambiguity), but from aspects of the phenomenon that make it difficult to understand. Complexity can, therefore, be viewed objectively (as part of the phenomenon itself) and subjectively, depending on how a particular individual views the situation [11]. Uncertainty arising from complexity can be reduced by improving the clarity and coherence of information.

This chapter examines current attitudes to uncertainty and its consequences in various areas. The chapter begins with a closer look at uncertainty and personality variables describing attitudes to uncertainty and how it is different from similar constructs. The focus will be on intolerance of uncertainty as the central concept present in the literature, which describes the core of the current approach to uncertainty. The chapter will also examine its differences from the two most similar variables—intolerance of ambiguity and risk perception. The role of intolerance of uncertainty is firstly described at a very general level as the predisposition of a particular way of information processing. It is further explored by looking at its position in anxiety disorders and other mental disorders. The next part examines the role of attitudes to uncertainty in well-being (subjectively assessed decision outcomes), mainly focusing on perceived stress. This is followed by discussing the consequences of attitudes to uncertainty in different domains, such as work or school. The final parts present the main conclusions and propose challenges that future research should address.

2 Uncertainty and Attitudes to Uncertainty

Our decisions are affected by multiple factors that can be classified into three main categories—decision features, situational factors, and individual differences [12]. Decision features include the different characteristics of decisions (how they are described or ordered), while situational factors include time pressure or the social context. Although these factors have been widely acknowledged, they do not have a uniform effect on all people. Indeed, individual differences can affect how people react to decision features and situational factors. While some features are quite general (e.g., decision-making styles or thinking styles), there are also ones that can have an effect in each decision situation, at least as it is related to uncertainty. Koerner et al. [2] have described intolerance of uncertainty, negative risk orientation, trait curiosity, indecisiveness, perceived constraints, self-oriented and socially prescribed perfectionism, intolerance of ambiguity, the need for predictability, and the need for order and structure as similar concepts. Similarly, Rosen et al. [13] have proposed intolerance of ambiguity, uncertainty orientation, and the need for cognitive closure as being very close concepts. There will be three concepts—intolerance of uncertainty, intolerance of ambiguity, and risk perception—discussed in more detail due to their prevalence in the current research.

Uncertainty tends to be perceived negatively and aversively [4] with most people seeking to eliminate or reduce it [13–15]. It can be a significant stressor in a person's life and has negative psychological and physiological consequences for an individual [16]. Individuals may experience negative emotions such as anger, frustration, feelings of helplessness or injustice in situations of uncertainty. These emotions may be experienced in the present moment as well as being anticipated in the future [10]. Uncertainty can exacerbate the negative effects of aversive events. Increased aversiveness, coupled with negative attitudes towards uncertainty, can lead to distorted expectations and misjudgments of contingencies [16].

Intolerance of uncertainty (henceforth: IU) can be defined as the cognitive, emotional and behavioral response to uncertainty in everyday situations [17]. Other definitions have described it as the predisposition to react negatively to an uncertain event or situation, independent of its probability of occurring or associated consequences [18] or a dispositional characteristic that represents the tendency to fear unpredictable and uncertain situations in the future [19]. It involves the belief that feeling uncertain is undesirable and that uncertain, unexpected situations and events are harmful and should be avoided [19]. It can also be viewed as a cognitive schema through which an individual perceives the world and which influences the way they perceive, interpret, and respond to uncertain situations [20]. Another definition has stated that it is an individual's dispositional incapacity to endure the aversive response triggered by the perceived absence of salient, critical, or sufficient information and is sustained by the associated perception of uncertainty [14]. According to Evoy et al. [21], IU is also the capacity of an individual to tolerate the consciously perceived fear of the unknown and the associated activation of the defensive motivational system. Similarly, IU can be viewed as stemming from negative beliefs about uncertainty with its implications and the fear of the unknown at its core [4]. Moreover, IU, together with intolerance of ambiguity (IA), can also be considered as a prominent lower-order uncertainty-specific construct behind the higher-order distress-tolerance [22].

Research has shown that IU is not a unidimensional construct and that more factors can be identified behind it. This depends on the theoretical perspective of the study and the results of the psychometric evaluation of the used measures. The first major factor configuration came with the introduction of the Intolerance of Uncertainty Scale [23]. It has four [19] (initially five) factors (uncertainty leads to the inability to act; uncertainty is stressful and upsetting; unexpected events are negative and should be avoided; being uncertain is unfair) but as a result of significant factors overlapping [24], the total score has been predominantly used. The most straightforward and currently used distinction has been provided by Carleton et al. [23] and distinguishes two factors in the revision of the previous scale—prospective anxiety and inhibitory anxiety. The former refers to anxiety about future events [23], which is characterized by a person's desire for predictability. This desire is triggered by anxious worry [25]. High prospective anxiety is associated with the use of planning or information-seeking strategies, i.e., seeking to reduce uncertainty [25, 26]. Inhibitory anxiety is more likely to be associated with the inability to act in uncertain situations [19]. It is characterized by paralysis in situations of uncertainty, impairs a person's functioning [27], and inhibits action [23]. High inhibitory anxiety leads to the use of avoidance strategies such as postponing decisions [25].

In the context of uncertainty, the term uncertainty orientation can be encountered. Within this concept, it is possible to distinguish two categories of people—uncertainty-oriented (motivated to resolve uncertainty) and certainty-oriented (trying to avoid uncertainty). While intolerance of uncertainty focuses on the psychological consequences of uncertainty, uncertainty orientation is concerned with the motivation to resolve or avoid uncertainty. At the same time, (in)certainty-oriented individuals do not feel threatened by uncertainty, while individuals with high intolerance of uncertainty do [24].

Another concept that is very close to uncertainty is ambiguity. This can be defined as the presence of more than one way of interpreting an event [28], which can also be a source of uncertainty [11]. It is based on vague, inconsistent, contradictory, or incomplete information [11, 29]. Based on the psychometric approach to measuring (in)tolerance of ambiguity, Lauriola et al. [29] have shown that attitude towards ambiguity is also a multidimensional construct with more components. These are (a) discomfort with ambiguity (unpleasant feelings related to experiencing ambiguity in interpersonal relationships, social or work situations); (b) moral absolutism (a black-and-white view of the world, inability to tolerate the presence of both positive and negative features within a single object); and (c) the need for complexity and novelty (a more tolerant and open-minded view of complex and novel situations) [29]. The common feature of both "intolerancies" is the tendency to react negatively to a situation of threat at the cognitive, affective and behavioral levels with the main difference being temporal orientation. While intolerance of ambiguity is associated with the current situation of vague and ambiguous features, intolerance of uncertainty refers to uncertainty towards the future [24, 30].

Another important distinction can be made between risk and uncertainty. In the case of risk, the probability of an event occurring is at least known to some extent, while in the case of uncertainty it is not [31]. The most frequent distinction between uncertainty, ambiguity and risk can be illustrated by the well-known Ellsberg paradox [32]. Participants in his experiments are provided with two bags of red and black marbles, where the ratio of red and black marbles is unknown for the first bag but is known for the second (50 red marbles, 50 black marbles). Participants are asked to bet on one of the bags that they think will produce a red (or black) ball. Most choose the second bag, even though both bags have an equal probability [32]. This can be explained by the desire to avoid uncertainty/ambiguity (first bag, unknown probability) in favor of risk (second bag, known probability).

While this approach views risk purely (and simplified) as "uncertainty with known probabilities", its general view in lay perception and psychological literature is different. The concept of risk is perceived somewhat negatively and is associated with something threatening and undesirable [33]. People often talk about risk behavior, risk groups, risk factors, or a risk environment as potentially harmful and should be avoided. Yet, risk can also bring some positive outcomes, at least from the person's perception, such as pleasure or new experiences [34].

The level of risk can be objectively assessed based on the likelihood and consequences of a particular event. However, social science is more interested in subjective perceptions of risk. In addition, the same person may adopt different attitudes in different situations because their perception of risk varies concerning the area to which the risk relates [33]. This may be in the area of health or safety (e.g., seat belt use, smoking), the social domain (e.g., confronting a family member), or the domain of ethical decisions (e.g., cheating in an exam [35]). Consequently, risk perception is understood as an assessment of how risky a certain behavior is [36]. It is assumed that if a person perceives a certain behavior as risky, they are less likely to engage in that behavior. Conversely, people who engage in risky behaviors underestimate the risk because they do not perceive the behavior as risky [36].

3 Consequences of Attitudes to Uncertainty

While the descriptive approach in decision-making tells us how people come to their decisions, the normative approach describes how these decisions and their outcomes should look. The domain of decision outcomes is quite heterogeneous with the quality and usefulness of any outcome being assessed by multiple criteria. Some studies use objective criteria that may be utilized in well-defined contexts such as experimental tasks, work productivity and school achievement. Other approaches are more focused on subjective assessment, such as that of well-being (perceived stress, worry, life satisfaction) and mental health (mainly anxiety

Table 1 Correlates of attitudes to uncertainty (domain and variables)

Information search	Mental disorders	Well-being	Educational and professional outcomes
Tendency to make threatening interpretations	Anxiety disorders	Stress	Burnout (and subsequent turnover intentions)
Reassurance seeking	Depression	Fear	Work behavior
Lower confidence	Obsessive–compulsive disorder		Entrepreneurial mindset
Preference for immediate reward	Eating disorders		Career decision-making Clinical reasoning

disorders). The general domain of information processing is first outlined in presenting the consequences of attitudes to uncertainty. Next, the role of attitudes to uncertainty in anxiety disorders and similar disorders will be presented, followed by other well-being indicators and COVID-19-related variables. The last part will be devoted to outcomes of attitudes to uncertainty in other real-life domains such as work and school. Table 1 provides a structure of the associations that will be described further.

4 Attitudes to Uncertainty and Information Processing

As attitudes to uncertainty and mainly IU describe how people approach information, more precisely perceiving information about future states as unsatisfactory and thus threatening [20], we start by describing the associations of IU to information search and processing. The primary claim is that individuals high in IU try to reduce perceived uncertainty. Dugas et al. [20] have shown that participants high in IU recall a higher proportion of words related to uncertainty and also report more concern about ambiguous situations. It is also interesting that the tendency to make threatening interpretations of ambiguous situations was related more to IU than to worry, anxiety, or depression. Similarly, individuals higher in IU tend to have a lower perceptual threshold of ambiguity (perceiving quite "certain" situations as unclear) [37] and make threatening interpretations of ambiguous situations [38]. Reuman et al. [38] have even used the term uncertainty-based reasoning when an individual perceives a situation as anxiety-provoking (and more likely to result in safety behavior) when uncertain aspects of the situation are apparent.

Similarly, IU was found to be related to threat-related reassurance seeking in Australian adults [39]. Self-reported IU is related to behavior associated with reducing uncertainty (e.g., taking health brochures, requesting information packages) [40], with ambiguity having a moderating role (weaker relationships with higher ambiguity) [37]. Conversely, IU was unrelated to the time taken to decide on an undergraduate or clinical sample [41]. High IU participants also yield higher online threat expectancy, post-experimental covariation estimates between cue and threat, and negative mood for target cues [42]. Surprisingly, the findings indicate that ambiguity may be a more powerful trigger for biased threat appraisal and negative affect than uncertainty [42]. One study found that the role of IU in information searching depends on the relevance of the threat. While individuals low and high in IU did not differ in information-seeking behavior in the high-relevance condition, the latter evidenced more information-seeking in the low-relevance condition [43]. IU was also positively associated with greater delay discounting behavior with a preference for smaller immediate rewards over larger distal rewards and was negatively associated with willingness to wait [44]. Similarly, in an experience-sampling study, IU was positively related to indecisiveness and more problematic safety behavior [45].

Carleton et al. [4] have used three behavioral measures of decision-making—the Wisconsin Card Sorting Game, the Risky Gains Task, and the Modified Iowa Gambling Task, with reports of weak but significant correlations with IUin each case. In contrast to previous studies, the results were found to be mixed when examining the partial indicators such as response latency, time spent evaluating feedback and sensitivity to negative outcomes. Additionally, the results of this study have indicated the usefulness of distinguishing inhibitory and prospective anxiety when these were not uniformly related to the target variables. Indeed, performance on both the executive functioning and risk-gaining tasks was inversely associated with both inhibitory and prospective anxiety, whereas the magnitude of the relationship was greater for prospective anxiety than inhibitory anxiety [4]. Similarly, the strength of the relationship was greater for inhibitory anxiety than for prospective anxiety, although the direction of the relationships of both subscales with anticipating temporally uncertain aversive shock was the same [46]. In another prospective study, IU was associated with distress about harm befalling oneself and perceived importance of the decision (solo condition), whereas inhibitory IU was associated with distress about harm befalling others (partner condition) [47]. In the situation of uncertainty induction, inhibitory IU positively predicted the use of under-engagement strategies of responses (aimed to avoid future uncertain situations) and negatively predicted the use of over-engagement ones (aiming to seek and gain certainty) [48].

The behavioral outcomes of IU have also been found to be manifested at the physiological level. The association of IU with heightened reactivity to uncertainty is reflected in greater activity of the anterior insula and amygdala, alterations in neural responses to rewards and errors evident in event-related potentials, a mixed pattern of startle responses to uncertain threat, and deficiencies in safety learning indexed by startle and skin conductance responding [49]. Moreover, IU has been found to be related to anticipatory physiological responding, skin conductance and corrugator supercilii brain area activity [50]. However, previous results have not been completely consistent, with another study reporting no association with skin conductance or heart rate, a negative correlation with typing speed and no relationship with error ratio in a typing task [51].

The past research on behavioral outcomes of IU in information processing can be summarized in five points [52]—(a) the preference for immediately available rewards, even when they are less probable or less valuable; (b) lower confidence in making high-risk decisions as well as a lower likelihood of changing a decision despite receiving new information; (c) a higher likelihood of seeking additional information to increase certainty; (d) a higher likelihood of increasing certainty by behaving, reacting or deciding more slowly; e) a higher likelihood of being distressed by uncertainty [52].

5 Attitudes to Uncertainty and Mental Health

The initial research on IU primarily focused on the relationship with worry and generalized anxiety disorder [17, 37]. However, it has since been found to be associated with a wide range of emotional disorder symptoms, suggesting that it is transdiagnostic in nature [5, 53, 54]. The importance of IU in anxiety research is emphasized by the development of an Intolerance of Uncertainty Model of generalized anxiety disorder, with this trait being positioned as the main factor in disorder development. Moreover, uncertainty is positioned as a central feature of anxiety-related experience [14] and IU appears to be a robust vulnerability factor associated with numerous psychological disorders [54]. Indeed, IU is viewed as a cognitive feature of anxiety, worry and mood disorders [14, 21].

IU is currently viewed as being related to multiple cognitive vulnerability factors in various disorders [54] and as being a transdiagnostic risk and maintaining factor and change mechanism [55]. A growing number of studies have explored the associations between IU and psychological disorders, their symptoms and subclinical values, which allowed to provide a more general and robust meta-analytic view. A meta-analysis from 2011 reported a robust moderate relationship of IU with general anxiety disorder, major depressive disorder and obsessive–compulsive disorder [5]. Other meta-analysis, which included 181 studies, found IU to be moderately associated with symptoms of generalized anxiety disorder, social anxiety disorder, panic disorder, agoraphobia, obsessive–compulsive disorder, depression, and eating disorders ($r = 0.51$), although heterogeneity was high. The random effect estimates ranged from 0.40 (eating disorders) and 0.42 (obsessive–compulsive disorder) to 0.50 (depression and social anxiety) and 0.57 (general anxiety disorder) [55]. Similar results have also been found in a meta-analysis focusing on children and adolescents, where IU accounted for 36% of the variance in anxiety and 40% in worry when the proportion of the sample from a clinical population had no moderator effect [56]. A positive association between IU and anxiety has also been confirmed in autistic people [57], while a further meta-analysis found IU related to eating disorders with an IU score higher in both anorexia nervosa and bulimia nervosa in comparison to the control group [58].

Empirical evidence has suggested that disorder-specific IU mediates the effect of trait IU on disorder-specific symptoms to some degree [59]. Moreover, trait IU is associated with disorder-specific cognitive vulnerabilities (e.g., negative metacognition in generalized anxiety disorder or agoraphobic cognition in panic disorder) [59]. As such, there may be different types of uncertainty and associated types of

IU with them [60]. It is also necessary to add that the associations of other mental disorders with IU are not only the result of the association with anxiety, although there is an independent relationship [61]. The positive relationship between IU and worry is mediated by negative problem orientation (inhibitory anxiety) and positive beliefs about worry (prospective anxiety), highlighting the value of distinguishing between these two factors [62]. IU, and inhibitory anxiety in particular, can be perceived as a contributor to fear and avoidance symptoms of panic attacks [13, 63]. It has also been found to be positively related to compulsive exercise [64]. IU is also associated with internet addictions, although it has been proposed that it is the mediator of more general variable future expectations [65]. On the other hand, current, past and non-opioid users do not differ in IU [66]. In addition, the effect of intolerance of uncertainty on mental health can be hypothesized as being mediated by difficulties in emotion regulation [67].

Given the role of dispositional IU in anxiety disorders, there have been numerous attempts to manipulate it to cause changes in anxiety. One meta-analysis has shown that psychological treatment yielded a large significant within-group effect size from pre-treatment to both post-treatment and follow-up for IU, as well as for worry, anxiety, and depression [3]. The effect was higher when treatment (Cognitive-Behavioral Therapy, CBT) directly targeted IU in comparison to general CBT. Moreover, a higher change in IU was associated with the amount of time spent directly targeting it [3]. A more recent meta-analysis examined the effect of randomized controlled trials on IU and the severity of anxiety-related symptoms (operationalized as differences between groups). There were even larger improvements in IU by comorbid depression and using certain treatment approaches [68]. A greater effect was associated with longer treatment duration and more intervention hours. IU can also be reduced in clinical samples (e.g., anorexia nervosa) using a group-based intervention [69].

In summary, IU has been found to be related to numerous psychological disorders, which are mostly anxiety-related. It seems that high IU individuals appraise uncertain/ambiguous situations as threatening and thus experience worry and anxiety, engaging in excessive information-seeking behavior in such situations [43]. The need to not only consider general, but disorder-specific IU has been justified [60]. On the other hand, there is still room for further investigation into the relationship of IU with emotional disorders, such as its conceptual foundations, development, assessment, or behavioral manifestation [52].

6 Attitudes to Uncertainty and Well-Being

While previous research has mainly focused on the role of IU in mental disorders, studies have also shown its associations with similar symptoms in non-clinical samples. Indeed, relationships have also been found concerning well-being indicators such as perceived stress or life satisfaction. Zvolensky [70] has provided a very broad framework for this relationship with IU, that is positioned as one of the five dimensions of distress tolerance. This trait was the only one to be confirmed empirically [22].

IU has been found to have a negative effect on the mental health of higher education students in addition to their fear, academic activities, and career aspects [71]. IU was also found to be positively related to COVID-19-related stress and traumatic stress symptoms [72]. Another study found IU to be associated with cancer-related, general, and genetic testing distress one month after receiving an uninformative genetic test. Distress was highest among patients with the highest perceived risk and IU [73]. Inhibitory, but not prospective, IU was found to be a predictor of post-traumatic stress symptoms in a military, non-clinical sample [74]. Similarly, inhibitory IU was also a more significant predictor of post-traumatic stress symptoms in students [75] or participants with a heterogeneous trauma history [76, 77]. However, the opposite pattern of IU factors was found in a veteran sample [78]. Numerous studies have also explored the role of IU in COVID-19-related symptoms and well-being indicators, confirming its role as a vulnerability factor in multiple negative reactions (higher stress, anxiety, fear) [79–82].

7 Attitudes to Uncertainty and Outcomes in Professional and Educational Areas

Previous research on attitudes to uncertainty has also shown the effect on "real-life" outcomes in domains such as work and school. In one of these studies [83], IU was found to affect Korean transplantation coordinators' turnover intention indirectly. In particular, IU leads to higher burnout that is subsequently associated with higher turnover intentions. The "positive" effect of IU on turnover intentions was only present in high grit. Similarly, IU was related to higher burnout in Italian general practitioners [84] and mental health professionals in Australia [85].

In another study on Israeli working adults [86], IU was positively associated with cyberloafing during the COVID-19 pandemic. Its effect, however, seems to have been mediated by perceived distress. More generally, ambiguity has been identified as one of the factors impacting virtual negative work behavior [87]. In another work-related study, IU was found to have an effect on the entrepreneurial mindset of South Korean university students, although this effect depended on the IU subscale. In the case of prospective anxiety, it was positive, while it was negative in terms of inhibitory anxiety [88]. Overall, IU was found to be negatively associated with career maturity (a construct consisting of career choice competencies and career choice attitudes) [89], proactive career behaviors (e.g., voluntary participation in further education or training), life goals (career, relationship and entertainment goals) and visions about the future (hope, optimism, pessimism) in Turkish university students [90]. In addition, a recent study has reported that IU mediates the effect of social class on career adaptability. In particular, social class is said to positively impact career adaptability through prospective anxiety and negatively impact career adaptability through inhibitory anxiety [91]. On the other

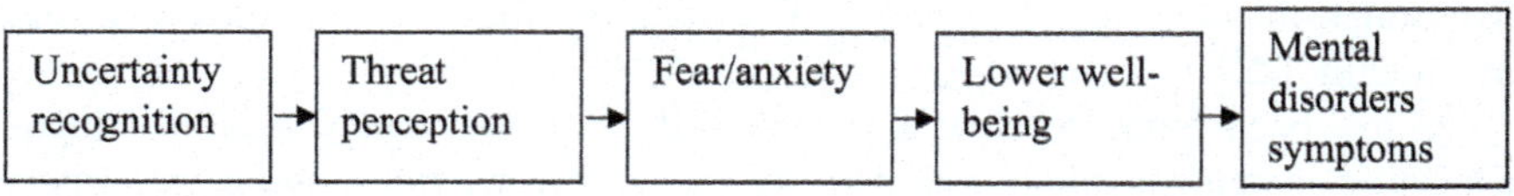

Fig. 1 Proposed process of the effect of intolerance of uncertainty on its outcomes

hand, two factors of the first version of the IU scale (self-referent implications and spoiled everything) were only very weakly related to GPA [90].

In a systematic review, Brun et al. [92] reported that IU affects the clinical reasoning of health professionals in several ways. It was found to be related to information withholding, more referrals to peers, less use of new types of medical interventions, and higher levels of overtesting. It was also found to be associated with more antibiotic prescriptions and can influence abortion recommendations in some cases. It may be expected that a similar pattern of avoidant behavior would be observed in other domains as well.

8 Conclusion

While current research on attitudes to uncertainty and their outcomes has gained the most attention with regard to anxiety disorders, it has also expanded to areas of mental health and well-being, as well as information processing and real-life outcomes. A number of studies have indicated the association of attitudes to uncertainty with maladaptive patterns of actual behavior that have consequences on mental health. The effect of IU on real-life outcomes seems to go from higher uncertainty recognition in people with a higher score of IU to threat perception, bringing anxiety and fear. These negative emotions result in lower well-being and higher stress. In vulnerable individuals, this can be followed by symptoms of mental disorders (Fig. 1).

Although the role of IU has been extensively studied in past decades, some issues remain unclear and present a challenge for future research. One of them is the factor structure of IU. While the two dominant factors—prospective and inhibitory anxiety—have been confirmed to be differently associated with various variables, other factor solutions should also be considered. Similarly, the used measures lack a robust theoretical background [93] and are mainly driven by clinical experience. From a practical point of view, more attention should be devoted to the manifestation in different disorders or to its associations with distinct clusters of symptoms given that IU is transdiagnostic and disorder-specific [53]. The real-life "objective" consequences of IU also seem to have been under-investigated, and its role in occupational, educational, or relationship domains should be explored in more detail.

Quote from the author:
Attitudes to uncertainty affect how we see the challenges the world brings and how we deal with them.

Acknowledgements This work was supported by the Research and Development Support Agency under contract No. APVV–19-0284.

References

1. Andersen S, Fountain J, Harrison GW, Rutström EE (2014) Estimating subjective probabilities. J Risk Uncertain 48(3):207–229. https://doi.org/10.1007/s11166-014-9194-z
2. Koerner N, Mejia T, Kusec A (2017) What's in a name? Intolerance of uncertainty, other uncertainty-relevant constructs, and their differential relations to worry and generalized anxiety disorder. Cogn Behav Ther 46(2):141–161. https://doi.org/10.1080/16506073.2016.1211172
3. Wilson EJ, Abbott MJ, Norton AR (2023) The impact of psychological treatment on intolerance of uncertainty in generalized anxiety disorder: a systematic review and meta-analysis. J Anxiety Disord 97:102729. https://doi.org/10.1016/j.janxdis.2023.102729
4. Carleton RN, Duranceau S, Shulman EP, Zerff M, Gonzales J, Mishra S (2016) Self-reported intolerance of uncertainty and behavioural decisions. J Behav Ther Exp Psychiatry 51:58–65. https://doi.org/10.1016/J.JBTEP.2015.12.004
5. Gentes EL, Ruscio AM (2011) A meta-analysis of the relation of intolerance of uncertainty to symptoms of generalized anxiety disorder, major depressive disorder, and obsessive–compulsive disorder. Clin Psychol Rev 31(6):923–933. https://doi.org/10.1016/j.cpr.2011.05.001
6. van den Bos K, Lind EA (2002) Uncertainty management by means of fairness judgments. In: Advances in experimental social psychology, vol 34. Academic Press, San Diego, CA, US, pp 1–60. https://doi.org/10.1016/S0065-2601(02)80003-X
7. Keren G, Gerritsen LEM (1999) On the robustness and possible accounts of ambiguity aversion. Acta Physiol (Oxf) 103(1–2):149–172. https://doi.org/10.1016/S0001-6918(99)00034-7
8. Gu Y, Gu S, Lei Y, Li H (2020) From uncertainty to anxiety: how uncertainty fuels anxiety in a process mediated by intolerance of uncertainty. Neural Plast 2020(1):8866386. https://doi.org/10.1155/2020/8866386
9. Wilkinson DJ (2006) Types of ambiguity. In: Wilkinson DJ (ed) The ambiguity advantage: what great leaders are great at. Palgrave Macmillan UK, London, pp 42–66. https://doi.org/10.1057/9780230597891_3
10. Freeston M, Tiplady A, Mawn L, Bottesi G, Thwaites S (2020) Towards a model of uncertainty distress in the context of coronavirus (COVID-19). Cogn Behav Therap 13:e31. https://doi.org/10.1017/S1754470X2000029X
11. Han PKJ, Klein WMP, Arora NK (2011) Varieties of uncertainty in health care: a conceptual taxonomy. Med Decis Making 31(6):828–838. https://doi.org/10.1177/0272989x11393976
12. Appelt KC, Milch KF, Handgraaf MJJ, Weber EU (2011) The decision making individual differences inventory and guidelines for the study of individual differences in judgment and decision-making research. Judgement Decis Mak 6(3):252–262
13. Carleton RN, Fetzner MG, Hackl JL, McEvoy P (2013) Intolerance of uncertainty as a contributor to fear and avoidance symptoms of panic attacks. Cogn Behav Ther 42(4):328–341. https://doi.org/10.1080/16506073.2013.792100
14. Carleton RN (2016) Into the unknown: A review and synthesis of contemporary models involving uncertainty. J Anxiety Disorders 39. https://doi.org/10.1016/j.janxdis.2016.02.007
15. Lovallo D, Kahneman D (2000) Living with uncertainty: attractiveness and resolution timing. J Behav Decis Mak 13(2):179–190. https://doi.org/10.1002/(SICI)1099-0771(200004/06)13:2%3c179::AID-BDM332%3e3.0.CO;2-J

16. Grupe DW, Nitschke JB (2011) Uncertainty is associated with biased expectancies and heightened responses to aversion. Emotion 11(2):413–424. https://doi.org/10.1037/a0022583
17. Freeston MH, Rhéaume J, Letarte H, Dugas MJ, Ladouceur R (1994) Why do people worry? Pers Individ Differ 17(6):791–802. https://doi.org/10.1016/0191-8869(94)90048-5
18. Ladouceur R, Gosselin P, Dugas MJ (2000) Experimental manipulation of intolerance of uncertainty: a study of a theoretical model of worry. Behav Res Ther 38(9):933–941. https://doi.org/10.1016/S0005-7967(99)00133-3
19. Buhr K, Dugas MJ (2002) The intolerance of uncertainty scale: psychometric properties of the English version. Behav Res Ther 40(8):931–945. https://doi.org/10.1016/S0005-7967(01)00092-4
20. Dugas MJ, Hedayati M, Karavidas A, Buhr K, Francis K, Phillips NA (2005) Intolerance of uncertainty and information processing: evidence of biased recall and interpretations. Cogn Ther Res 29(1):57–70. https://doi.org/10.1007/s10608-005-1648-9
21. McEvoy PM, Carleton RN, Correa K, Shankman SA, Shihata S (2019) Intolerance of uncertainty. In: The Cambridge handbook of anxiety and related disorders. Cambridge University Press, New York, NY, US, pp 196–226. https://doi.org/10.1017/9781108140416.008
22. Bardeen JR, Fergus TA, Orcutt HK (2013) Testing a hierarchical model of distress tolerance. J Psychopathol Behav Assess 35(4):495–505. https://doi.org/10.1007/s10862-013-9359-0
23. Carleton RN, Norton MAPJ, Asmundson GJG (2007) Fearing the unknown: a short version of the intolerance of uncertainty scale. J Anxiety Disord 21(1):105–117. https://doi.org/10.1016/j.janxdis.2006.03.014
24. Rosen NO, Ivanova E, Knäuper B (2014) Differentiating intolerance of uncertainty from three related but distinct constructs. Anxiety Stress Coping 27(1):55–73. https://doi.org/10.1080/10615806.2013.815743
25. Hong RY, Lee SSM (2015) Further clarifying prospective and inhibitory intolerance of uncertainty: factorial and construct validity of test scores from the intolerance of uncertainty scale. Psychol Assess 27(2):605–620. https://doi.org/10.1037/pas0000074
26. Birrell J, Meares K, Wilkinson A, Freeston M (2011) Toward a definition of intolerance of uncertainty: a review of factor analytical studies of the intolerance of uncertainty scale. Clin Psychol Rev 31(7):1198–1208. https://doi.org/10.1016/j.cpr.2011.07.009
27. Berenbaum H, Bredemeier K, Thompson RJ (2008) Intolerance of uncertainty: exploring its dimensionality and associations with need for cognitive closure, psychopathology, and personality. J Anxiety Disord 22(1):117–125. https://doi.org/10.1016/j.janxdis.2007.01.004
28. Furnham A, Ribchester T (1995) Tolerance of ambiguity: a review of the concept, its measurement and applications. Curr Psychol 14(3):179–199. https://doi.org/10.1007/BF02686907
29. Lauriola M, Foschi R, Mosca O, Weller J (2016) Attitude toward ambiguity: empirically robust factors in self-report personality scales. Assessment 23(3):353–373. https://doi.org/10.1177/1073191115577188
30. Grenier S, Barrette A-M, Ladouceur R (2005) Intolerance of uncertainty and intolerance of ambiguity: similarities and differences. Pers Individ Differ 39(3):593–600. https://doi.org/10.1016/J.PAID.2005.02.014
31. Gigerenzer G (2007) Gut feelings: the intelligence of the unconscious. Penguin Books, London
32. Ellsberg D (1961) Risk, ambiguity, and the Savage axioms. Q J Econ 75(4):643–669
33. Blais A-R, Weber EU (2006) A domain-specific risk-taking (DOSPERT) scale for adult populations. Judgment Decis Mak 1(1):33–47
34. Zimmermann G (2010) Risk perception, emotion regulation and impulsivity as predictors of risk behaviours among adolescents in Switzerland. J Youth Stud 13(1):83–99. https://doi.org/10.1080/13676260903173488
35. Weber EU, Blais A-R, Betz NE (2002) A domain-specific risk-attitude scale: measuring risk perceptions and risk behaviors. J Behav Decis Mak 15(August):263–290. https://doi.org/10.1002/bdm.414
36. Bonem EM, Ellsworth PC, Gonzalez R (2015) Age differences in risk: perceptions, intentions and domains. J Behav Decis Mak 28(4):317–330. https://doi.org/10.1002/bdm.1848

37. Ladouceur R, Talbot F, Dugas MJ (1997) Behavioral expressions of intolerance of uncertainty in worry: experimental findings. Behav Modif 21(3):355–371. https://doi.org/10.1177/01454455970213006
38. Reuman L, Jacoby RJ, Fabricant LE, Herring B, Abramowitz JS (2015) Uncertainty as an anxiety cue at high and low levels of threat. J Behav Ther Exp Psychiatry 47:111–119. https://doi.org/10.1016/j.jbtep.2014.12.002
39. Clark GI, Rock AJ, Clark LH, Murray-Lyon K (2020) Adult attachment, worry and reassurance seeking: investigating the role of intolerance of uncertainty. Clin Psychol 24(3):294–305. https://doi.org/10.1111/cp.12218
40. Rosen NO et al (2010) The impact of intolerance of uncertainty on anxiety after receiving an informational intervention about HPV: a randomised controlled study. Psychol Health 25(6):651–668. https://doi.org/10.1080/08870440902822913
41. Jacoby RJ, Abramowitz JS, Buck BE, Fabricant LE (2014) How is the beads task related to intolerance of uncertainty in anxiety disorders? J Anxiety Disord 28(6):495–503. https://doi.org/10.1016/j.janxdis.2014.05.005
42. Chen JT-H, Lovibond PF (2016) Intolerance of uncertainty is associated with increased threat appraisal and negative affect under ambiguity but not uncertainty. Behav Ther 47(1):42–53. https://doi.org/10.1016/j.beth.2015.09.004
43. Bartoszek G, Ranney RM, Curanovic I, Costello SJ, Behar E (2022) Intolerance of uncertainty and information-seeking behavior: experimental manipulation of threat relevance. Behav Res Ther 154:104125. https://doi.org/10.1016/j.brat.2022.104125
44. Luhmann CC, Ishida K, Hajcak G (2011) Intolerance of uncertainty and decisions about delayed, probabilistic rewards. Behav Ther 42(3):378–386. https://doi.org/10.1016/j.beth.2010.09.002
45. Appel H, Krasko J, Luhmann M, Gerlach AL (2024) Intolerance of uncertainty predicts indecisiveness and safety behavior in real-life decision making: results from an experience sampling study. J Res Pers 110:104490. https://doi.org/10.1016/j.jrp.2024.104490
46. Nelson BD, Shankman SA (2011) Does intolerance of uncertainty predict anticipatory startle responses to uncertain threat? Int J Psychophysiol 81(2):107–115. https://doi.org/10.1016/j.ijpsycho.2011.05.003
47. Jacoby RJ, Reuman L, Blakey SM, Hartsock J, Abramowitz JS (2019) "What if I make a mistake?": examining uncertainty-related distress when decisions may harm oneself vs. others. J Obsessive-Compulsive Relat Disord 20:50–58. https://doi.org/10.1016/j.jocrd.2017.11.003
48. Bottesi G, Carraro E, Martignon A, Cerea S, Ghisi M (2019) "I'm uncertain: what should i do?": an investigation of behavioral responses to everyday life uncertain situations. J Cogn Ther 12(1):55–72. https://doi.org/10.1007/s41811-019-00040-y
49. Tanovic E, Gee DG, Joormann J (2018) Intolerance of uncertainty: neural and psychophysiological correlates of the perception of uncertainty as threatening. Clin Psychol Rev 60:87–99. https://doi.org/10.1016/j.cpr.2018.01.001
50. Morriss J (2019) What do I do now? Intolerance of uncertainty is associated with discrete patterns of anticipatory physiological responding to different contexts. Psychophysiology 56(9):e13396. https://doi.org/10.1111/psyp.13396
51. Thibodeau MA, Carleton RN, Gómez-Pérez L, Asmundson GJG (2013) "What if I make a mistake?": intolerance of uncertainty is associated with poor behavioral performance. J Nerv Ment Dis 201(9):760–766. https://doi.org/10.1097/NMD.0b013e3182a21298
52. Shihata S, McEvoy PM, Mullan BA, Carleton RN (2016) Intolerance of uncertainty in emotional disorders: what uncertainties remain? J Anxiety Disord 41:115–124. https://doi.org/10.1016/j.janxdis.2016.05.001
53. Mahoney AEJ, McEvoy PM (2012) A transdiagnostic examination of intolerance of uncertainty across anxiety and depressive disorders. Cogn Behav Ther 41(3):212–222. https://doi.org/10.1080/16506073.2011.622130
54. Hong RY, Cheung MW-L (2015) The structure of cognitive vulnerabilities to depression and anxiety: evidence for a common core etiologic process based on a meta-analytic review. Clin Psychol Sci 3(6):892–912. https://doi.org/10.1177/2167702614553789

55. McEvoy PM, Hyett MP, Shihata S, Price JE, Strachan L (2019) The impact of methodological and measurement factors on transdiagnostic associations with intolerance of uncertainty: a meta-analysis. Clin Psychol Rev 73:101778. https://doi.org/10.1016/j.cpr.2019.101778
56. Osmanağaoğlu N, Creswell C, Dodd HF (2018) Intolerance of Uncertainty, anxiety, and worry in children and adolescents: a meta-analysis. J Affect Disord 225:80–90. https://doi.org/10.1016/j.jad.2017.07.035
57. Jenkinson R, Milne E, Thompson A (2020) The relationship between intolerance of uncertainty and anxiety in autism: a systematic literature review and meta-analysis. Autism 24(8):1933–1944. https://doi.org/10.1177/1362361320932437
58. Brown M, Robinson L, Campione GC, Wuensch K, Hildebrandt T, Micali N (2017) Intolerance of uncertainty in eating disorders: a systematic review and meta-analysis. Eur Eat Disord Rev 25(5):329–343. https://doi.org/10.1002/erv.2523
59. Shihata S, McEvoy PM, Mullan BA (2017) Pathways from uncertainty to anxiety: an evaluation of a hierarchical model of trait and disorder-specific intolerance of uncertainty on anxiety disorder symptoms. J Anxiety Disord 45:72–79. https://doi.org/10.1016/j.janxdis.2016.12.001
60. Mahoney AEJ, McEvoy PM (2012) Trait versus situation-specific intolerance of uncertainty in a clinical sample with anxiety and depressive disorders. Cogn Behav Ther 41(1):26–39. https://doi.org/10.1080/16506073.2011.622131
61. Jensen D, Cohen JN, Mennin DS, Fresco DM, Heimberg RG (2016) Clarifying the unique associations among intolerance of uncertainty, anxiety, and depression. Cogn Behav Ther 45(6):431–444. https://doi.org/10.1080/16506073.2016.1197308
62. Groves D, Clark LH, Clark GI (2020) The relationship between worry and intolerance of uncertainty subtypes. Aust Psychol 55(2):132–142. https://doi.org/10.1111/ap.12421
63. Carleton RN, Duranceau S, Freeston MH, Boelen PA, McCabe RE, Antony MM (2014) "But it might be a heart attack": intolerance of uncertainty and panic disorder symptoms. J Anxiety Disord 28(5):463–470. https://doi.org/10.1016/j.janxdis.2014.04.006
64. Scharmer C, Martinez K, Gorrell S, Reilly EE, Donahue JM, Anderson DA (2020) Eating disorder pathology and compulsive exercise during the COVID-19 public health emergency: examining risk associated with COVID-19 anxiety and intolerance of uncertainty. Int J Eat Disord 53(12):2049–2054. https://doi.org/10.1002/eat.23395
65. Du G, Lyu H (2021) Future expectations and internet addiction among adolescents: the roles of intolerance of uncertainty and perceived social support. Front Psychiatry 12. https://doi.org/10.3389/fpsyt.2021.727106
66. Danışman M, İspir GZ (2024) Decision-making styles, magical thinking, and intolerance of uncertainty in opioid use disorder. Ind J Psychiatry 66(6):545. https://doi.org/10.4103/indianjpsychiatry.indianjpsychiatry_630_23
67. Gao Z-H, Li J (2024) Intolerance of uncertainty and mental health in China "Post-pandemic" age: the mediating role of difficulties in emotion regulation. PLOS ONE 19(2):e0298044. https://doi.org/10.1371/journal.pone.0298044
68. Miller ML, McGuire JF (2023) Targeting intolerance of uncertainty in treatment: a meta-analysis of therapeutic effects, treatment moderators, and underlying mechanisms. J Affect Disord 341:283–295. https://doi.org/10.1016/j.jad.2023.08.132
69. Sternheim L, Harrison A (2018) The acceptability, feasibility and possible benefits of a group-based intervention targeting intolerance of uncertainty in adolescent inpatients with anorexia nervosa. Cogent Psychol 5(1):1441594. https://doi.org/10.1080/23311908.2018.1441594
70. Zvolensky MJ, Vujanovic AA, Bernstein A, Leyro T (2010) Distress tolerance: theory, measurement, and relations to psychopathology. Curr Dir Psychol Sci 19(6):406–410. https://doi.org/10.1177/0963721410388642
71. Lucchese VC, Cas ARD, Vasconcellos SJL (2023) Intolerance of uncertainty in students during the COVID-19 pandemic: scoping review. PTP 25(2). https://doi.org/10.5935/1980-6906/ePTPPE15294.en

72. Paluszek MM, Asmundson AJN, Landry CA, McKay D, Taylor S, Asmundson GJG (2021) Effects of anxiety sensitivity, disgust, and intolerance of uncertainty on the COVID stress syndrome: a longitudinal assessment of transdiagnostic constructs and the behavioural immune system. Cogn Behav Ther 50(3):191–203. https://doi.org/10.1080/16506073.2021.1877339
73. O'Neill SC et al (2006) Tolerance for uncertainty and perceived risk among women receiving uninformative BRCA1/2 test results. Am J Med Genet C Semin Med Genet 142C(4):251–259. https://doi.org/10.1002/ajmg.c.30104
74. Hromova HM (2022) Interrelation between intolerance of uncertainty and the time perspective profile in the military. Curr Issues Personal Psychol 10(4):321–332. https://doi.org/10.5114/cipp.2021.111984
75. Boelen PA (2019) Intolerance of uncertainty predicts analogue posttraumatic stress following adverse life events. Anxiety Stress Coping 32(5):498–504. https://doi.org/10.1080/10615806.2019.1623881
76. Bardeen JR, Fergus TA, Orcutt HK (2017) Examining the specific dimensions of distress tolerance that prospectively predict perceived stress. Cogn Behav Ther 46(3):211–223. https://doi.org/10.1080/16506073.2016.1233454
77. Fetzner MG, Horswill SC, Boelen PA, Carleton RN (2013) Intolerance of uncertainty and PTSD symptoms: exploring the construct relationship in a community sample with a heterogeneous trauma history. Cogn Ther Res 37(4):725–734. https://doi.org/10.1007/s10608-013-9531-6
78. Raines AM, Oglesby ME, Walton JL, True G, Franklin CL (2019) Intolerance of uncertainty and DSM-5 PTSD symptoms: associations among a treatment seeking veteran sample. J Anxiety Disord 62:61–67. https://doi.org/10.1016/j.janxdis.2018.12.002
79. Jacoby RJ, Szkutak A, Shin J, Lerner J, Wilhelm S (2023) Feeling uncertain despite knowing the risk: patients with OCD (but not controls) experience known and unknown probabilistic decisions as similarly distressing and uncertain. J Obsessive-Compul Relat Disord 39:100842. https://doi.org/10.1016/j.jocrd.2023.100842
80. Bavclar J et al (2023) Intolerance of uncertainty and reactions to the COVID-19 pandemic. J Gen Psychol 150(2):143–170. https://doi.org/10.1080/00221309.2021.1922346
81. del-Valle MV et al (2022) Intolerance of COVID-19-related uncertainty and depressive and anxiety symptoms during the pandemic: a longitudinal study in Argentina. J Anxiety Disord 86:102531. https://doi.org/10.1016/j.janxdis.2022.102531
82. Satici B, Saricali M, Satici SA, Griffiths MD (2022) Intolerance of uncertainty and mental wellbeing: Serial Mediation by rumination and fear of COVID-19. Int J Ment Health Addict 20(5):2731–2742. https://doi.org/10.1007/s11469-020-00305-0
83. Lee S, Jeon KO, Kim H, Chung EK (2020) Effects of intolerance of uncertainty on turnover intention in transplantation coordinators: the roles of burnout and grit. Korean J Transplant 34(4):265–271. https://doi.org/10.4285/kjt.20.0044
84. Di Monte C, Monaco S, Mariani R, Di Trani M (2020) From resilience to Burnout: psychological features of Italian general practitioners during COVID-19 emergency. Front Psychol 11. https://doi.org/10.3389/fpsyg.2020.567201
85. Malouf P, Quinlan E, Mohi S (2023) Predicting burnout in Australian mental health professionals: uncertainty tolerance, impostorism and psychological inflexibility. Clin Psychol 27(2):186–195. https://doi.org/10.1080/13284207.2022.2163159
86. Reizer A, Galperin BL, Chavan M, Behl A, Pereira V (2022) Examining the relationship between fear of COVID-19, intolerance for uncertainty, and cyberloafing: a mediational model. J Bus Res 145:660–670. https://doi.org/10.1016/j.jbusres.2022.03.037
87. Keating DJ, Cullen-Lester KL, Meuser JD (2024) Virtual work conditions impact negative work behaviors via ambiguity, anonymity, and (un)accountability: an integrative review. J Appl Psychol 109(2):169–201. https://doi.org/10.1037/apl0001126
88. Lee A, Jung E (2021) The mediating role of entrepreneurial mindset between intolerance of uncertainty and career adaptability. Sustainability 13(13), Art no 13. https://doi.org/10.3390/su13137099

89. Chen Q, Lu L, Huang H, Fang Y (2023) Fear of COVID-19 and the career maturity of Chinese international high school students: the mediating effect of the intolerance of uncertainty. J Health Psychol 28(5):450–461. https://doi.org/10.1177/13591053221116678
90. Doğanülkü HA (2024) Life goals and proactive career behaviors: the mediating role of visions about the future and the moderating role of intolerance of uncertainty. Curr Psychol 43(1):72–84. https://doi.org/10.1007/s12144-022-03236-2
91. Wang R, Lin X, Wang L (2022) The paradoxical effects of social class on career adaptability: the role of intolerance of uncertainty. Front Psychol 13. https://doi.org/10.3389/fpsyg.2022.1064603
92. Brun C, Zerhouni O, Akinyemi A, Houtin L, Monvoisin R, Pinsault N (2023) Impact of uncertainty intolerance on clinical reasoning: a scoping review of the 21st-century literature. J Eval Clin Pract 29(3):539–553. https://doi.org/10.1111/jep.13756
93. Lin Y, Mason D, Zhong S, Hirsch C, Happé F (2024) How is "intolerance of uncertainty" (IU) measured? A systematic review of assessment tools for IU and the psychometric properties of IU questionnaires. Clin Psychol Sci Pract 31(2):179–204. https://doi.org/10.1037/cps0000199

Jozef Bavolar is an associate professor of psychology at the Pavol Jozef Safarik University in Košice, Slovakia. He graduated in Psychology in 2005 and received his Ph.D. in Social Psychology and Psychology of Work in 2009. His research focuses on individual differences in decision-making, mainly decision-making styles, decision-making competence, indecisiveness and the intolerance of uncertainty. He studies mainly their associations with real-life decision outcomes. He publishes in a variety of general psychological journals and journals focusing on judgment and decision-making.

Decision Making in Health Care: "If Only I'd Known then What I Know Now"

9

Maka Tsulukidze, Stuart W. Grande, and Ingrid Skop

Abstract

The chapter provides an in-depth exploration of medical decision making (MDM), provides a review of the evolution of shared decision making (SDM) from its origins grounded in the early 20th-century concept of informed consent to its current status. Among the many complexities surrounding SDM implementation, particular attention is given to the key actors of SDM—patients and physicians. The same complexities are also considered in the light of new technological advances powered by artificial intelligence, which offers immense opportunities but also presents numerous challenges. Using empirical evidence, the chapter offers the analysis of power dynamics characteristic to MDM and the fluid nature of SDM, discusses different ways of SDM implementation and the need for more developed approach to its application in clinical practice. Finally, as educators, clinicians and researchers, we also offer our personal views informed by our practice and experience. We hope that this thorough examination of medical decision making, SDM's theoretical foundations, practical challenges, and potential future developments will be an important resource for both researchers and practitioners for understanding, improving and implementing patient-centered care.

M. Tsulukidze (✉)
Department of Health Sciences, Florida Gulf Coast University, Fort Myers, FL, USA
e-mail: mtsulukidze@fgcu.edu

S. W. Grande
Public Health Administration and Policy, School of Public Health, University of Minnesota, Minneapolis, MN, USA
e-mail: sgrande@umn.edu

I. Skop
Charlotte Lozier Institute, Arlington, VA, USA
e-mail: lskop@lozierinstitute.org

© The Author(s), under exclusive license to Springer Nature Switzerland AG 2025

T. McElroy (ed.), *Decision-Making in Life and Work*, Integrated Science 36,
https://doi.org/10.1007/978-3-031-88415-3_9

Graphical Abstract

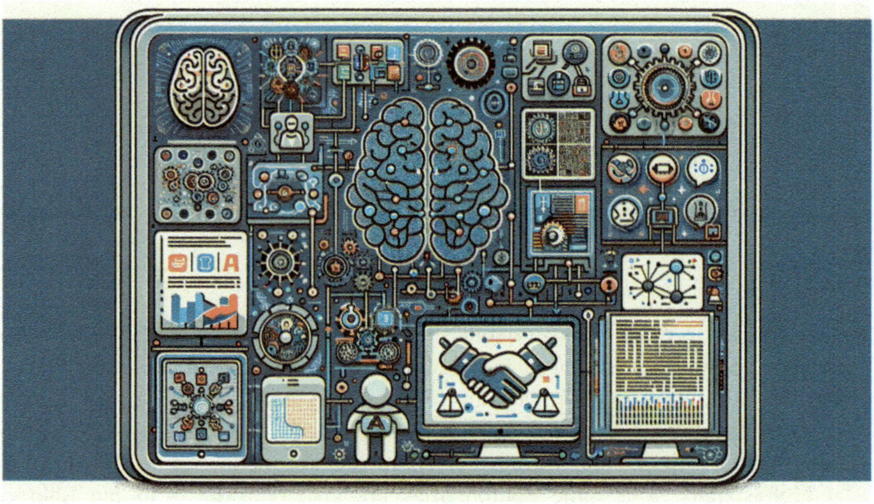

Keywords

Medical decision making • Shared decision making • Historical perspectives and evolution of medical decision making • Future directions in medical decision making

When I [MT] shared Kerry O'Connell's profoundly and deeply personal story [1] of becoming disabled at the age of 48 due to arm surgery and hospital-acquired infection with my students, the energy in the class shifted. My students, future clinicians and health care professionals instantly realized the story was too important and carried real life implications for their future practice. Many patients do not get to tell their stories so eloquently or at all. The suffering, however, is equally immense and often irreparable. Robust medical decision making, practiced responsibly and professionally, can alleviate such suffering. As Schwartz and Bergus correctly noted, "decision making is a key activity, perhaps the most important activity, in the practice of healthcare".[2]

The latest advances in medical sciences, research and clinical breakthroughs have revolutionized and changed health care as we once knew it [3]. However, application of this knowledge into clinical practice, to individual patients, is not always at a pace with scientific developments. Rethinking medical decision making is arguably one of the biggest challenges in the lack of wider spread of medical progress.

1 Medical Decision Making: History and Evolution

According to Schwartz and Bergus, "Medical decision science is a field that encompasses several related pursuits. As a normative endeavor, it proposes standards for ideal decision making. As a descriptive endeavor, it seeks to explain how physicians and patients routinely make decisions, and has identified both barriers to, and facilitators of, effective decision making. As a prescriptive endeavor, it seeks to develop tools that can guide physicians, their patients, and health care policymakers to make good decisions in practice." [2]

Historically, physicians often paternalistically guided patients toward acceptance of their preferred recommendations when there were several treatment alternatives available. Patients accepted such guidance, under the belief that the "doctor knows best". Yet, increasingly, it became recognized that patients often regretted blindly accepting their physician's recommendation, as they belatedly realized that other options had been available that they may have preferred to choose.

The term "informed consent" was legally recognized in the US for the first time in 1914 in the case law on *Schloendorff v. Society of New York Hospitals* [4]. The court decided in favor of Mrs. Schioendorff who, as a competent patient, had consented to abdominal examination under anesthesia but was not informed about the tumor, which was removed by the surgeon without informing her about the possible adverse outcomes [5, 6].

Informed consent has become a standard recommendation for quality medical care, balancing the ethical principles of beneficence, non-maleficence, autonomy and justice. The American Medical Association's Principles of Medical Ethics defines informed consent as: "The process of informed consent occurs when communication between a patient and physician results in the patient's authorization or agreement to undergo a specific medical intervention. In seeking a patient's informed consent (or the consent of the patient's surrogate if the patient lacks decision-making capacity or declines to participate in making decisions), physicians should: (a) Assess the patient's ability to understand relevant medical information and the implications of treatment alternatives and to make an independent, voluntary decision. (b) Present relevant information accurately and sensitively, in keeping with the patient's preferences for receiving medical information. The physician should include information about: (i) the diagnosis (when known); (ii) the nature and purpose of recommended interventions; (iii) the burdens, risks, and expected benefits of all options, including forgoing treatment. (c) Document the informed consent conversation and the patient's (or surrogate's) decision in the medical record in some manner. When the patient/surrogate has provided specific written consent, the consent form should be included in the record." [7].

The judicial system has clarified the components of informed consent further. In the Canterbury v Spence case, the DC Circuit Court of Appeals, documented that when obtaining informed consent, a physician must disclose: (1) condition being treated; (2) nature and character of the proposed treatment or surgical procedure; (3) anticipated results; (4) recognized possible alternative forms of treatment; and (5) recognized serious possible risks, complications, and anticipated benefits involved in the treatment or surgical procedure, as well as the recognized possible alternative forms of treatment, including non-treatment [8].

To ensure informed consent is provided, clinicians are legally and ethically obligated to help patients understand the risks, benefits, and alternatives of a given clinical treatment or intervention. An important part of the consent is that the patient is competent to make a voluntary decision about the proposed treatment, procedure or intervention and, based on the clinician's assessment, understands the offered plan of treatment.

In the U.S. the required standard for informed consent is determined by the state based on the following three legal approaches to adequate informed consent: (1) Subjective standard: *What would this patient need to know and understand to make an informed decision?* (2) Reasonable patient standard: What would the average patient need to know to be an informed participant in the decision? (3) Reasonable physician standard: What would a typical physician say about this procedure? [Shah P, Thornton I, Turrin D, et al. Informed Consent. [Updated 2023 Jun 5]. In: StatPearls [Internet]. Treasure Island (FL): StatPearls Publishing; 2024 Jan-. Available from: https://www.ncbi.nlm.nih.gov/books/NBK430827/].

Most states use the "reasonable patient standard" as it focuses on what a typical patient would need to know to understand the medical decision [Shah P, Thornton I, Turrin D, et al. Informed Consent. [Updated 2023 Jun 5]. In: StatPearls [Internet]. Treasure Island (FL): StatPearls Publishing; 2024 Jan-. Available from: https://www.ncbi.nlm.nih.gov/books/NBK430827/].

Currently, informed consent is applied in all areas of medical care to ensure legal compliance and protection of both clinicians and patients and ethically support patient's autonomy. However, inadequacy of informed consent has become increasingly obvious with time. First, "informed consent defines and protects only a very narrow version of patient autonomy, essentially a veto power over a short list of potential interventions offered by a clinician." [Tonelli, M. R., & Sullivan, M. D. (2019). Person-centred shared decision making. Journal of evaluation in clinical practice, 25(6), 1057–1062.]

"The legal doctrine of informed consent doctrine focuses only on disclosure, not on patient understanding. The physician's duty is only to "deliver" certain information to the patient, not to ensure that the patient receives and appreciates it. Despite its name, "informed consent" fails to assure that the patient's consent is "informed." [Pope, T. M. (2019). Informed consent requires understanding: complete disclosure is not enough. The American Journal of Bioethics, 19(5), 27–28.]

In addition, consent forms fall short of providing information to guide decision making even for patients with higher levels of education. In research examining the comprehension, many of the 68% of the participants with education beyond high school, still missed comprehension questions and found the consent forms difficult to read. [Manta, C. J., Ortiz, J., Moulton, B. W., & Sonnad, S. S. (2021).

From the patient perspective, consent forms fall short of providing information to guide decision making. Journal of patient safety, 17(3), e149–e154.]

An alarming 88% of patients who had signed informed consent for elective cardiac catheterization held fundamentally mistaken beliefs about the potential benefits of the procedure. [Rothberg, M. B., S. K. Sivalingam, R. Kleppel, et al. 2015. Informed decision making for percutaneous coronary intervention for stable coronary disease. JAMA Internal Medicine. 175(7): 1199–1206. doi: https://doi.org/10.1001/jamainternmed.2015.1657.]

When examining understanding of informed consent forms, a study found that patients with both high and low literacy expressed a desire for extra educational information provided in addition to consent forms and listed video, diagram or illustration as helpful tools facilitating understanding of medical information. [Manta, C. J., Ortiz, J., Moulton, B. W., & Sonnad, S. S. (2021). From the patient perspective, consent forms fall short of providing information to guide decision making. Journal of patient safety, 17(3), e149–e154.]

Recognizing that informed consent "strives not to interfere with patient autonomy but does nothing to promote patient autonomy" [Tonelli, M. R., & Sullivan, M. D. (2019). Person-centred shared decision making. Journal of evaluation in clinical practice, 25(6), 1057–1062.] and provides little if any support to guarantee patient informed decision making, it was proposed to "rethink informed consent" [King JS, Moulton BW. Rethinking informed consent: the case for shared medical decision-making. Am J Law Med. 2006;32(4):429–501. doi: https://doi.org/10.1177/009885880603200401PubMedGoogle ScholarCrossref] and introduce "shared medical decision-making" (SDM) [9].

The proposed shared medical decision-making principles extended the concept of informed consent beyond simple information transfer [9] to identify patient's informed preferences based on deliberate and in-depth discussion between patient and clinician.

The SDM model, published in 1997 [Charles C, Gafni A, Whelan T. Shared decision-making in the medical encounter: what does it mean? (or it takes at least two to tango). Soc Sci Med. 1997;44:681–92.] consisted of four key elements: (1) decision making involves at least two participants: a clinician and a patient; (2) both parties share information; (3) both parties seek agreement about the preferred treatment; (4) clinician and patient must reach an informed, mutually agreed decision about the best available treatment option.

In 1982, the President's Commission for the Study of Ethical Problems in Medicine and Biomedical and Behavioral Research issued a report that proposed further advancement of informed consent and then stated, "The ethical foundation of informed consent can be traced to the promotion of 2 values: personal well-being and self-determination. To ensure that these values are respected and enhanced, the Commission finds that patients... must have all relevant information regarding their condition and alternative treatments, including possible benefits, risks, costs, other consequences, and significant uncertainties.... Ethically valid consent is a process of shared decision making based upon mutual respect and participation, not a ritual to be equated with reciting the contents of a form..." [10]

In 2011 Salzburg statement on SDM Formally acknowledged patients' role of in making medical decision making and called "on patients and clinicians to work together to be coproducers of health." [11]. In current medical practice, SDM is almost universally accepted in the United States and globally based on both ethical considerations and empirical evidence. It represents a pivotal shift in the healthcare landscape, epitomizing a movement towards more collaborative, patient-centered care. Over the past 25 years, SDM has evolved significantly, informed by a confluence of historical, sociological, and clinical insights that underscore the importance of the patient's role in healthcare decisions. Policymakers, payers and health systems are expected to advocate for engaging patients in shared decision making as a standard of care for clinical practice.

2 So, What is SDM?

> "Listen to your patient; he is telling you the diagnosis" a quote attributed to Sir William Osler, a 19th-century physician and chair of Clinical Medicine at the University of Pennsylvania, emphasizes the importance of carefully listening to a patient while taking a medical history. Adapted to SDM it might say, "Listen to your patient; he is telling you the ***preference*** diagnosis."

SDM, as we know it today, emerged as a response to nearly half a century of observed power imbalances at the point of care, advocating for a model where patients are not merely passive recipients of care but active, informed participants. It recognizes that while patients may lack the professional expertise to diagnose or treat ailments, they hold invaluable personal insights and preferences that are crucial to the care process. This has a special relevance where uncertainty or complexity may be high.

Certain medical conditions for which there is no clear "correct" clinical option, can have a number of preference-sensitive decisions available. If a patient's preferences, values and needs are not identified correctly, if they do not understand the available options, treatments and interventions received by the patient may prove unwanted, unnecessary, and even unsafe. Such value-discordant care has been described as preference misdiagnosis [12]. However, unlike clinical misdiagnosis, it is 'silent' [12], undetected and unrecognized in current health care systems.

Answering the question, what is SDM requires a certain amount of story-telling and reflection. Focusing on the sociocultural dimensions of medicine and healthcare, medical anthropologists uncover the underlying factors that drive patient-provider interactions. Mary-Joe DelVecchio Good, has extensively explored the narratives and practices of medicine and healthcare, how medical knowledge and authority are constructed and how these constructions influence patient-provider relationships [13]. For example, her research on the cultural dimensions of pain and suffering reveals how cultural beliefs and expectations shape the ways in which patients express symptoms and how providers interpret and respond to these expressions [14].

Sienna Craig has contributed significantly to our understanding of how traditional and biomedical healthcare systems intersect and interact. Her work in the Himalayan regions, where traditional Tibetan medicine coexists with Western biomedical practices, offers a compelling case study of how diverse healthcare paradigms can influence patient-provider relationships [15]. Craig's research highlights the importance of understanding cultural context in healthcare delivery and the potential for integrative approaches that respect and incorporate multiple healing traditions.

Medical anthropology also emphasizes the role of power dynamics in patient-provider relationships. The field explores how social hierarchies, institutional structures, and cultural norms can create imbalances in these interactions, affecting the quality and equity of healthcare. By bringing attention to these issues, medical anthropologists advocate for more inclusive and responsive healthcare systems that acknowledge and address the diverse needs and perspectives of patients. In addition to advocacy, medical anthropology has leveraged methods like ethnography to help researchers examine patient-provider relationships in detail, capturing both the interactions but also the unique qualities of human communication and expression.

Equally important to the development of SDM is the sociological perspective. Medical sociologists have examined the illness behavior as well as the influence of highly structured medical environments like hospitals or insurance markets on human behavior and trust [16], the sick role as it pertains to the various expressions of human illness and response to illness [17]. This work laid the foundation for deeper discussions into ways people experience pain as well as how systems have evolved to treat people who are seen as sick, ill, diseased, or otherwise outside the normal responses of human behavior. The impact of this work remains salient today as we hear more about patients altering their behaviors to gain access to different treatments or patients changing their expression of pain to demand more from their provider [18].

Collectively, this work suggests that patient engagement in healthcare is significantly shaped by human interactions with, and perceptions of, the healthcare system. This line of thought is further expanded by Eliot Freidson, who delves into the profession of medicine, examining how the medical field's structure, with its defined barriers of entry, membership, and knowledge authority, inherently positions patients at a disadvantage when engaging with healthcare systems [19].

The historical roots of SDM are also deeply entwined with the seminal work of pioneers such as Moira Stewart, Debra Roter, Ian McWhinney, and Joel Leventhal, who were some of the first physicians and researchers to extensively explore patient-provider relationships [20]. Their collective research underscores the transformative potential of patient engagement in healthcare conversations, advocating for a model of care that respects and integrates the patient's voice as a fundamental component of clinical decision-making.

For understanding "What is Shared Decision Making" we should also highlight the contributions of Wennberg and Gittlesohn's work on small area variation in

Vermont in the late 1960's. At the about the same time McWhinney was building a reputation for family medicine in Canada, Wennberg and Gittlesohn were examining differences in various patient outcomes across the state of Vermont, with an interest in trying to first describe then explore practice variation [21]. They would later discover and identify three areas of practice variation: effective, preference-sensitive, and supply-sensitive care. Based on the idea that preference-sensitive care demonstrated how outcomes could vary based on what mattered to the patient rather than what the evidence claimed, gave strong support for the notion that a patient's views were important. While there are deeper explorations of these concepts elsewhere [22], even a short introduction is valuable to help seat SDM within this broader foundation, which essentially states that medicine is about control and power. Navigating the system, from a patient perspective, is about who has the power and how medical decisions are made and executed.

This backdrop of power imbalance sets the stage for understanding the emergence and necessity of SDM. Ivan Illich's critique in "Limits of Medicine", provides a poignant commentary on the potential detriments of medical intervention, emphasizing the importance of re-evaluating the role of medicine in society and, by extension, the role of the patient in healthcare [23].

In the 1990s emerged a new model of thinking that was guided in many ways by these earlier challenges to medical practice, and particular the accepted model of paternalism. Emanuel and Emanuel's seminal work characterizing a "new" model of deliberative describing the patient-provider relationship [24]. This new model conceived a standard of practice that involved informing, sharing, and eliciting patient values as central to the decision making practice. It was around this time in the early 1990s that SDM started to take shape.

At its core, SDM is characterized by several foundational elements agreed upon by many experts in the field, despite varying definitions. These elements include establishing a partnership between patient and provider, presenting treatment options along with their evidence base, eliciting patient preferences, and integrating these preferences with the evidence to guide care decisions. This model not only respects the patient's experiential authority but also acknowledges their right and ability to understand and weigh the risks and benefits of different treatment paths.

When it comes to answering the question WHAT IS SDM, however, a specific, universally adopted definition is lacking [H. Bomhof-Roordink, F. R. Gärtner, A. M. Stiggelbout, A. H. Pieterse. Key components of shared decision making models: a systematic review BMJ Open, 9 (2019), Article e031763] and many studies do not identify the SDM framework or a specific definition of SDM used in their research [Makoul G, Clayman ML. An integrative model of shared decision making in medical encounters. Patient Education and Counseling, 2006; 60: 301–312. Moumjid N, Gafni A, Bremond A, Carrere MO. Shared decision making in the medical encounter: are we all talking about the same thing? Medical Decision Making, 2007; 27: 539–546]. Aptly titling their research as "Shared decision making in the medical encounter: are we all talking about the same thing?" Moumjid and colleagues [Moumjid N, Gafni A, Bremond A, Carrere MO. Shared decision making in the medical encounter: are we all talking about the same thing? Medical Decision Making, 2007; 27: 539–546] expressed the domain-wide confusion clearly and succinctly.

In broad terms, SDM is conceptualized as the process through which patients and clinicians collaboratively make a decision considering patient's medical situation and views and preferences [25, 26]. Therefore, the best available clinical evidence on risks and expected outcomes of a certain treatment or intervention is balanced against patient preferences and values.

To further address the lack of unified SDM model, Makoul and Clayman proposed a set of essential elements to form an integrative model of SDM (e.g., Define and/or explain the problem, Discuss pros/cons, Patient values/preferences, Make or explicitly defer decision) [26]. However, consistency and uniformity are lacking with regard to the elements of SDM models. A systematic review of key components of shared decision making models identified 40 SDM models with significant heterogeneity around their essential elements despite some similarities between the models. [Bomhof-Roordink, H., Gärtner, F. R., Stiggelbout, A. M., & Pieterse, A. H. (2019). Key components of shared decision making models: a systematic review. BMJ open, 9(12), e031763.]

This analysis showed that identifying patient preferences about the available treatment options was one of the central concepts of SDM. *Describe treatment options* was the most prominent component across 40 models included in the review. *Make the decision*, *Patient preferences*, *Tailor information*, *Deliberate*, *Create choice awareness*, and *Learn about the patient* were present in >50% of models.

This heterogeneity extends to the definition of key actors in a decision making process. Ten models out of 40 included in the same systematic review, identified healthcare professional as a sole actor, 37 of the 40 models identified one or more actors, including both patient and clinician. [Bomhof-Roordink, H., Gärtner, F. R., Stiggelbout, A. M., & Pieterse, A. H. (2019). Key components of shared decision making models: a systematic review. BMJ open, 9(12), e031763.]

Although key actors may vary based on medical condition and may sometimes include other clinicians and patient family members, the absence of consensus on who should participate in SDM and to what extent reflects a significant divide in views on how SDM should be practiced in clinical care.

3 Patients as SDM Actors

The fact that SDM models diverge on whether patient is a key actor in medical decision making [Bomhof-Roordink, H., Gärtner, F. R., Stiggelbout, A. M., & Pieterse, A. H. (2019). Key components of shared decision making models: a systematic review. BMJ open, 9(12), e031763.] points to the fact that immense amount of work needs to be done to bridge the gap between the existing reality and the ethical imperative of engaging patients as full partners in their own care.

Characterizing patients' perspectives on shared decision making encompass topics such as whether decisions were shared, their perceptions of physician behaviors [27], impact of SDM on patient satisfaction [28], and whether or not they want to share decisions with their physicians [27].

Characterizing patients' perspectives on shared decision making, existing research explores whether decisions were shared between patient and clinician [Kremer H., Ironson G., Schneiderman N,. Hautzinger M. _It_s my body_: does patient involvement in decision making reduce decisional conflict? Medical Decision Making, 2007; 27: 522–532. Swenson SL, Buell S, Zettler P, White M, Ruston DC, Lo B. Patient-centered communication: do patients really prefer it? Journal of General Internal Medicine, 2004; 19: 1069–1079.], patients' perceptions of physician behaviors (Burt, J., Abel, G., Elliott, M. N., Elmore, N., Newbould, J., Davey, A., Llanwarne, N., Maramba, I., Paddison, C., Campbell, J., & Roland, M. (2018). The Evaluation of Physicians' Communication Skills From Multiple Perspectives. Annals of Family Medicine, 16(4), 330–337. https://doi.org/https://doi.org/10.1370/afm.2241), impact of SDM on patient satisfaction (Shepherd, J., Gurney, S., & Patel, H. P. (2022). Shared decision making and personalized care support planning: pillars of integrated care for older people. Clinics in Integrated Care, 12, 100,097. https://doi.org/https://doi.org/10.1016/j.intcar.2022.100097) and whether or not patients want to share decisions with their physicians (Burton, D., Blundell, N., Jones, M., Fraser, A., & Elwyn, G. (2010). Shared decision-making in cardiology: do patients want it and do doctors provide it? Patient Education and Counseling, 80(2), 173–179.)

There is an argument that some patients may not value the invitation to be actively involved in their own care as a decision-maker, as they feel uncomfortable, uncertain or incapable of navigating the decision complexities. In most cases, it has been shown that patients are scared or fear asking questions of their providers/doctors assuming their care may be at risk if they ask the wrong question or they are seen as disruptive [29]. There is even less data on how vulnerable populations, such as elderly patients, patients with language barrier and lower literacy levels perceive their engagement in clinical care. As a result, there are refuted and debatable findings that cite these groups as less engaged in the decision-making process. What we know based on current evidence is that marginalized populations in the community are not included in the development of SDM-based tools or approaches, on average. Diversity of cultural backgrounds is also referenced as a contributor in making it difficult for some patients to engage with their provider [30].

However, evidence is quite strong on SDM benefits experienced by patients particularly around perceived confidence, satisfaction, and reduced conflict [31]. When SDM is supported by a clinical decision tool or decision aid, evidence shows an increased patient knowledge of available treatment options and outcomes and reduced decisional conflict. Research also shows that patients who watched a video describing the procedure, risks, benefits and alternatives had significantly improved comprehension compared to patients who only received information verbally and the comprehension benefits were even greater in patients with low literacy levels

[Manta, C. J., Ortiz, J., Moulton, B. W., & Sonnad, S. S. (2021). From the patient perspective, consent forms fall short of providing information to guide decision making. Journal of patient safety, 17(3), e149–e154.]

Temporal trends in patient behavior are also important to note. A systematic review of patient preferences for shared decisions showed that the majority of patients preferred sharing decisions in 71% of studies dated 2000 and later, compared to only 50% of studies dated before 2000. [Chewning B, Bylund CL, Shah B, Arora NK, Gueguen JA, Makoul G. Patient preferences for shared decisions: a systematic review. Patient Educ Couns 2012;86:9–18.] This growing interest indicates that patient non-involvement may not be a lack of desire but instead a lack of awareness, as well as necessary knowledge, on how to engage in SDM. [Légaré, F., & Thompson-Leduc, P. (2014). Twelve myths about shared decision making. Patient education and counseling, 96(3), 281–286.]

The role of SDM in enhancing patient understanding and reducing decisional conflict can be transformative [32]. Patients that are actively involved in the decision-making process, are empowered to understand complex treatment options and make informed choices that resonate with their personal and cultural values. Participating in medical decision making can also increase patient satisfaction and confidence in treatment decisions [33, 34]. When patients feel heard and respected, their trust in the healthcare system and their providers deepens, fostering a therapeutic alliance that is essential for effective care. This is particularly vital for marginalized populations who may have experienced historical mistrust and discrimination within healthcare settings.

King [35] extends this discussion by highlighting the role of SDM in augmenting patient knowledge. As patients engage more deeply with their care options, they become more knowledgeable about their conditions and the potential pathways for treatment. This increased knowledge not only bolsters their confidence but also equips them to advocate for themselves more effectively, a critical skill for navigating complex healthcare systems.

The future of SDM will depend on the interests of patients in being engaged in their clinical decision making. Entire movements have started in favor of patients taking more of a role in their care and treatment. Montori's book "Why we revolt" [36] is a call for a patient revolution of sorts [37], citing the industrial mechanism of current medicine and the consequences for patient care. He and others make the case that patients' face severe barriers to effective care in medicine, be they knowledge of practice, awareness of treatment options and risks, as well as basic health literacy. Under this premise, any method or tool to help frame patients' perspectives is needed and necessary.

This premise is of particular importance as it relates to vulnerable populations, which includes but are not limited to racial and ethnic minorities, economically disadvantaged and low-income groups, and elderly. Existing research indicates that vulnerable populations face barriers to participation in SDM, which exacerbates disparities in decision quality [Grabinski, V. F., Myckatyn, T. M., Lee, C. N., Philpott-Streiff, S. E., & Politi, M. C. (2018). Importance of shared decision-making for vulnerable populations: examples from postmastectomy breast reconstruction. Health Equity, 2(1), 234–238.]

> I thought [only] people in Hollywood had reconstructed breasts. I was a poor, old, Black lady, and it didn't matter whether I had one breast… I gladly would have talked with somebody if somebody had approached me with the details about it… I guess they felt like a woman my age, you don't matter. You're gettin' ready to die anyway. You may as well die with one breast. [Hasak JM, Myckatyn TM, Grabinski VF, et al. Stakeholders' perspectives on postmastectomy breast reconstruction: recognizing ways to improve shared decision making. Plast Reconstr Surg Glob Open. 2017;5:e1569.]

This powerful and poignant comment, made by a breast cancer survivor recounting her experiences with postmastectomy breast reconstruction, reveals an internal defeat that vulnerable patients feel due to multiple layers of decisional blocks: self doubt, lack of communication, lack of resources, perceived discourgment from external groups …

Vulnerable population-specific barriers to SDM may include distrust in the healthcare system, heightened power differentials with their providers and implicit biases held by providers [Grabinski, V. F., Myckatyn, T. M., Lee, C. N., Philpott-Streiff, S. E., & Politi, M. C. (2018). Importance of shared decision-making for vulnerable populations: examples from postmastectomy breast reconstruction. Health Equity, 2(1), 234–238.; Peek, M. E., Odoms-Young, A., Quinn, M. T., Gorawara-Bhat, R., Wilson, S. C., & Chin, M. H. (2010). Race and shared decision-making: perspectives of African-Americans with diabetes. Social science & medicine, 71(1), 1–9.] As a consequence, patients who experience physician bias, discrimination, and cultural discordance are often less willing to share health information and less willing to adhere to physician treatment recommendations. [Peek, M. E., Odoms-Young, A., Quinn, M. T., Gorawara-Bhat, R., Wilson, S. C., & Chin, M. H. (2010). Race and shared decision-making: perspectives of African-Americans with diabetes. Social science & medicine, 71(1), 1–9.]

As we think about implementation of SDM and its vital role in the elimination or mitigation of multiple biases and barriers at the point of care, we identify several important strategies that apply to both general and vulnerable populations, yet to a different degree. In Table 1 provided later in this chapter, we summarize the barriers to SDM implementation, identified in the literature, and potential strategies to address them.

As healthcare continues to evolve, the principles of SDM offer a roadmap for creating a system that truly meets the needs of all patients, irrespective of their background or social standing. As patients with multiple vulnerabilities receive additional support to meaningfully engage with providers [Grabinski, V. F., Myckatyn, T. M., Lee, C. N., Philpott-Streiff, S. E., & Politi, M. C. (2018). Importance of shared decision-making for vulnerable populations: examples from postmastectomy breast reconstruction. Health Equity, 2(1), 234–238.], SDM may become a model of care that shows a real potential to moderate health disparities experienced by vulnerable populations, mitigate disparities through the basic principles

Table 1 Major barriers and potential strategies for engaging patients in shared decision-making

Barrier	Potential Strategy
Individual physician level	
Lack of provider training in SDM	Increased SDM training in medical education
Provider biases and assumptions	Training for addressing biases, implicit bias training, understanding different forms of patient vulnerability
Time constraints	Finding opportunities and place in clinical care to insert SDM implementation strategies, e.g., providing decision aids prior to consultation, consider special needs of vulnerable patients, including more time needed to convey the information, and design SDM implementation accordingly
Individual patient level	
Lack of patient participation and willingness to be engaged in SDM	Patient education about the value of SDM, and the ways in which they can actively participate in their own care
Lack of including patients, especially from vulnerable groups, in SDM research	Training for researchers to identify gaps in patient inclusion methodologies, increased funding to study SDM for vulnerable populations, understanding concerns that vulnerable patients face, including access to and cost of health care
Health system level	
Numerous decision aid shortcomings, including difficulties to keep the decision aids up to date, challenges related to certification of decision aids, limited applicability of decision aids to vulnerable patients	Establish a national body responsible for certifying decision aids, place special emphasis on developing aids for vulnerable populations, seek feedback from a diverse group of patients during decision aid development, considering different reading levels, different formats, such as graphical or pictorial presentation of information
Lack of SDM mandates for preference-sensitive care	Identify preference-sensitive procedures and mandate SDM while providing care

(continued)

Table 1 (continued)

Barrier	Potential Strategy
Policy level	
Lack of provider reimbursement for practicing SDM	Establish and promote policies for provider reimbursement, championing SDM as a critical component of healthcare reform, wider implementation of CMS reimbursement plans for SDM focused clinical encounters
Lack of legislation and policy initiatives to promote the adoption of SDM	Develop and promote legal process that supports a reasonable-patient–centered standard through shared decision making to support SDM implementation, replicate the legislative model implemented by Washington State incorporating shared decision making in the informed consent and promoting the use of decision aids in addition to or as an alternative to standard informed consent documents

of preference elicitation, partnership, reduction of decisional conflict, increased knowledge, and improved decisional self-efficacy [Durand, M.-A., Carpenter, L., Dolan, H., Bravo, P., Mann, M., Bunn, F., & Elwyn, G. (2014). Do interventions designed to support shared decision-making reduce health inequalities? A systematic review and meta-analysis. PloS One, 9(4), e94670]. In this light, SDM is not just a clinical strategy but a moral imperative, paving the way for a more just and compassionate healthcare landscape.

4 Clinicians as SDM Actors

Similar divergence of views exists about physician's role and desire to practice SDM. A systematic review of barriers to SDM found that many physicians report lack of applicability of SDM due to patient characteristics and the clinical situation. [Légaré, France, et al., "Barriers and Facilitators to Implementing Shared Decision Making in Clinical Practice: Update of a Systematic Review of Health Professionals' Perceptions," Patient Education and Counseling, Vol. 73, No. 3 (August 2008)] Moreover, doctors may hold biases toward patients who they feel are not able to understand complex medical information. [van Ryn M, Burke J. The effect of patient race and socio-economic status on physicians' perceptions of patients. Soc Sci Med. 2000;50(6):813–28.]

Physicians usually have more detailed knowledge of potential medical interventions than the patient. The physician's challenge is to distill this knowledge into an appropriate discussion, taking into account the patient's educational and cultural background and emotional factors, while remaining aware of their own intrinsic bias that may cloud the discussion. Physicians and their patients are prone to subjective interpretations of objective data based on past experiences. This is a human reality and should be acknowledged and addressed.

The physician's training background and prior experience may predispose them toward a certain intervention. If surgical and medical options are both available, a surgeon may be inclined to promote surgery, whereas an internal medicine specialist may emphasize the medical option. Many allopathic medical schools (conferring M.D. or medical doctorate) emphasize the role of medications or surgical procedures as a primary intervention (and indeed, many patients also expect such an intervention to be offered as a first-line therapy). Conversely, those receiving their education from an osteopathic medical school (conferring D.O. or doctor of osteopathy) often emphasize a more holistic approach, recommending lifestyle changes or non-pharmaceutical interventions as a first step.

Personality characteristics may influence a physician's ability to infer a patient's desires. A lack of good clinical communication skills or inability to build rapport may lead to a failure to connect or understand a patient's preferences. Physicians, often by personality and training, have usually become confident in quickly making decisions and acting upon them decisively. They may hesitate to admit the role of uncertainty in predicting future outcomes, leading to erroneous expectations by a patient.

Similarly, misunderstood cultural factors may influence a physician's discussion. Some cultures demonstrate extreme deference to a physician's recommendations and members of these groups may be uncomfortable being asked to choose between options. Other patriarchal societies may fail to empower women to make their own decisions. A perceived imbalance of power of physician over the patient, as may occur in the setting of patients with illegal residency status, may also result in failure of the patient to disclose their true desires [38].

Politically charged, ideologically driven medical interventions, similarly, may be approached differently by physicians compared to less divisive issues, based on their own inherent biases. As a practicing obstetrician (IS), I would like to highlight induced abortion, perhaps the most divisive of all medical interventions, which can serve as an example of injecting provider biases into patient care and a major failure of providing a care that identifies and meets women's preferences. I have seen paternalistic pressure exerted upon patients I have cared for many times. A young Black woman hospitalized with hyperemesis and extreme anxiety regarding her pregnancy told me how, her mother and physician overruled her objections during her previous pregnancy as a teenager to encourage and perform a late second trimester abortion, when she wanted to give birth to her child. In her current pregnancy, she was again being pressured toward another late abortion. Additionally, her boyfriend was adding to her coercion by forcing her to leave their shared apartment until the pregnancy ended. She tearfully expressed her desire to give birth to her child and expressed her belief that the circumstances were causing her to experience anxiety and depression. Unfortunately, she was not an exception. Greater than 95% of elective terminations of pregnancy are performed for social or financial, not medical, reasons [39]. We need to seek ways, which would enable us to support these women in voicing their preferences and making the decisions that align with their desires, values and beliefs.

In a highly contentious and ideologically driven issue, such as abortion, provider bias can take various forms. For example, in a survey conducted by the Collaborative for Reproductive Equity [https://core.wisc.edu/2021/12/06/cores-survey-of-doctors-highlights-widespread-support-for-abortion-access/], more than three-quarters of doctors expressed support for abortion access, and more than nine in ten expressed concern about abortion restrictions. This "pro-choice" bias may compel providers to see abortion as the best option their patient, motivating them to promote this option as a preferred available alternative.

Conversely, some organizations have expressed concern that vulnerable women may receive deceptive counseling in order to influence their decision in a crisis pregnancy, alleging that pro-life advocates may overstate the risks of abortion, downplay the impact of pregnancy and childbirth or otherwise emotionally manipulate women, causing them to continue a pregnancy rather than obtain an abortion [https://www.acog.org/advocacy/abortion-is-essential/trending-issues/issue-brief-crisis-pregnancy-centers; https://www.politico.com/newsletters/women-rule/2024/07/12/the-latest-abortion-battle-crisis-pregnancy-centers-00167783].

In discordance with the provider biases described above, survey data demonstrate that 60% of women with a history of abortion would have preferred to give birth to the child, if they had more support or an improved financial situation [40]. Additionally, abundant data demonstrate that the subset of women who choose to abort a pregnancy that is wanted or meaningful to them, are at much higher risk of mental health complications [Reardon, DC. 'The abortion and mental health controversy: A comprehensive literature review of common ground agreements, disagreements, actionable recommendations, and research opportunities', SAGE Open Medicine, https://doi.org/https://doi.org/10.1177/2050312118807624.], particularly anxiety, depression, substance and alcohol abuse, and self-harm [Coleman, PK 'Abortion and mental health: quantitative synthesis and analysis of research published 1995–2009', Br J Psychiatry, 2011;199:180–186.], potentially leading to "deaths of despair" from suicide, overdoses, accidents or even homicide from destructive intimate partner relationships [Gissler M, Kauppila R, Merilainen J, et al. Pregnancy Associated Deaths in Finland 1987–1994. Acta Obstetricia et Gynecologica Scandinavica. 1997;76:651–657; Gissler M, Berg C, Bouvier-Colle MH, et al. Injury deaths, suicides and homicides associated with pregnancy, Finland 1987–2000. Eur J of Public Health. 2005;15(5):459–463; Gissler M, Berg C, Bouvier-Colle MH, et al. Pregnancy Associated Mortality After Birth, Spontaneous Abortion or Induced Abortion in Finland. 1987–2000. American Journal Obstetrics and Gynecology. 2004;190(2):422–427; Gissler M, Berg C, et al. Methods of Identifying Pregnancy Related Deaths: Population Based Data From Finland. 1987–2000. Pediatrics and Perinatology Epidemiology. 2004;18(6):448–55; Gissler M, et al. Suicides after pregnancy in Finland, 1987–94. Register linkage study. British Medical Journal. 1996;313:14,311,434; Karalis E, et al. Decreasing mortality during pregnancy and for a year after while mortality after termination of pregnancy remains high: a population based register study of pregnancy associated deaths in Finland 2001–2012. British Journal Obstet Gynecol. 2017;124:1115–1121.]

Additionally complicating the counseling regarding this important decision is the separation of elective abortion providers from the standard medical community. Only 7–14% of practicing obstetrician/gynecologists say they would perform an elective abortion when requested by a patient [Desai S, Jones R, Castle K. Estimating abortion provision and abortion referrals among United States obstetricians and gynecologists in private practice. *Contraception* 2018;97:297–302; Stuhlberg DB, Dude AM, Dahlquist I, Curlin FA. Abortion provision among practicing obstetrician-gynecologists. *Obstet Gynecol* 2011;118(3):609–614. doi: https://doi.org/10.1097/AOG.0b013e31822ad973.], so most abortions are not performed by the patient's established physician, but in clinics dedicated to abortion provision. Planned Parenthood's most recent annual report demonstrates that patients are assumed to have already chosen abortion when they present for counseling, as 97.2% of their pregnancy services are abortions (0.7% miscarriage management, 1.6% referrals for prenatal care and 0.5% adoption counseling) [41]. Conversely, when abortion-vulnerable women present to a crisis pregnancy center for an ultrasound, 78% will continue their pregnancies [42].

The failure to provide a neutral discussion of the risks, benefits and alternatives of all the available options to women considering abortion, demonstrates that true informed consent counseling may not be provided to most women facing problematic pregnancies. Additionally, the failure of the U.S. to mandate and enforce abortion complication reporting, leaves providers unable to accurately counsel women on the true incidence of complications following abortion [43].

5 SDM Adoption and Implementation

5.1 Current State of SDM Adoption: Barriers and Potential Strategies ...

Health care systems and policy makers in the US and globally have recognized the importance of adopting SDM. The reality, however, in 2024 is that U.S. health systems are not built or purposed to support interventions like SDM. Despite fairly strong evidence that practicing SDM is associated with better treatment plans that reflect patients' goals, views and preferences [Stiggelbout A. M., Pieterse A. H., De Haes J. C. J. M. Shared decision making: concepts, evidence, and practice. Patient Educ Couns 2015;98:1172–9. https://doi.org/10.1016/j.pec.2015.06.022.] has a positive effect on patient outcomes and reduces costs, [Stiggelbout A. M., Van Der Weijden T., De Wit M. P. T., Frosch D., Légaré F., Montori V. M., et al. Shared decision making: really putting patients at the centre of healthcare. BMJ 2012;344:e256. Stacey D., Légaré F., Lewis K., Barry M. J., Bennett C. L., Eden K. B., et al. Decision aids for people facing health treatment or screening decisions. Cochrane Database Syst Rev 2017;4:CD001431.] its adoption has been limited. Several key reasons can be identified as constraints for the implementation of SDM in clinical practice.

For example, a recent review found that hospital settings were not conducive to promoting SDM conversations (lack of privacy). Other notable barriers included: knowledge (awareness of something), skills (proficiency acquired through practice), professional roles (perceptions of roles and behaviors in work setting), beliefs about capabilities (does the clinician want to do SDM), optimism (confidence to achieve goals), and beliefs about consequences (fear of negative outcome) [Waddell, A., Lennox, A., Spassova, G., & Bragge, P. (2021). Barriers and facilitators to shared decision-making in hospitals from policy to practice: a systematic review. Implementation Science: IS, 16(1), 74. https://doi.org/10.1186/s13012-021-01142-y].

At an ***individual physician and patient level***, the most often cited challenges to expanding the use of SDM include lack of clinician training, time constraints in clinical settings, lack of patient participation and willingness to be engaged in SDM.

At a health *system level*, one of the most often cited barriers is decision aid shortcomings.

International Patient Decision Aid Standards (IPDAS) Collaboration defines patient decision aids as "tools designed to help people participate in decision making about health care options. They provide information about the options, and help patients to construct, clarify, and communicate the personal values they associate with the different features of the options. They provide structured guidance in the steps of decision making". [International Patient Decision Aid Standards Collaboration http://ipdas.ohri.ca/what.html].

These evidence-based patient education tools are designed in various formats, including printed materials with graphics, photographs, diagrams, videos and website-based interactive programs and provide actionable information on the risks, benefits and burdens of treatment options, and help patients identify and communicate their preferences.

Research, including more than 130 randomized controlled studies, shows that patient decision aids provide balanced, evidence-based information about treatment options and help patients gain significant knowledge and a more accurate understanding of relevant clinical evidence [44]. Despite numerous attempts and body of research aimed at developing and testing decision aids, challenges for effective implementation of patient decision aids in clinical practice remain. These include fast paced clinical advances, such as new treatment options and new information on risks and complications, which makes it difficult to keep the decision aids up to date. Certification of decision aids is another challenge as there is currently no national body responsible for certifying decision aids which would ensure quality of aids.

Policy level barriers include lack of provider reimbursement for practicing SDM, lack of legal process that supports a reasonable-patient–centered standard through shared decision making.

However, the growing interest in SDM transcends the clinical domain, gaining significant traction among policymakers, health foundations, and non-profit organizations within the United States. This momentum is fueled by a growing recognition of SDM's potential to not only enhance patient care but also to streamline healthcare delivery in a manner that is both cost-effective and patient-centered [45]. Government bodies and influential health organizations are increasingly championing SDM as a critical component of healthcare reform, emphasizing its role in improving health outcomes and patient satisfaction, while also addressing the pressing need to curb escalating healthcare costs [46].

For instance, the Agency for Healthcare Research and Quality (AHRQ), a pivotal entity within the U.S. Department of Health and Human Services, has been at the forefront of promoting SDM. AHRQ's funding of research and development of decision aids is a testament to the government's commitment to embedding SDM within the fabric of healthcare provision (AHRQ). AHRQ provides a collection of tools and training resources to support the implementation of shared decision making in practice. Its SHARE Approach [Agency for Healthcare Research and Quality. "The SHARE Approach." Accessed May 25, 2024, from https://www.ahrq.gov/health-literacy/professional-training/shared-decision/index.html], a five-step process for shared decision-making includes exploring and comparing the benefits, harms, and risks of each option through meaningful dialogue about what matters most to the patient, ensuring that healthcare decisions are informed by both the best available evidence and the patients' personal values and preferences. This initiative underscores a strategic shift towards patient-centered care, recognizing the invaluable role of patient engagement in achieving optimal health outcomes.

Furthermore, leading health foundations and non-profits are amplifying this shift towards SDM. Organizations such as the Robert Wood Johnson Foundation and the Patient-Centered Outcomes Research Institute (PCORI) have dedicated substantial resources to advancing SDM research and practice. Their grants and programs support innovative projects that seek to integrate SDM into routine care, particularly for underserved and marginalized communities. These efforts reflect a broader trend among health philanthropies to prioritize initiatives that empower patients and make the healthcare system more responsive to their needs.

Similarly, the Centers for Medicare & Medicaid Services (CMS) has introduced specific reimbursement plans for SDM focused clinical encounters, to support patient participation in decision making and higher quality informed consent. Rare exceptions promoting SDM exist at a state policy level as well. For example, Washington State has enacted legislation incorporating shared decision making in the informed consent and promoting the use of decision aids in addition to or as an alternative to standard informed consent documents [Spatz, E. S., Krumholz, H. M., & Moulton, B. W. (2016). The new era of informed consent: getting to a reasonable-patient standard through shared decision making. *Jama, 315*(19), 2063–2064.].

However, all these advances have not been sufficient to make a significant progress towards implementation and a wider adoption of SDM. All key factors should align to ensure that SDM takes place as "the right process, with the right partners, at the right time and place." [47]

5.2 New Models for Advancing SDM Implementation

New technological capabilities such as mobile health (mHealth), the application of digital tools through smartphones or computers, have emerged as useful resources to help patients improve access and engagement in health care. While incredible advantages such as the ability of digital tools to quickly upload the most up-to-date evidence available, or their ease of application through the ubiquitous use of smartphones by clinicians and patients are weighed against the concerns that these advances only benefit the most empowered and privileged in society, thus further promoting a rise in health disparities [Abbasgholizadeh Rahimi, S., Menear, M., Robitaille, H., & Légaré, F. (2017). Are mobile health applications useful for supporting shared decision making in diagnostic and treatment decisions? Global Health Action, 10(sup3), 1,332,259. https://doi.org/10.1080/16549716.2017.1332259], the hopes of SDM advancement through mHealth applications remains high. Studies have shown that mHealth promotes greater participation of patients in SDM [Omboni S., Caserini M., Coronetti C. Telemedicine and m-Health in hypertension management: technologies, applications and clinical evidence. *High Blood Press Cardiovasc Prev.* 2016;23:187–196.] and increased patient satisfaction with patient–provider interactions [Borgelt L. M., Liston R., Giacomini K., et al. Evaluation of shared decision making between patients and providers to improve menopause health outcomes. *Menopause.* 2015;22:1399.]

Telehealth has also shown promise to advance implementation of SDM [Ozanne, E. M., Noseworthy, P. A., Cameron, K. A., Schmidt, M., Cavanaugh, K., Pershing, M. L., Guzman, A., Sivly, A., & Fagerlin, A. (2020). Shared decision making in the era of telehealth: Implications for practice and research. MDM Policy & Practice, 5(2), 2,381,468,320,976,364. https://doi.org/10.1177/2381468320976364]. In addition to improving access by creating a capacity for the "remote SDM", i.e., provision of SDM via telemedicine the application of telehealth can also promote productive patient-clinician interactions and patient-centered care [Hartasanchez, S. A., Heen, A. F., Kunneman, M., Garcia-Bautista, A., Hargraves, I. G., Prokop, L. J., … & Montori, V. M. (2022). Remote shared decision making through telemedicine: a systematic review of the literature. Patient education and counseling, 105(2), 356–365.] In addition, e-health technologies offer opportunities for sharing telehealth visits between primary and specialist providers thus making both provider-to-provider communication and patient-provider communication more effective and efficient.

Research has also shown the immense value of integrating SDM systems within electronic medical records [Davis, S., Roudsari, A., Raworth, R., Courtney, K. L., & MacKay, L. (2017). Shared decision-making using personal health record technology: a scoping review at the crossroads. Journal of the American Medical Informatics Association: JAMIA, 24(4), 857–866. https://doi.org/10.1093/jamia/ocw172]. However, knowledge around best practices is limited and much like telehealth and mHealth innovations, these advances come with a great deal of both skepticism and optimism.

5.3 New Era of Medical Decision Making: "Dr. AI Will See You Soon"

> Imagine waking up tomorrow to find a new continent that has emerged overnight. Not a landmass — a digital realm populated by 100 billion AI agents, and a billion robots. All ready to go. Welcome to AI Atlantis: a seemingly far-off utopia which is actually a vivid reality unfolding before our very eyes.
>
> Emad Mostaque, How to Think About AI, White Paper. Accessed September 2, 2024. https://x.com/SchellingAI/status/1818600200232927721

No technological innovation has entered the health care at a speed, magnitude and force like Artificial Intelligence. In 1996, when Professor Coiera predicted that "providing information on health will no longer be the exclusive remit of health care professionals. The quantity of information on the Internet will continue to grow over the next few years, as will the proportion of people with access to it" [*Coiera, E. (1996). The Internet's challenge to health care provision. Bmj, 312(7022), 3–4.*], the extent to which his prediction would prove to be true was not fully recognized. Yet, the speed of this digital revolution is light years slower compared to the advances made by AI. Since early 2000s, as patients first started to use a newly launched search engine Google, it took several years before 'Dr. Google', a term describing the use of the internet for searching health information, became an invisible actor in every health care setting globally. However, "Dr. AI will see you soon" already appears as a headline or title in numerous publications from major media sources such as the Economist, the Wall Street Journal, CNBS to scientific publications [*Bohr, A., & Memarzadeh, K. (2022). Chapter 2-the rise of artificial intelligence in healthcare applications. Artificial intelligence in healthcare, 25–60.; Bunz, M., & Braghieri, M. (2022). The AI doctor will see you now: assessing the framing of AI in news coverage. AI & society, 37(1), 9–22.*].

In a recent article, evolution of AI in health care was aptly described as "A Changing Actor Landscape" [*Mogo, E. August 21, 2024. Dr. AI Will See You Now, JSTOR Daily,* https://daily.jstor.org/dr-ai-will-see-you-now/, *accessed September 1, 2024*]. Indeed, cases of extraordinary clinical performance of AI applications in various health care settings and technologies which outperform or are on par with humans already exist *[Bohr, A., & Memarzadeh, K. (2022). Chapter 2-the rise of artificial intelligence in healthcare applications. Artificial intelligence in healthcare,* 25–60.*; Davenport T, Kalakota R. The potential for artificial intelligence in healthcare. Future Healthc J. 2019;6(2):94–8.*] The Babylon AI powered Triage and Diagnostic System identified the condition modelled by a clinical vignette with accuracy comparable to human doctors (in terms of precision and recall) and the triage advice recommended by the AI System was, on average, safer than that of human doctors [*Razzaki, S., Baker, A., Perov, Y., Middleton, K., Baxter, J., Mullarkey, D., … & Johri, S. (2018). A comparative study of artificial intelligence and human doctors for the purpose of triage and diagnosis. arXiv preprint arXiv:1806.10698.*], ChatGPT, the popular chatbot from OpenAI diagnosed a case missed by 17 doctors over 3 years [*Holohan M., Sept. 11, 2023, A boy saw 17 doctors over 3 years for chronic pain. ChatGPT found the diagnosis. TODAY,* https://www.today.com/health/mom-chatgpt-diagnosis-pain-rcna101843, *Accessed September 2, 2024*]; ChatGPT also performed as well as, and in some cases better than a trained doctor in clinical reasoning when suggesting likely diagnoses for patients assessed in emergency departments [*Rodman, A., Buckley, T. A., Manrai, A. K., & Morgan, D. J. (2023). Artificial intelligence vs clinician performance in estimating probabilities of diagnoses before and after testing. JAMA Network Open, 6(12), e2347075-e2347075.; Ten Berg, H., van Bakel, B., van de Wouw, L., Jie, K. E., Schipper, A., Jansen, H., … & Kurstjens, S. (2024). ChatGPT and generating a differential diagnosis early in an emergency department presentation. Annals of Emergency Medicine, 83(1), 83–86.*]

When Nature published the article "Google AI has better bedside manner than human doctors—and makes better diagnoses" [*Lenharo, M. (2024). Google AI has better bedside manner than human doctors—and makes better diagnoses. Nature, 625(7996), 643–644.*], many in medical community criticized the publication as a non-peer reviewed article promoting 'AI hype. However, the fact remains that AI can be trained using large language model (LLM) on vast text quantities, trillions of words (tokens) [*Emad Mostaque, How to Think About AI,*

White Paper. Accessed September 2, 2024. https://x.com/SchellingAI/status/1818600200232927721] and to make it even more extraordinary, AI models can adapt based on context [*Emad Mostaque, How to Think About AI, White Paper. Accessed September 2, 2024.* https://x.com/SchellingAI/status/1818600200232927721] and do this enormously faster than humans. Therefore, there is a potential for AI models to learn empathy through the process of context-based training [*Diamandis, P. (Host), Aug 13, 2024, Moonshots with Peter Diamandis, Episode EP114, Humanoid Robots, the Job Market & Mass Automation - The Current State of AI w/ Emad Mostaquel, accssed on September 2, 2024* https://www.youtube.com/watch?v=8LRBhT3TLr8&t=1868s]. In fact, this AI model already exists in a form of companion robots for social and emotional stimulation, assisting patients with stress or depression by connecting emotionally and providing social interaction [Bohr, A., & Memarzadeh, K. (2020). The rise of artificial intelligence in healthcare applications. In *Artificial Intelligence in healthcare* (pp. 25–60). Academic Press.]

As bold as this statement may sound, we can realistically expect to see the accounts of AI outperforming humans in the area of patient-catered care as well. Consider this example: a patient consulting a plastic surgeon for a postmastectomy breast reconstruction recalls "...he [plastic surgeon] said, 'Well, you're old anyhow, so what difference does it make?' Isn't that cruel? Those were the words." *[Hasak JM, Myckatyn TM, Grabinski VF,* et al. *Stakeholders' perspectives on postmastectomy breast reconstruction: recognizing ways to improve shared decision making. Plast Reconstr Surg Glob Open. 2017;5:e1569.]* It is hard to imagine a similar scenario in AI-led SDM. The challenge will be training AI properly but once this is achieved, AI models can be fully integrated in the area of patient communication, preference diagnosis and helping patients navigate the decision-making process.

As AI evolves to be a reliable and trustworthy actor in health care delivery, patients will respond accordingly. While at the time of this writing, in summer 2024, the key areas of AI applications involve diagnosis, medical imaging, treatment recommendations, patient monitoring, and administrative activities, [*Davenport T, Kalakota R. The potential for artificial intelligence in healthcare. Future Healthc J. 2019;6(2):94–8.*] the biggest change, with a potential to reshape the health care delivery will be related to the changing role of the patient in health care system to the active and even equal partner [Žaliauskaitė, M. (2021). Role of ruler or intruder? Patient's right to autonomy in the age of innovation and technologies. *AI & SOCIETY*, *36*(2), 573–583.]. To a certain extent, this has already happened as a result of technological developments in healthcare and their key benefit to date—ubiquitous access to health information. It has empowered patients, transformed their views and redefined the roles of physician, patient and their relationship. [Tsulukidze, M., Grande, S. W., Thompson, R., Rudd, K., & Elwyn, G. (2015). Patients covertly recording clinical encounters: threat or opportunity? A qualitative analysis of online texts. *PloS one*, *10*(5), e0125824.; Žaliauskaitė, M. (2021). Role of ruler or intruder? Patient's right to autonomy in the age of innovation and technologies. *AI & SOCIETY*, *36*(2), 573–583.]. In fact, patients in their new role as consumers and principal users of innovative technologies [Žaliauskaitė, M. (2021). Role of ruler or intruder? Patient's right to autonomy in the age of innovation and technologies. *AI & SOCIETY*, *36*(2), 573–583; Huckvale, K., Wang, C. J., Majeed, A., & Car, J. (2019). Digital health at fifteen: more human (more needed). BMC medicine, 17, 1–4.] have sought ways to acquire and assert more power in the health care process by taking control, and 'ownership' of the clinical encounter [Tsulukidze, M., Grande, S. W., Thompson, R., Rudd, K., & Elwyn, G. (2015). Patients covertly recording clinical encounters: threat or opportunity? A qualitative analysis of online texts. *PloS one*, *10*(5), e0125824]. It is very likely that the AI-powered agency will significantly enhance this patient-consumer movement. It will enable patients to act as main drivers in medical decision making in ways we may not fully envision at this time.

So far, evidence of the use of AI in SDM is in its infancy and given the novelty of AI tools, we cannot fully assess their impact on the patient centered care [*Sauerbrei, A., Kerasidou, A., Lucivero, F., & Hallowell, N. (2023). The impact of artificial intelligence on the person-centred, doctor-patient relationship: some problems and solutions. BMC Medical Informatics and Decision Making, 23(1), 73.*] Research exploring the understanding and attitudes of patients regarding AI in the context of risk prediction for shared clinical decision-making after knee replacement surgery, "highlighted the potential of AI to enable patients to develop realistic expectations and equip them with personalized risk information to discuss in shared decision-making conversations with the surgeon." [*Gould, D. J., Dowsey, M. M., Glanville-Hearst, M., Spelman, T., Bailey, J. A., Choong, P. F., & Bunzli, S. (2023). Patients' Views on AI for Risk Prediction in Shared Decision-Making for Knee Replacement Surgery: Qualitative Interview Study. Journal of Medical Internet Research, 25, e43632.]* A review of literature on application of AI in SDM noted a lack of emphasis on patients' values and preferences, lack of clarity about different aspects of AI interventions and recommended to strengthen and standardize the use of AI in different steps of SDM, and evaluate its impact on various decisions, populations, and settings. *[Abbasgholizadeh Rahimi, S., Cwintal, M., Huang, Y., Ghadiri, P., Grad, R., Poenaru, D., ... & Pluye, P. (2022). Application of artificial intelligence in shared decision making: scoping review. JMIR Medical Informatics, 10(8), e36199.]*

The greatest challenge to implementing AI models in various health care domains, however, is not the technological capabilities but rather ensuring their adoption in clinical practice. [*Bohr, A., & Memarzadeh, K. (2020). The rise of artificial intelligence in healthcare applications. In Artificial Intelligence in healthcare (pp. 25–60). Academic Press.]* As with any innovation, and more so with something of this immense potential and power, the multifaceted efforts will need to cover policy and regulations, clinical training and patient education, reimbursement and payer systems, integration with the existing technologies, such as electronic health records. At this time, questions drastically outnumber the answers. We do not yet know how key ethical and legal challenges, such as safety and transparency, data protection and privacy, liability issues will be addressed. The impact of the large-scale deployment of AI on the doctor-patient relationship is also unclear and difficult to predict [*Sauerbrei, A., Kerasidou, A., Lucivero, F., & Hallowell, N. (2023). The impact of artificial intelligence on the person-centred, doctor-patient relationship: some problems and solutions. BMC Medical Informatics and Decision Making, 23(1), 73.; Niel O, Bastard P. Artificial intelligence in nephrology: core concepts, clinical applications, and perspectives*]. At the same time, some skepticism has been expressed that AI may further dehumanize the practice of medicine [*Sauerbrei, A., Kerasidou, A., Lucivero, F., & Hallowell, N. (2023). The impact of artificial intelligence on the person-centred, doctor-patient relationship: some problems and solutions. BMC Medical Informatics and Decision Making, 23(1), 73.; Sparrow, R., & Hatherley, J. (2020). High hopes for "Deep Medicine"? AI, economics, and the future of care. Hastings Center Report, 50(1), 14–17.*] and lead back to paternalism by way of imposing choices justified based on its superior analytical capabilities [*Sauerbrei, A., Kerasidou, A., Lucivero, F., & Hallowell, N. (2023). The impact of artificial intelligence on the person-centred, doctor-patient relationship: some problems and solutions. BMC Medical Informatics and Decision Making, 23(1), 73.]*

One thing is clear, neither AI skeptics nor optimists have a full understanding of the possibilities that may emerge in the AI-powered health care. To realize the tremendous potential of AI and transform healthcare for the better, all stakeholders including AI developers, clinicians, patients, ethicists, and legislators, policymakers must engage in a constructive discussion on how to implemented AI successfully in clinical practice [Gerke, S., Minssen, T., & Cohen, G. (2020). Ethical and legal challenges of artificial intelligence-driven healthcare. In Artificial intelligence in healthcare (pp. 295–336). Academic Press.]

Much like Professor Coiera's prediction 28 years ago [*Coiera, E. (1996). The Internet's challenge to health care provision. Bmj, 312(7022), 3–4.*], the AI's implications also extend far beyond the technology itself, this time at a bigger scale, however.

> The AI revolution is like a tsunami, and the water is already receding from the shores. Some are still playing on the beach, oblivious. Others are running for the high ground. But the smart ones? They're already there.
>
> Emad Mostaque, How to Think About AI, White Paper. Accessed September 2, 2024. https://x.com/SchellingAI/status/1818600200232927721

We must act now decisively to make sure that we are among the 'smart ones'.

6 Conclusions

6.1 "What Would You Do, If You Were Me?"

The question, patients ask their clinicians all too often and the question often not welcomed by clinicians [Tonelli, M. R., & Sullivan, M. D. (2019). Person-centred shared decision making. Journal of evaluation in clinical practice, 25(6),

1057–1062.]. This question is also actively discouraged by many SDM proponents based on the argument that it may introduce physician bias. However, merit of this question needs careful exploration before it is rejected.

Predominantly accepted SDM models are quite prescriptive in nature, offering 'stages' that physicians are expected to progress through as they engage their patients in SDM, such as problem definition, presenting options with risks and benefits, eliciting patients' values and preferences and agreeing on or deferring a decision [Lown, B. A., Hanson, J. L., & Clark, W. D. (2009). Mutual influence in shared decision making: a collaborative study of patients and physicians. Health Expectations, 12(2), 160–174] However, such models do not account for patient—physician interaction as "non-linear process for shared decisions, in which the outcome of a shared decision cannot be predicted with certainty at the start of any one encounter" [Suchman AL. A new theoretical foundation for relationship-centered care. Complex responsive processes of relating. Journal of General Internal Medicine, 2006; 21 (Suppl. 1): S40–S44.]. Accepting that SDM is a fluid and dynamic process not merely sequential steps or 'stages' [Lown, B. A., Hanson, J. L., & Clark, W. D. (2009). Mutual influence in shared decision making: a collaborative study of patients and physicians. Health Expectations, 12(2), 160–174] also invites consideration of interpersonal dynamics between and bidirectional influence of these dynamics on SDM actors. We think these are important gaps and potential contributors to slow SDM adoption.

6.2 "A Patient in the Clinic; A Person in the World": Humanity and Personhood in Health Care

In addition to lack of recognizing a dynamic nature of the shared decision-making process, narrowly conceptualizing and operationalizing decision making also excludes important influences of decision partners involved in the decision making process, and limits the scope and context in which SDM should develop.

"What would you do, if you were me?"—reflects a complex and intricate nature of clinical care, the fact that humans on opposite sides of the decision are navigating immense uncertainties. In a more nuanced SDM model this question is not only encouraged but proposed as a question to be asked even when patient doesn't ask it. Asking "What would I do if I were in their place?" re-introduces humanity into the clinical encounter, and "acknowledges clinicians as persons, requiring them to understand patients as persons" [Tonelli, M. R., & Sullivan, M. D. (2019). Person-centred shared decision making. Journal of evaluation in clinical practice, 25(6), 1057–1062.].

Excellent studies, such as "A patient in the clinic; a person in the world": why shared decision making needs to center on the person rather than the medical encounter" [Clayman, M. L., Gulbrandsen, P., & Morris, M. A. (2017). A patient in the clinic; a person in the world. Why shared decision making needs to center on the person rather than the medical encounter. Patient education and counseling, 100(3), 600–604.] have found that sharing decision involve physician–patient relationship that go beyond developing a partnership around a decision, and includes "self-expression, humour, expressing empathy and compassion, staying in connection, taking time, trusting the patient, adopting a non-judgemental attitude towards the patient and respecting his or her intelligence, culture, circumstances and style". [Lown, B. A., Hanson, J. L., & Clark, W. D. (2009). Mutual influence in shared decision making: a collaborative study of patients and physicians. Health Expectations, 12(2), 160–174].

Knowing the patient as a person allows clinician to contextualize care and patient to trust the care they receive.

In a study of mutual influence of patients and physicians, patients and physicians who facilitate SDM were described as "actively seeking a personal connection with each other". [Lown, B. A., Hanson, J. L., & Clark, W. D. (2009). Mutual influence in shared decision making: a collaborative study of patients and physicians. Health Expectations, 12(2), 160–174].

Recognizing the bidirectional nature of SDM, also calls for examining how patients encourage or elicit shared decision-making behaviors in their physicians.

> Might patients play an active role in shaping physicians' approaches to a decision, just as physicians may play a role in shaping the degree of patient involvement in a decision? Might both persons change their minds in the course of a decision, or might both persons begin without a clear decision in mind, and shape a decision that neither could have conceptualized alone? [Lown, B. A., Hanson, J. L., & Clark, W. D. (2009). Mutual influence in shared decision making: a collaborative study of patients and physicians. Health Expectations, 12(2), 160–174]

Research has shown that indeed, patients can influence the communication behaviors of physicians by actively participating and may improve clinical outcomes. [Cegala DJ, McClure L, Marinelli TM, Post DM. The effects of communication skills training on patients_participation during medical interviews. Patient Education and Counseling, 2000; 41: 209–222.]

"The keys to more frequent occurrence of successful shared decision making may therefore lie equally in the hands of both patients and physicians." [Lown, B. A., Hanson, J. L., & Clark, W. D. (2009). Mutual influence in shared decision making: a collaborative study of patients and physicians. Health Expectations, 12(2), 160–174].

At its core, SDM is deeply rooted in the historical evolution of patient-provider communication, which has increasingly recognized the value of trust, empathy, and mutual respect. These elements are essential for meaningful conversations about treatment options, particularly for marginalized populations who may have different health beliefs, practices, and experiences of stigma and discrimination. SDM can be a powerful vehicle for engaging minoritized patients [48]48, historically sidelined in their care, often due to systemic barriers and prejudices that pervade healthcare systems, thereby contributing to the reduction of overall health disparities. By acknowledging and integrating diverse patient voices into the care process, SDM helps to mitigate the social and cultural distances that can impede effective healthcare delivery.

6.3 "Meta-process" of Making Decisions About Decision Making: Power Dynamics and Control

> We sort of realized that there's a meta-process that has to occur first, which is the physician and the patient have to agree about how the decision is going to get made. Some patients will want the physician basically to lead them and say, "This is what you should do". And other patients basically would have the physician explain the alternatives to them. So the first step actually is to come together and create a decision about how the process will be implemented. [Patient statement, Lown, B. A., Hanson, J. L., & Clark, W. D. (2009). Mutual influence in shared decision making: a collaborative study of patients and physicians. Health Expectations, 12(2), 160–174]

The importance of equality in physician patient relationship has been long recognized. Research published in 1959 suggested that patients who played an active role in making treatment decisions, more successfully adhered to prescribed drug regimen [Menzel H, Coleman J, Katz E. Dimensions of being modern in medical practice. J Chronic Dis 1959;9:20–40]. Such active participation was described as "helping the doctor by trying to figure out problems together".

"The doctor should consider the patient as an equal partner-as intelligent as himself-and give the patient a chance to help the doctor by trying to figure out problems together. The patient should have the freedom and the chance to say what he thinks about a certain therapeutic approach." [Menzel H, Coleman J, Katz E. Dimensions of being modern in medical practice. JChronic Dis 1959;9:20–40].

This stresses the fact that in addition to being a dynamic and fluid process, SDM is also a synergistic progression in which roles of its actors are mutually negotiated and information sharing is adjusted based on the level of trust, respect and shared responsibility.

These roles and responsibilities sometimes shift, control may be equally shared, or shift so "either the physician or the patient may assume, defer or share control in decision making",

"'I've…[made] a decision that my doctor absolutely hated. And, I think, the best thing he did was actually expressed that. He said, "Today you are saying no. Can we agree to talk about it tomorrow?" And I said, "Well, we can agree to talk about it an hour from now, two hours from now, a day from now, but it's not going to change my mind". Well, surprisingly, I changed my mind'. "Lown, B. A., Hanson, J. L., & Clark, W. D. (2009). Mutual influence in shared decision making: a collaborative study of patients and physicians. Health Expectations, 12(2), 160–174].

The concept of SDM is built on the foundational principle of partnership. This model transcends the traditional, paternalistic approach to healthcare, where decisions were predominantly made by healthcare providers, often leaving patients passive and disengaged. SDM, in contrast, fosters a collaborative environment where the knowledge and expertise of the provider converge with the values, preferences, and lived experiences of the patient.

6.4 Path Forward: "Precision SDM"

In the era of precision medicine that accounts for individual differences in patients' genes, environments, and lifestyles, we should consider ways in which SDM is tailored to individual patients based on their personality profile, style and preference for communication, values and preferences, in other words, develop and apply "precision SDM".

"Rather than assuming all patients under all circumstances require the same degree or kind of SDM, [..] provider DM style be tailored to patients' preferences, disease state, personality, motivation, and psychological need satisfaction (i.e., autonomy, competence, and relatedness). This means that while all SDM should be patient centered, it does not always have to be patient driven. Under some circumstances, a more provider-driven approach may be more practical, ethical, and efficacious." [Resnicow, K., Catley, D., Goggin, K., Hawley, S., & Williams, G. C. (2022). Shared decision making in health care: theoretical perspectives for why it works and for whom. Medical Decision Making, 42(6), 755–764.]

Practicing precision SDM will be even more accessible and feasible as AI-enabled SDM tools become available. The potential to process the vast quantity of information, including patient-reported outcomes and feedback loops will facilitate precision SDM and inform decision-making, promote SDM in treatment encounters, offering unprecedented opportunities for enhanced decision-making, personalized care and streamlined care processes. *[Kaplan, F. L. (2024). Beyond Shared Decision-Making: Integrating Coproduction, Learning Health Systems, Artificial Intelligence, and Workforce Development for Patient-Centered Care.*]

However, more research is necessary to support the utility of "precision SDM" [Resnicow, K., Catley, D., Goggin, K., Hawley, S., & Williams, G. C. (2022). Shared decision making in health care: theoretical perspectives for why it works and for whom. Medical Decision Making, 42(6), 755–764.] and integrate the human and bidirectional nature of SDM, as described above, in all SDM conversations. Rethinking how SDM is conceptualized and implemented can accelerate its adoption and benefit all parties involved. Broader and correct understanding of SDM, will "ensure that the patient and the health professional take part in SDM with the right information, using the right tools, in the right manner, in the right setting, and at the right time". [47]

SDM is still in its infancy despite the proliferation of literature, policy and practice attention given to SDM and a rapidly growing body of research on its characteristics, benefits and implementation recommendations. Gaps in understanding and agreement on what SDM constitutes must be filled if we want to achieve its effective implementation, assessment, and wide adoption. Future studies should focus on expanding the understanding of roles, responsibilities, and input of key SDM actors, including technology-enabled tools and identifying the key drivers of clinical culture change, which is required to truly empower patients and physicians as main SDM actors and implementers.

6.4.1 Core Messages

- "Decision making is a key activity, perhaps the most important activity, in the practice of healthcare".[2]
- Since early 1900s patient autonomy and agency in medical decision making has been protected by legally mandated informed consent.
- Since early 2000s shared decision-making principles extended the concept of informed consent beyond simple information transfer signaling a pivotal shift in health care and epitomizing a movement towards more collaborative, patient-centered care.
- While shared decision-making is now considered to be an essential part of health care, it is not yet fully developed as a theory and model of clinical practice.
- System wide adoption and implementation of SDM requires clear understanding of how SDM works, for whom, when and where.

Glossary

Medical Decision Making involves the observation and collection of information (the art of medicine); coalescing this information to best determine syndromes, diseases, and treatments (the science of medicine); and, finally, making decisions with the patient, factoring in cultural, economic, interpersonal, religious, and ethical variables that influence patient participation or adherence (the ethics of medicine) [*Bostic, J. Q., & Rho, Y. (2006). Target-symptom psychopharmacology: between the forest and the trees. Child and adolescent psychiatric clinics of North America, 15(1), 289–302.*]

Shared Decision Making—(SDM) while there is debate on exact definitions, but experts agree, that SDM is a collaborative process where patients and providers work together to make care decisions. This process ensures that patient voice is central. SDM involves a provider giving evidence-based information about treatment options, risks and benefits while a patient provides their values, preferences and circumstances. SDM seeks concensus for optimal treatment based on an integration of these clinical and patient perspectives (Charles et al. 1997).

Decision Analysis this describes a systematic and visual approach to evaluating different healthcare options using decision trees, probabilities, Markov models, and outcomes to assist clinical/medical decision making by providing likely consequences to different outcomes based on discrete choices

•

Patient-Centered Care may be best characterized by Leventein's work from the 1980s focused on the Patient-Centered Clinical Method, which integrates patient "understanding of illness experience" with the clinicians "differential diagnosis" [49]. Ultimately, this approach contributed to ways of characterizing the relationship between patients and their providers.

Decision Aids A decision aid can be defined or described as a tool or resource designed to assist individuals in making informed decisions about their healthcare treatment. These work by providing clear, evidence-based information about the different options available and the associated outcomes, benefits, and risks. Decision aids often include structured approaches to clarify the values and preferences of the decision-maker, facilitating a more deliberate and informed choice (Stacey et al. 2017).

Decision Support Tools These tools are like decision aids yet are primarily focused on platforms aligned with clinical decision support systems (CDSS) engineered to improve personalization of care and treatment [50, 51]. Examples of these tools might be a digital platform to educate patients on different genetic variations on breast cancer and guide clinicians around the most up-to-date evidence on treatment, risks, and outcomes for breast cancer treatment.

Patient Engagement tools These types of tools are generally referring to resources with a goal of involving patients in their healthcare treatment decisions. These could focus on increasing patient knowledge, improving health literacy, or increasing patient motivation to actively work to support their own health. Examples include online platforms, digital applications, short videos and other communication tools to promote patient engagement and interaction with their care [52].

Preference-sensitive care some health conditions have multiple evidence-based options available for treatment requiring preference-sensitive trade-off decisions based on individual patient's values or preferences in deciding the best course of action. Shared decision-making (SDM) is often indicated in preference-sensitive situations.

References

1. O'Connell K (2012) If only I'd known then what I know now. Health Aff 31(8):1848–1848
2. Schwartz A, Bergus G (2008) Medical decision making: a physician's guide. Cambridge University Press, Cambridge
3. Collins FS, Varmus H (2015) A new initiative on precision medicine. N Engl J Med 372(9):793–795
4. Schloendorff v. Society of New York Hospitals. In: Landmark Cases in Bioethics (1914)

5. Faden RR, Beauchamp TL (1986) A history and theory of informed consent. Oxford University Press
6. Maclean A (2009) Autonomy, informed consent and medical law: a relational challenge. Cambridge University Press
7. American Medical Association. Opinion 2.1.1 Informed Consent
8. Canterbury v. Spence, 464 F.2d 772 (D.C. Cir. 1972)
9. King JS, Moulton BW (2006) Rethinking informed consent: the case for shared medical decision-making. Am J Law Med 32(4):429–501. https://doi.org/10.1177/009885880603200401
10. (1982) President's Commission for the Study of Ethical Problems in Medicine and Biomedical and Behavioral Research. Making health care decisions. The ethical and legal implications of informed consent in the patient-practitioner relationship. Washington, DC: The Commission
11. Salzburg Global Seminar (2011) Salzburg statement on shared decision making. BMJ 342:d1745. https://doi.org/10.1136/bmj.d1745
12. Mulley AG, Trimble C, Elwyn G (2012) Stop the silent misdiagnosis: patients' preferences matter. BMJ 345:e6572. https://doi.org/10.1136/bmj.e6572
13. DelVecchio Good MJ (2007) The medical imaginary and the biotechnical embrace. In: Subjectivity: ethnographic investigations. University of California Press, pp 362–380
14. Good BJ, Good MJD (1981) The meaning of symptoms: a cultural hermeneutic model for clinical practice. In: The relevance of social science for medicine. Springer Netherlands, pp 165–196
15. Craig SR (2012) Healing elements: efficacy and the social ecologies of Tibetan medicine. University of California Press
16. Mechanic D (1989) Medical sociology: some tensions among theory, method, and substance. J Health Soc Behav 30(2):147. https://doi.org/10.2307/2137012
17. Parsons T (1951) The social system. Routledge
18. Parsons T (1975) The sick role and the role of the physician reconsidered. Milbank Memorial Fund Quarterly Health Soc 53(3):257. https://doi.org/10.2307/3349493
19. Freidson E (1970) Profession of medicine: a study of the sociology of applied knowledge. Harper & Row
20. Grande SW, Elwyn G (2009) Evolution of decision making models in healthcare and formalizing shared decision making. In: Elwyn G, Edwards A (eds) Shared decision making in health care: achieving evidence-based patient choice, 2nd ed. Oxford University Press
21. Wennberg JE, Gittelsohn A (1973) Small area variations in health care delivery: a population-based health information system can guide planning and regulatory decision-making. Science 182(4117):1102–1108. https://doi.org/10.1126/science.182.4117.1102
22. Wennberg JE (2010) Tracking medicine: a researcher's quest to understand health care. Oxford University Press
23. Illich I (1976) Medical nemesis: the expropriation of health. Pantheon Books
24. Emanuel EJ, Emanuel LL (1992) Four models of the physician-patient relationship. JAMA 267(16):2221–2226
25. Charles C, Gafni A, Whelan T (1997) Shared decision-making in the medical encounter: what does it mean? (or it takes at least two to tango). Soc Sci Med 44(5):681–692. https://doi.org/10.1016/s0277-9536(96)00221-3
26. Makoul G, Clayman ML (2006) An integrative model of shared decision making in medical encounters. Patient Educ Couns 60(3):301–312. https://doi.org/10.1016/j.pec.2005.06.010
27. Burton D, Blundell N, Jones M, Fraser A, Elwyn G (2010) Shared decision-making in cardiology: do patients want it and do doctors provide it? Patient Educ Couns 80(2):173–179. https://doi.org/10.1016/j.pec.2009.10.013
28. Shepherd HL, Matthews M, Becker L et al (2022) The impact of shared decision-making on healthcare: a systematic review and meta-analysis. BMC Health Serv Res 22(1):820. Published 2022 Jun 22. https://doi.org/10.1186/s12913-022-08197-7
29. Frosch DL, May SG, Rendle KA, Tietbohl C, Elwyn G, Authoritarian physicians and patients' fear of being labeled 'difficult' among key obstacles to shared decision making

30. Pérez-Stable EJ, El-Toukhy S (2018) Communicating with diverse patients: How patient and clinician factors affect disparities. Patient Educ Couns 101(12):2186–2194. https://doi.org/10.1016/j.pec.2018.08.021
31. Stacey D, Légaré F, Lewis K et al (2017) Decision aids for people facing health treatment or screening decisions. Cochrane Database Syst Rev 4(4):CD001431. Published 2017 Apr 12. https://doi.org/10.1002/14651858.CD001431.pub5
32. Zisman-Ilani Y, Barnett E, Harik J, Pavlo AJ, O'Connell M (2017) Expanding the concept of shared decision making for mental health: systematic search and scoping review of interventions. Ment Health Rev J 22(3):191–213. https://doi.org/10.1108/MHRJ-01-2017-0002
33. Castaneda-Guarderas A, Glassberg J, Grudzen CR, Ngai KM, Samuels-Kalow ME, Shelton E, Wall SP, Richardson LD (2016) Shared decision making with vulnerable populations in the emergency department. Acad Emerg Med 23(12):1410–1416. https://doi.org/10.1111/acem.13134
34. Durand MA, Carpenter L, Dolan H et al (2014) Do interventions designed to support shared decision-making reduce health inequalities? A systematic review and meta-analysis. PLoS One 9(4):e94670. Published 2014 Apr 15. https://doi.org/10.1371/journal.pone.0094670
35. King J, Eckman MH, Moulton BW (2011) The potential of shared decision making to reduce health disparities. J Law Med Ethics 39(Suppl 1):30–33. https://doi.org/10.1111/j.1748-720X.2011.00561.x
36. Montori VM (2020) Why we revolt: a patient revolution for careful and kind care. The patient revolution
37. Richards T, Montori VM, Godlee F, Lapsley P, Paul D (2013) Let the patient revolution begin. BMJ 346:f2614. https://doi.org/10.1136/bmj.f2614
38. Jagosh J, Boudreau JD, Steinert Y, MacDonald ME, Ingram L (2011) The importance of physician listening from the patients' perspective: enhancing diagnosis, healing, and the doctor-patient relationship. Patient Educ Couns 85(3):369–374. https://doi.org/10.1016/j.pec.2011.01.028
39. Jones RK, Jerman J (2017) Population group abortion rates and lifetime incidence of abortion: United States, 2008–2014. Am J Public Health 107(12):1904–1909. https://doi.org/10.2105/AJPH.2017.304042
40. Coleman PK, Reardon DC, Strahan T, Cougle JR (2005) The psychology of abortion: a review and suggestions for future research. Psychol Health 20(2):237–271. https://doi.org/10.1080/08870440412331337084
41. Planned Parenthood Federation of America. Planned Parenthood 2018–2019 Annual Report. https://www.plannedparenthood.org/uploads/filer_public/2e/da/2eda3f50-82aa-4ddb-acce-c2854c4ea80b/2018-2019_annual_report.pdf Accessed 27 Aug 2023
42. Glessner TD (2015) The relationship between abortion and fetal tissue research. Issues Law Med 30(2):189–220
43. Stroud L, Hester DM (2016) Abortion reportinG in the United States: an assessment of state laws and the systematic underreporting of abortion deaths. Charlotte Lozier Institute. February 2016. https://lozierinstitute.org/wp-content/uploads/2016/01/State-Abortion-Reporting-Underreporting_OP2014.pdf. Accessed 27 Aug 2023
44. Pope TM (2017) Certified patient decision aids: solving persistent problems with informed consent law. J Law Med Ethics 45(1):12–40. https://doi.org/10.1177/1073110517703097
45. Veroff D, Marr A, Wennberg DE (2013) Enhanced support for shared decision making reduced costs of care for patients with preference-sensitive conditions. Health Aff (Millwood). 32(2):285–293. https://doi.org/10.1377/hlthaff.2011.0941
46. Oshima Lee E, Emanuel EJ (2013) Shared decision making to improve care and reduce costs. N Engl J Med 368(1):6–8. https://doi.org/10.1056/NEJMp1209500
47. Dobler CC, Midthun DE, Montori VM (2017) Quality of shared decision making in lung cancer screening: the right process, with the right partners, at the right time and place. Mayo Clin Proc 92:1612–1616. https://doi.org/10.1016/j.mayocp.2017.08.010

48. Peek ME, Odoms-Young A, Quinn MT, Gorawara-Bhat R, Wilson SC, Chin MH (2010) Race and shared decision-making: perspectives of African-Americans with diabetes. Soc Sci Med 71(1):1–9. https://doi.org/10.1016/j.socscimed.2010.03.014
49. Levenstein JH, McKracken EC, McWhinney IR, et al (1986) The patient-centred clinical method. 1. A model for the doctor-patient interaction in family medicine. Fam Pract 3:24–30
50. Agarwal S, Glenton C, Tamrat T, Henschke N, Maayan N, Fønhus MS, Mehl GL, Lewin S (2021) Decision-support tools via mobile devices to improve quality of care in primary healthcare settings. Cochrane Database Syst Rev 7. https://doi.org/10.1002/14651858.CD012944.pub2
51. Liu J, Wyatt JC, Altman DG (2006) Decision tools in health care: focus on the problem, not the solution. BMC Med Inform Decis Mak 6(1). https://doi.org/10.1186/1472-6947-6-4
52. Clavel N, Paquette J, Dumez V, Del Grande C, Ghadiri DPS, Pomey M-P, Normandin L (2021) Patient engagement in care: a scoping review of recently validated tools assessing patients' and healthcare professionals' preferences and experience. Health Expect: Int J Public Participation Health Care Health Policy 24(6):1924–1935. https://doi.org/10.1111/hex.13344

Maka Tsulukidze, M.D., Ph.D. is an Associate Professor at Florida Gulf Coast University, Marieb College of Health and Human Services. Dr. Tsulukidze earned a Ph.D. in Health Sciences Research from University of North Carolina at Charlotte, M.D. from Tbilisi Medical Academy and completed Postdoctoral Fellowship at Dartmouth Center for Health Care Delivery Science. She has authored studies published in the Archives of Surgery, Patient Education and Counseling, PLOS One, American Journal of Alzheimer's Disease and Other Dementias, Educational Gerontology, and other peer-reviewed journals.

Stuart W. Grande is an Associate Professor who teaches and conducts health services research in the Division of Health Policy and Management at the University of Minnesota, School of Public Health. He holds a Ph.D. in health behavior and an MPA in public management from Indiana University. He completed a postdoctoral fellowship in Shared Decision Making (SDM) under the mentorship of Glyn Elwyn at The Dartmouth Institute (TDI) in Hanover, NH. Professor Grande's current research considers the collaborative benefits of new technologies on enhancing patient-provider communication and reducing bias at the point of care.

Ingrid Skop, M.D., F.A.C.O.G. has been a practicing board-certified obstetrician-gynecologist in San Antonio, Texas for over 30 years. She received her Bachelor of Science in physiology from Oklahoma State University, her medical doctorate from Washington University School of Medicine, and completed her residency in obstetrics and gynecology at the University of Texas Health Science Center at San Antonio. She practices with OB Hospitalist Group, is Vice President and Director of Medical Affairs for the Charlotte Lozier Institute and serves as a member of the Texas Maternal Morbidity and Mortality Review Committee. She has published peer reviewed articles focusing on abortion safety, abortion drugs and maternal mortality, and has testified to state and federal legislatures and judicial courts.

Neuroscience Evidence and Legal Decision-Making

10

Alissa Anderson Call, Liz Wilson, and Twila Wingrove

Most juries feel that most mental patients are really faking… if you show them the X-ray, they're convinced.

Dr. Bernard Diamond

Abstract

The introduction of neuroscience evidence in the courtroom has grown since the early twentieth century. Neuroscience evidence encompasses various forms of genetic (e.g., MAOA tests) and brain data (e.g., fMRI, CT, MRI, EEG, PET scans) which have been introduced in both criminal and civil cases to assess competency, influence sentencing, and substantiate personal injury claims. However, the effects of neuroscience evidence (e.g., fMRI scans) on legal judgments in criminal and civil cases are complex and have yet to be fully understood. This chapter explores the integration of neuroscience evidence into legal decision-making, briefly examining its historical context and application. We cover the controversy surrounding the neuroscience evidence bias and review the extant literature on the effects of neuroscience on legal decisions for both criminal and civil cases. Overall, the existing literature suggests a mitigating effect of neuroscience evidence for legal decisions for criminal cases (e.g., capital punishment) and increases damage awards for the plaintiff in civil

A. A. Call (✉)
Department of Psychology, Black Hills State University, Spearfish, SD, USA
e-mail: alissa.call@bhsu.edu

L. Wilson · T. Wingrove
Department of Psychology, Appalachian State University, Boone, NC, USA
e-mail: wilsonea6@appstate.edu

T. Wingrove
e-mail: wingroveta@appstate.edu

© The Author(s), under exclusive license to Springer Nature Switzerland AG 2025

T. McElroy (ed.), *Decision-Making in Life and Work*, Integrated Science 36,
https://doi.org/10.1007/978-3-031-88415-3_10

cases (e.g., personal injury). Factors such as defendant dangerousness may also moderate the effects of neuroscience evidence in criminal cases. Importantly, neuroimages do not appear to dramatically influence legal judgments over and above expert explanations of a neuroscientific fashion and no stark contrasts between forms of neuroimages were identified. We conclude with recommendations for future research endeavors, emphasizing the need to understand the impact of defendant and juror characteristics and the framing of neuroscience evidence on legal outcomes.

Keywords

Neuroscience • Juror decision-making • Neuroimages • Expert evidence • Criminal law • Civil law • Legal decisions

1 Introduction

As Dr. Diamond suggests, the effects of neuroscience evidence presentation in court *may* be quite convincing. However, do the results of real cases support this claim? Take for example, the case of Brian Dugan. Throughout the early 1980s, Brian Dugan kidnapped, raped, and murdered three people: two young girls and one woman [1]. In 2009, researchers scanned Dugan's brain with a functional magnetic resonance imaging (fMRI) and his scans showed distinct defects in his brain, suggesting he was unable to control his impulses—at least that is how one neuroscientist interpreted it. While the neuroscientific expert testified on these fMRI scans for Dugan's defense at trial, the scans were not actually shown to the jury because the prosecution thought it would be too damning to their case. The jury deliberated for less than one hour, returning ten votes for the death penalty with two jurors abstaining—meaning Dugan would be sentenced to life without parole. However, at the request of one juror for more time to review several trial testimony transcripts, the judge granted the jury additional deliberation time. The following day, the jury unanimously agreed to sentence Dugan to death for the murders of the three females.

While the ultimate case outcome was not in Dugan's favor, many legal experts agree that the neuroscientific evidence weakened the prosecution's case during the sentencing phase [1]. Moreover, Dugan's case was innovative and controversial, being the first to admit fMRI scans as evidence in a courtroom. Dugan committed these murders over 25 years ago meaning that these current fMRI scans were unlikely to represent a mental state of brain activity for crimes that happened decades ago, sparking significant controversy due to the long gap between the crimes and the brain scans.

But is it worthwhile for the defense to invest in neuroscientific experts and/or to present neuroimages at trial? Should prosecutors be afraid of showing these images, like in the Dugan case? Further, does the extant empirical research suggest neuroscience evidence produces a "seductive allure" that influences legal decisions? In this chapter, we discuss various forms of neuroscience evidence, the history of its presence in court, and types of cases that incorporate this form of evidence.[1] Next, we discuss the controversy surrounding the neuroscience evidence bias from non-legal research contexts and then review the existing research on whether these types of evidence produce a seductive allure on fact finders and, thus, influence their legal decisions for both criminal and civil cases. And finally, we conclude with recommendations for future research endeavors.

2 A Brief History of Neuroscience in the Courtroom

Neuroscience evidence refers to information and research findings gleaned from the field of neuroscience that are presented as evidence in legal cases. This evidence can encompass various types of data and research related to the brain, such as brain imaging (e.g., fMRI, PET, CT), studies on brain structure and function, and research on the neural basis of behavior, cognition, and emotions. More recently, behavioral genetic evidence (e.g., MAOA genotype tests) has been introduced at trial in both criminal [5] and civil cases [6]. Neuroscience evidence is used in a wide range of legal contexts to help establish or challenge aspects of a case. Courts evaluate neuroscience evidence based on relevance to the case, reliability of the information, and if it meets the current standards of practice (i.e., the *Daubert* standard). Typical delivery of neuroscience evidence in the courtroom is via expert witness testimony and may include the presentation of neuroimages (e.g., fMRI scans). These experts interpret and explain complex scientific data to make the evidence accessible to judges, attorneys, and juries.

In addition to the fMRI scans discussed above, a variety of forms of neuroscience information have been introduced as evidence in court dating back to the early twentieth century. The electroencephalography (EEG) was first introduced in the courtroom in the 1940s in a case involving an epileptic defendant [7]. During this time, some lawyers sought to present EEG evidence to protect their clients whose rights were allegedly denied by laws, while others used this type of evidence to identify neural markers of violent behavior [7].

In 1981, the attempted assassination of President Ronald Reagan by John Hinckley led to one of the most infamous criminal cases involving the presentation of neuroscience in the courtroom. Hinckley's defense team sought to present computed tomography (CT) scans to support their argument that he suffered from schizophrenia, and therefore, should be found not guilty by reason of insanity (NGRI). Despite the prosecution's objections to introduce Hinckley's CT scans during trial, the presiding judge allowed them to be presented. Consequently, the jury found Hinckley NGRI [8].

[1] Despite being beyond the scope of this chapter, it is worth noting that there are works that discuss the potential ethical considerations surrounding the presentation of neuroscience in court contexts [2–4].

Around a decade later, positron emission tomography (PET) scans made their way into the courtroom. Herbert Weinstein, a 65-year-old New York City advertising executive, was charged with second-degree murder for strangling his wife and throwing her body out of the window of their 12th-floor apartment. Weinstein readily admitted to the alleged crime and showed little remorse for his actions, leading his defense team to call for PET scans to investigate the causes of his behavior. The result of the PET scans indicated an abnormal cyst nestled in Weinstein's arachnoid membrane, suggesting his brain function was disrupted. His team then claimed Weinstein did not have the mental state to be found criminally responsible. Subsequently, Weinstein plead guilty to a lesser charge of manslaughter [9].

A year after the Weinstein case, the rules for what could be introduced in federal trials changed substantially from the "Frye standard" to the "Daubert standard" as a result of *Daubert v. Merrell Dow Pharmaceuticals, Inc.* [10]. The *Frye* standard allowed evidence to be admitted to trial if the method by which that evidence was obtained was generally accepted by the relevant scientific community. *Frye* was regarded as a more conservative standard that helped to keep pseudoscience from being presented during trial. However, in the 1993 *Daubert* trial, the Supreme Court replaced the *Frye* standard with a standard known as the *Federal Rules of Evidence (FRE)*, which offered a more liberal position towards allowing evidence to be presented in trial—now known more commonly as the *Daubert* standard. Under *Daubert*, a judge can allow expert evidence to be presented in court if it meets a set of criteria including whether the information can help the fact finders understand the evidence or to determine a fact in issue, is data-driven, the product of reliable methods, and applies to the case. Thus, *Daubert* has allowed for more neuroscience expert evidence and neuroimaging to be introduced in courts whether the expert's interpretations are accepted by the general scientific community—or not.

Take for example, the case of Grady Nelson, who in 2005 killed his wife by stabbing her 61 times and then stabbed and raped his 11-year-old stepdaughter [11]. Nelson's defense was one of the first to present neuroscience testimony regarding quantitative electroencephalogram (qEEG) scans of his brain, arguing that these scans revealed abnormalities in Nelson's left frontal lobe and that this impairment affected his ability to control his violent impulses. Standard electroencephalogram (EEG) tests record electrical activity in the brain by placing electrodes on the skull while qEEG tests translate EEG results into colorful images of the brain to detect abnormalities. Although a neuroscientist for the prosecution criticized the qEEG data as flawed, two jurors cited the evidence as a factor in their decision to vote against the death penalty, resulting in a life sentence without parole. This case demonstrates that even when the interpretation of neuroscientific evidence is not agreed upon by the scientific community from which it stems, it can still be persuasive in the eyes of legal fact-finders.

Behavioral genetic evidence has also made its appearance in the courtroom. In 1995, a convicted murderer petitioned to have his death sentence overturned because the trial court had refused to pay for a test of his monoamine oxidase A

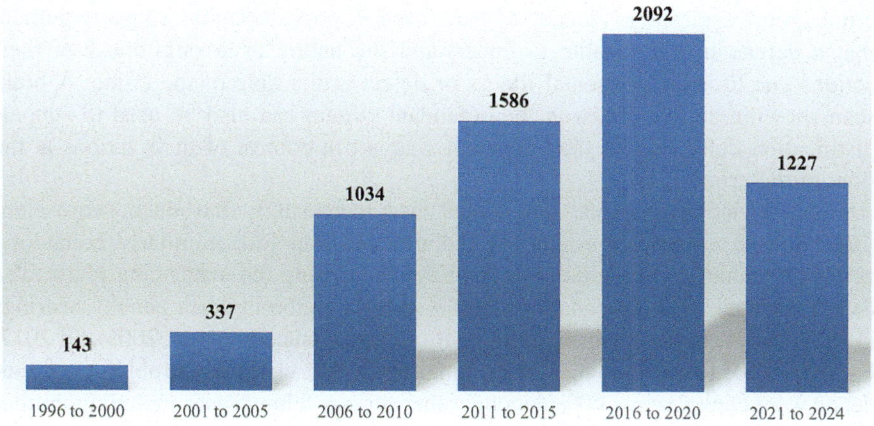

Fig. 10.1 Number of Law Review Articles Published on "Neuroscience" (1996–2024). *Note* The number of law review articles published on "neuroscience" as indexed by Lexis Nexis has steadily increased over the past 28 years

(MAOA) allele [12]. MAOA is an enzyme that degrades monoamine neurotransmitters. Much research has suggested a link between low MAOA production and a history of childhood maltreatment to violent crime in male populations [13–16]. Consequently, some forensic investigators began to collect MAOA data in a variety of criminal cases [17] and this form of evidence has been successful in reducing prison sentences in both Italy [18] and the United States [19].

The use of neuroscience in the courts has not gone unnoticed by legal scholars. A quick search on Lexis Nexis indicates a substantial increase in law review articles published on "neuroscience" since 1996 starting at 143 (Fig. 10.1). In fact, the number of articles increased 15-fold by the year 2020, reaching heights of 2092 published works. While the number of articles has dwindled to 1227 in the years of 2021–2024, the current 2024 year has yet to conclude. These large numbers of published works suggest massive attention being paid to the introduction of neuroscientific evidence in the courtroom, but how often is this form of expert evidence presented? And in what contexts?

3 Case Contexts Where Neuroscience Evidence is Used

In criminal cases, neuroscience evidence can be introduced at multiple phases within a trial. In the pretrial phase, brain scans can be presented to support claims of competency, meaning the defendant has the mental capacity to stand trial, plead guilty, or to confess [20]. This is often done alongside a psychological evaluation of the defendant to assess cognitive function and is the second most common use of neuroscientific and behavioral genetic evidence used in criminal court [20].

Despite being less helpful in determining a defendant's responsibility for a crime, neuroscientific evidence has been used to prove insanity, a legal argument that a defendant was unable to understand the nature or wrongfulness of their actions due to a severe mental illness or defect at the time of the crime. A brain scan showing abnormalities in the defendant's brain can also be used to support justification defenses that the defendant was not in control of their actions at the time of the crime [21].

Post-conviction, criminal defendants also increasingly introduce neuroscientific evidence to mitigate sentencing recommendations [20]. Similarly, behavioral genetic evidence is most likely to be presented during the sentencing phase of a trial, to mitigate recommendations, and is most common in death penalty hearings [19]. However, in her analysis of judicial opinions issued between 2005 and 2012, Farahany [20] found that 40% of criminal cases that utilized neurobiological evidence were capital while 60% denoted other serious felony cases including murder (23%), drugs (13%), and assault charges (11%). It is important to note, however, that only a fraction of cases go to court and result in a guilty verdict. Further, only a segment of these convictions reaches appellate court and produce written opinions that can later be used for data analytic purposes.

In the United States, the most common type of neuroimaging presented in criminal court are magnetic resonance imaging (MRI; about 25% in both Denno [19], and Farahany [20]), CT scans (24% in Denno [19]; 23% in Farahany [20]), and EEG (18% in Farahany [20], and 25% in Denno [19]). More sophisticated forms of neuroimaging such as single-photon emission computed tomography (SPECT) or fMRI scans are less common [20].

In her review of 553 criminal cases including the presentation of neuroscientific evidence between the years of 1992 and 2012, Denno [22] observed that in nearly all cases, neuroscientific information was presented as mitigating evidence by the defense. Conversely, in only 7% of cases, was this material presented by the prosecution as aggravating evidence [22]. Many cases that utilize neuroscientific evidence include claims of ineffective counsel introduced during an appeal (i.e., *Strickland* claims; *Strickland v. Washington,* [23]). For example, *Strickland* claims appeared in 53% of cases reviewed by Denno [22], 45% of those reviewed by Farahany [20], and 27% of United States cases sampled by Gaudet and Marchant [24].

There is much less information available concerning the rates to which neuroscientific and behavioral genetic evidence are presented in civil cases. Nevertheless, neuroscience evidence can be used in personal injury claims to demonstrate a defendant has suffered from brain injuries (e.g., showing structural damage from an accident; Calvillo and Irimia [25]). Neuroscience evidence can play a crucial role in disability claims that demonstrate cognitive impairments invisible through traditional medical evaluations [26].

Legally, the presentation of neuroscience evidence in civil court faces similar issues as in criminal court. It is difficult to prove that a defendant suffered from or obtained symptoms caused by a neurological disorder at a certain point in time. For instance, in *Brent Boyd v. Bert Bell/Pete Rozelle NFL Players Retirement Plan*

[27], Brent Boyd, a former National Football League (NFL) player, retired and later claimed head trauma from his career in the NFL. In 1997, Boyd was denied disability benefits from the NFL. In 2000, he appealed this decision, citing former brain damage from a head injury sustained in 1980. The NFL Retirement Board reviewed the medical evidence, including brain scans and evaluations from several doctors, some of whom suggested Boyd's issues stemmed from head injuries obtained during his career in the NFL. In contrast, others argued that Boyd's problems were more likely linked to conditions unrelated to football (e.g., chronic pain, mental illness, and substance abuse). The board denied Boyd's claim for degenerative benefits, concluding that his disability did not arise from NFL-related activities [27]. Despite Boyd's claims being denied, the case played a significant role in ongoing discussions about brain injuries in professional football and other contact sports.

4 Neuroscience and Lay Views

Due to the rise in attention being paid to the introduction of neuroscientific evidence in the courtroom (Fig. 10.1), as well as increasing utility of neuroscience-based evidence, legal scholars and attorneys expressed concern about how fact finders might be persuaded by such evidence at trial. Questions of whether neuroscientific evidence—and particularly, neuroimages—may be unduly prejudicial in the courtroom soon arose. This apprehension has sparked much empirical work on how lay people and jurors attend to, interact with, and use neuroscientific and neuroimaging evidence to render decisions.

In a seminal series of (non-legal) studies on the persuasiveness of neuroscience, Weisberg et al. [28] conducted a study presenting a brief explanation of psychological phenomena to a sample of naïve adult participants, followed by a more elaborate description incorporating either a high- or low-quality explanation and either a neuroscientific component or not. Importantly, the neuroscientific component was irrelevant to the explanation of the psychological phenomena. Despite this irrelevance, participants rated the low-quality explanation as more satisfying when the neuroscience information was included versus when it was excluded. Weisberg et al. [28] then replicated these findings with a sample of university students enrolled in an introductory cognitive neuroscience class. Akin to the novice sample, students equipped with critical thinking skills to analyze neuroscience information judged both the low- and high-quality explanations as being more satisfying with the presence of the neuroscience explanation.

In a more recent study, Weisberg et al. [29] followed up with this work by investigating whether length of neuroscientific information explanations or use of neuroscientific jargon impacted satisfactory ratings yet observed nonsignificant results. In other words, neither length of explanation nor jargon affected the persuasiveness of the presented neuroscientific information. These authors coined this effect the "seductive allure" of neuroscience. Moreover, these findings have been replicated using larger samples of neuroscientifically naïve individuals [30] and have been expanded upon [31]. Taken together, findings from these studies have propelled additional research examining how neuroscientific evidence influences decisions in legal contexts.

Extending the work of Weisberg et al. [28], McCabe and Castel [32] examined the effects of neuroscientific information on lay views, paying particular interest to the influence of neuroimages. In a series of three studies, the authors presented summaries of reported research (both fictitious and real) that were rated by undergraduate student samples as more convincing when coupled with a brain image compared to identical articles containing no image, a bar graph, or a topographical map [32]. The authors concluded that brain images may be more persuasive than other depictions of brain activity because they provide a more tangible explanation for cognitive processes that is more reductionistic and thus, digestible for laypeople to comprehend. These "neuroimaging bias" findings have particular importance regarding the presentation of such images in court to support the defense or prosecution's claims, thus prompting additional research on how these images influence legal decision-making.

However, it is important to note that additional researchers have struggled to replicate these findings. For example, Gruber and Dickerson [33] assessed undergraduates' ratings of a news article when paired with a neuroimage, an artistic drawing of a human head, an image from a movie, or no image. The researchers reported no significant differences in ratings based on image pairing [33]. Additionally, Hook and Farah [34] were unable to replicate the seductive allure effect of neuroimages (i.e., fMRI scans) compared to stock photographs or bar charts in their study using a national participant sample. In a series of 10 experiments using larger participant samples, Michael et al. [30] were unable to replicate these findings. And finally, Schweitzer et al. [35] were unsuccessful in replicating McCabe and Castel's findings across a series of direct and conceptual studies.

Nevertheless, extant research [28, 29, 36, 37] suggests the presence of neuroscientific evidence influences lay perceptions and critiques of scientific claims, even when the information provided is irrelevant. Importantly, these non-legal findings may translate to other contexts such as interpreting the credibility of news articles as well as evidence presentation in criminal and civil courts. If the presence of neuroscience information is seductively alluring to a layperson, it is reasonable to expect the same effects will be found in a jury comprised of laypeople.

5 Neuroscience Evidence and Jurors' Decisions for Criminal Cases

The mock juror literature can be broken down into three categories as first defined by Aono and colleagues [38]. These categories include: (1) Neuroscientific expert testimony *without* neuroimages versus no expert testimony [39], (2) Neuroscientific expert testimony *with* neuroimages versus no expert testimony [40], and (3) Neuroscientific expert testimony *with* neuroimages versus *without* neuroimages [41]. We have also elected to incorporate a fourth category comparing various forms of neuroimages paired with neuroscientific expert testimony [42]. A full review of the research studies within each of these four categories is located in Table 10.1. The remainder of these sections will focus on the effects of neuroscience evidence presentation on mock jurors' legal judgments in criminal cases. Within each section, we have also organized our explanations of each study by legal judgment (e.g., guilty/NGRI, death sentence/life without parole, sentence length).

5.1 Neuroscientific Expert Testimony without Neuroimages

Regarding the first study category, neuroscientific expert testimony without neuroimages versus no expert testimony, none of the studies revealed significant effects on *verdict* decisions. Interestingly, only a handful of studies observed significant effects of neuroscientific expert evidence without images on mock juror *sentencing* decisions. For example, in their study, Greene and Cahill [44] investigated the impact of diagnostic evidence presentation and defendant dangerousness on mock jurors' sentencing recommendations (death or life in prison without parole). The defendant was convicted of first-degree murder of his cellmate. In their study, defense expert testimony described the defendant's disorder as only a psychosis diagnosis (without neuroscientific evidence) or a psychosis diagnosis with summaries of the defendant's neuropsychological tests and an expert's interpretation of these test results. They observed participants render a reduced sentence from the death penalty to life without parole but *only* for the defendant who was described as being high-risk compared to low-risk of future dangerousness. Compared to the neuroscientific expert condition, participants who read the diagnosis only information were nearly 12 times more likely to sentence the defendant to death. This suggests neuroscientific expert testimony has a mitigating effect on capital sentencing decisions, under certain contexts (i.e., defendant dangerousness).

Saks and colleagues [41] (Experiment 1) also found a mitigating effect of neuroscience evidence on sentencing recommendations (death or life in prison without parole). The authors presented a national sample of "death qualified" mock jurors with a defendant convicted of first-degree murder and was described as either healthy, diagnosed with psychopathy, or diagnosed with schizophrenia. In the neuroscientific evidence condition, the defense presented expert testimony from two independent neuroscientists, who affirmed (or disputed) the defendant's diagnosis after having reviewed his fMRI scans. The authors found a mitigating effect of neuroscientific evidence sans images when the defendant was portrayed as having a schizophrenia diagnosis compared to when he was described as healthy or diagnosed as a psychopath [41]. These findings further support the notion that general neuroscientific expert evidence can mitigate sentencing recommendations in capital cases when specific details about the defendant are also discussed.

Other study findings have indicated a mitigating effect of neuroscientific expert evidence regarding prison sentence recommendations. For example, Allen et al.

Table 10.1 Summary of key variables for reviewed studies

Study	Number of participants	Sample type	Legal judgment	Case type	Condition: no expert testimony	Condition: expert testimony without neuroimages	Additional independent variables	Effect on verdict	Effect on sentencing
No neuroscientific expert testimony versus neuroscientific expert testimony without neuroimages									
Allen et al. (2019)	330	Amazon Mechanical Turk (MTurk); U.S. adult mock judges	Sentence	Criminal; sexual assault	Diagnosis of impulse control disorder	Large tumor detected in the "impulse control" region of defendant's brain	Treatment/ dangerousness: treated + low risk versus untreatable + high risk	N/A	Yes: reduced prison sentence lengths
Appelbaum and Scurich (2014)	191	Knowledge Networks (KN) participants; mock jurors	Verdict (first degree manslaughter/ second-degree manslaughter/ first-degree); sentence murder/ second-degree murder)	Criminal; murder	Explanation of defendant's impulsive behavior: simple impulsive behavior	Explanation of defendant's impulsive behavior: childhood abuse versus genetic predisposition versus childhood abuse + genetic predisposition	Defendant race: white versus black	No: no change in verdict	Yes: increased prison sentence for childhood abuse and childhood abuse + genetic predisposition conditions

(continued)

Table 10.1 (continued)

Study	Number of participants	Sample type	Legal judgment	Case type	Condition: no expert testimony	Condition: expert testimony without neuroimages	Additional independent variables	Effect on verdict	Effect on sentencing
Aspinwall et al. (2012)	181	U.S. state trial judges	Sentence	Criminal; aggravated battery	Biomechanism absent: Psychiatrist testified defendant was a diagnosed psychopath	Biomechanism present: psychiatrist testified defendant was a diagnosed psychopath and neurobiologist testified to the biomechanism (i.e., low MAOA activity) contributing to the development of psychopathy	Presenting party: prosecutions versus defense	N/A	Yes: reduced prison sentence lengths
Costa et al. (2017)	279	Undergraduate students from two universities; mock jurors	Verdict (first-degree murder/ second-degree murder/ first-degree manslaughter/ second-degree manslaughter)	Criminal; murder	Explanations for crime: jealousy motive	Explanations for crime: child abuse versus genetic risk versus child abuse + genetic risk	N/A	No: no change in verdict	N/A

(continued)

Table 10.1 (continued)

Study	Number of participants	Sample type	Legal judgment	Case type	Condition: no expert testimony	Condition: expert testimony without neuroimages	Additional independent variables	Effect on verdict	Effect on sentencing
Cheung and Heine (2015)—Experiment 1[b]	132	MTurk U.S. participants; mock jurors	Verdict (first-degree murder/ second-degree murder/ manslaughter/ not guilty); sentence	Criminal; murder	No explanation given for defendant's behavior	Explanation for defendant's behavior: genetic versus environmental	N/A	No: no change in verdict	No: no change in sentence
Cheung and Heine (2015)—Experiment 2	165	Undergraduate students; mock jurors	Verdict (first-degree murder/ second-degree murder/ manslaughter/ not guilty); sentence	Criminal; murder	No explanation given for defendant's behavior	Explanation for defendant's behavior: genetic versus environmental	N/A	No: no change in verdict	No: no change in sentence

(continued)

Table 10.1 (continued)

Study	Number of participants	Sample type	Legal judgment	Case type	Condition: no expert testimony	Condition: expert testimony without neuroimages	Additional independent variables	Effect on verdict	Effect on sentencing
Cheung and Heine (2015)—Experiment 3	298	MTurk U.S. participants; mock jurors	Sentence	Criminal; murder	Explanation for defendant's behavior: environmental	Explanation for defendant's behavior: genetic versus defendant's gene created a stronger impulse to act violently [gene-impulse] versus defendant's gene decreased his ability to control his violent impulses [gene-control]	N/A	N/A	Yes: reduced sentence for all three gene conditions; additional significant reduction in sentence for gene-impulse versus genetic condition
Cheung and Heine (2015)—aggregate dataset	350	Undergraduates and MTurk participants; mock jurors	Verdict (first-degree murder/ second-degree murder/ manslaughter/ not guilty); sentence	Criminal; murder	Explanation for defendant's behavior: environmental	Explanation for defendant's behavior: genetic	N/A	No: no change in verdict	Yes: reduced sentence for genetic condition

(continued)

Table 10.1 (continued)

Study	Number of participants	Sample type	Legal judgment	Case type	Condition: no expert testimony	Condition: expert testimony without neuroimages	Additional independent variables	Effect on verdict	Effect on sentencing
Fuss, Dressing, and Briken (2015)	372	German court judges	Sentence	Criminal; aggravated battery	Biomechanism absent: Psychiatrist testified defendant was a diagnosed psychopath	Biomechanism present: Psychiatrist testified defendant was a diagnosed psychopath and neurobiologist testified to the biomechanism (i.e., low MAOA activity) contributing to the development of psychopathy	Presenting party: prosecutions versus defense	N/A	No: no change in sentence
Greene and Cahill (2012)	208	University students; "death qualified" mock jurors	Death/life sentence	Criminal; first degree murder	Psychosis diagnosis	Psychosis diagnosis + frontal brain damage	Defendant dangerousness: low versus high risk	N/A	Yes: reduced death sentence (only for high risk defendants)

(continued)

Table 10.1 (continued)

Study	Number of participants	Sample type	Legal judgment	Case type	Condition: no expert testimony	Condition: expert testimony without neuroimages	Additional independent variables	Effect on verdict	Effect on sentencing
LaDuke et al. (2018)	896	MTurk; U.S. adult mock jurors	Sentence	Criminal; burglary + aggravated assault	Facts of case: mock police report of the crime, defendant statement, defendant plea	Video footage of psychologist describing the defendant's neurological abnormalities, presents MRI or fMRI scans, described defendant as high risk for future violence	Structural [MRI] versus functional [fMRI] neuroimaging	N/A	No: no change in sentence
Marshall et al. (2017)—Experiment 2	400	MTurk, U.S. adult mock jurors	Sentence	Criminal; murder	Psychiatrist testified defendant had a lack of impulse control and feelings of guiltlessness, explained characteristics of psychopathy	Neuroscientist testified defendant exhibited under activation in amygdala and orbitofrontal cortex, explained characteristics of psychopathy	N/A	N/A	No: no change in sentence

(continued)

Table 10.1 (continued)

Study	Number of participants	Sample type	Legal judgment	Case type	Condition: no expert testimony	Condition: expert testimony without neuroimages	Additional independent variables	Effect on verdict	Effect on sentencing
Mowle et al. (2016)	419	Adult community members summoned for jury duty	Verdict (guilty/not guilty); sentence	Criminal; robbery + assault	Psychologist discusses defendant's mental disorder and traumatic brain injury	Psychologist discusses defendant's diagnosis and damage to prefrontal cortex leading to increased impulsivity	Diagnosis: schizophrenia versus psychopathy	No: no change in verdict	No: no change in sentence
Phalen, Salerno, and Schweitzer (2021)—Experiment 2	353	Prolific participants, mock jurors	Total damage award, pain and suffering award	Civil; personal injury	Pain measurement assessment technique: clinical battery	Pain measurement assessment technique: neuroimaging without visual	Pain type: emotional versus physical	N/A	N/A to sentencing; yes: increased damage awards
Saks et al. (2014)—Experiment 1	825	Survey Sampling International (SSI), U.S. adult sample, "death qualified" mock jurors	Death/life sentence	Criminal; frist degree murder	None	Mental disorder diagnosis based on fMRI scans	Diagnosis: schizophrenia versus psychopathy versus healthy	N/A	Yes: reduced death sentences (only for defendants with schizophrenia)

(continued)

Table 10.1 (continued)

Study	Number of participants	Sample type	Legal judgment	Case type	Condition: no expert testimony	Condition: expert testimony without neuroimages	Additional independent variables	Effect on verdict	Effect on sentencing
Schweitzer et al. (2011)—Experiment 1	237	SSI, U.S. adult mock jurors	Verdict (first-degree, second-degree, manslaughter, not guilty); sentence	Criminal; armed robbery + homicide	Defense attorney argues neurological defect prevents defendant from forming intent	Neuroscientist testifies defendant suffered from structural fontal lobe damage, cannot form intent, and was also prone to losing control when becoming angry	N/A	No: no change in verdict	No: no change in sentence
Schweitzer et al. (2011)—Experiment 2	294	SSI, U.S. adult mock jurors	Verdict (guilty/ not guilty); sentence	Criminal; robbery + assault	Defense attorney argues neurological defect prevents defendant from forming intent	Neuroscientist testifies defendant suffered from structural fontal lobe damage, cannot form intent, and was also prone to losing control when becoming angry	N/A	No: no change in verdict	No: no change in sentence

(continued)

Table 10.1 (continued)

Study	Number of participants	Sample type	Legal judgment	Case type	Condition: no expert testimony	Condition: expert testimony without neuroimages	Additional independent variables	Effect on verdict	Effect on sentencing
Schweitzer et al. (2011)—Experiment 3	512	SSI, U.S. adult mock jurors	Verdict (simple assault/ aggravated assault/not guilty); sentence	Criminal; assault	Defense attorney argues neurological defect prevents defendant from forming intent	Neuroscientist testifies defendant suffered from structural fontal lobe damage, cannot form intent, and was also prone to losing control when becoming angry	N/A	No: no change in verdict	No: no change in sentence
Schweitzer et al. (2011)—Experiment 4	433	SSI, U.S. adult mock jurors	Verdict (simple assault/ aggravated assault/not guilty); sentence	Criminal; assault	Defense attorney argues neurological defect prevents defendant from forming intent	Neuroscientist testifies defendant suffered from structural fontal lobe damage, cannot form intent, and was also prone to losing control when becoming angry	N/A	No: no change in verdict	No: no change in sentence

(continued)

Table 10.1 (continued)

Study	Number of participants	Sample type	Legal judgment	Case type	Condition: no expert testimony	Condition: expert testimony without neuroimages	Additional independent variables	Effect on verdict	Effect on sentencing
Schweitzer et al. (2011)—Meta-analysis	1374	Meta-analysis	Meta-analysis	Criminal; meta-analysis	Meta-analysis	Meta-analysis	N/A	No: no change in verdict	No: no change in sentence
No neuroscientific expert testimony versus neuroscientific expert testimony with neuroimages									
Study	Number of participants	Sample type	Legal judgment	Case type	Condition: no expert testimony	Condition: expert testimony with neuroimages	Additional independent variables	Effect on verdict	Effect on sentencing
Appelbaum et al. (2015)—Experiment 1	960	YouGov participants, U.S. adult mock jurors	Sentence	Criminal; murder	Defense attorney claimed defendant's behavior was an impulsive act	Psychiatrist's explanation of behavior: genetic [with genetic test results] versus neuro [with MRI scans], versus genetic + neuro [with images]	Heinousness of crime: low versus high; criminal history: none, non-violent; violent	N/A	No: no change in sentencing

(continued)

Table 10.1 (continued)

No neuroscientific expert testimony versus neuroscientific expert testimony with neuroimages									
Study	Number of participants	Sample type	Legal judgment	Case type	Condition: no expert testimony	Condition: expert testimony with neuroimages	Additional independent variables	Effect on verdict	Effect on sentencing
Appelbaum et al. (2015)—Experiment 3	763	YouGov participants, U.S. adult "death qualified" mock jurors	Death/life sentence	Criminal; first-degree murder	Defense attorney claimed defendant's behavior was an impulsive act	Psychiatrist's explanation of behavior: genetic [with genetic test results] versus neuro [with MRI scans], versus genetic + neuro [with images]	Heinousness of crime: low versus high; criminal history: none, non-violent, violent	N/A	Yes: reduced death sentences (only for the neuro with MRI scans condition versus no expert condition)
Greene and Cahill (2012)	208	University students; "death qualified" mock jurors	Death/life sentence	Criminal; first degree murder	Psychosis diagnosis	Psychosis diagnosis + MRI and PET scans	Defendant dangerousness: low versus high risk	N/A	Yes: reduced death sentence (only for high risk defendants)

(continued)

Table 10.1 (continued)

No neuroscientific expert testimony versus neuroscientific expert testimony with neuroimages									
Study	Number of participants	Sample type	Legal judgment	Case type	Condition: no expert testimony	Condition: expert testimony with neuroimages	Additional independent variables	Effect on verdict	Effect on sentencing
Gurley and Marcus (2008)	394	Introductory psychology university students, mock jurors	Verdict (Guilty/NGRI)	Criminal; murder	Psychologist and psychiatrist testify to defendant's mental diagnosis	Psychologist and psychiatrist testify to defendant's mental diagnosis and presented 4 MRI scans showing prefrontal cortex damage, implying difficulty with impulse control	Diagnosis: Schizophrenia (psychosis) versus psychopathy	Yes: reduced guilty verdicts	N/A
LaDuke et al. (2018)	896	MTurk; U.S. adult mock jurors	Sentence	Criminal; burglary + aggravated assault	Facts of case: mock police report of the crime, defendant statement, defendant plea	Expert + MRI or fMRI	Structural [MRI] versus functional [fMRI] neuroimaging	N/A	No: no change in sentence
Mowle et al. (2016)	419	Adult community members summoned for jury duty	Verdict (guilty/not guilty); sentence	Criminal; robbery + assault	Psychologist discusses defendant's mental disorder and traumatic brain injury	Psychologist testimony + brain scan	Diagnosis: schizophrenia versus psychopathy	No: no change in verdict	No: no change in sentence

(continued)

Table 10.1 (continued)

Study	Number of participants	Sample type	Legal judgment	Case type	Condition: no expert testimony	Condition: expert testimony with neuroimages	Additional independent variables	Effect on verdict	Effect on sentencing
No neuroscientific expert testimony versus neuroscientific expert testimony with neuroimages									
Phalen, Salerno, and Schweitzer (2021)—Experiment 2	353	Prolific participants, mock jurors	Total damage award, pain and suffering award	Civil; personal injury	Pain measurement assessment technique: clinical battery	Pain measurement assessment technique: neuroimaging + visual	Pain type: emotional versus physical	N/A	N/A to sentencing; yes: increased damage awards
Saks et al. (2014)—Experiment 1	825	U.S. adult sample, "death qualified" mock jurors	Death/life sentence	Criminal; first-degree murder	No neuroscientific evidence	Neuroscientific expert + fMRI	Diagnosis: Psychopathy versus schizophrenia versus healthy	N/A	No: no change in sentence
Schweitzer and Saks (2011)	1170	SSI, U.S. adult mock jurors	Verdict (Guilty/NGRI or GMBI)	Criminal; assault	Defense attorney claimed defendant suffered from mental disorder causing irrational and uncontrollable actions. Family testified to defendant suffering neglect and abuse as a child	Neurologist testified to MRI showing damage to frontal lobe, could cause defendant to act irrationally and uncontrollably, presented MRI	N/A	Yes: reduced guilty verdicts	N/A

(continued)

Table 10.1 (continued)

No neuroscientific expert testimony versus neuroscientific expert testimony with neuroimages									
Study	Number of participants	Sample type	Legal judgment	Case type	Condition: no expert testimony	Condition: expert testimony with neuroimages	Additional independent variables	Effect on verdict	Effect on sentencing
Schweitzer et al. (2011)—Experiment 1	237	SSI, U.S. adult mock jurors	Verdict (first-degree, second-degree, manslaughter, not guilty); sentence	Criminal; armed robbery + homicide	Defense attorney argues neurological defect prevents defendant from forming intent	Expert + fMRI	N/A	No: no change in verdict	No: no change in sentence
Schweitzer et al. (2011)—Experiment 2	294	SSI, U.S. adult mock jurors	Verdict (guilty/not guilty); sentence	Criminal; robbery + assault	Defense attorney argues neurological defect prevents defendant from forming intent	Expert + fMRI	N/A	No: no change in verdict	No: no change in sentence
Schweitzer et al. (2011)—Experiment 3	512	SSI, U.S. adult mock jurors	Verdict (simple assault/aggravated assault/not guilty); sentence	Criminal; assault	Defense attorney argues neurological defect prevents defendant from forming intent	Expert + fMRI and MRI	N/A	No: no change in verdict	No: no change in sentence
Schweitzer et al. (2011)—Experiment 4	433	SSI, U.S. adult mock jurors	Verdict (simple assault/aggravated assault/not guilty); sentence	Criminal; assault	Defense attorney argues neurological defect prevents defendant from forming intent	Expert + MRI	N/A	Yes: reduced verdict severity	No: no change in sentence

(continued)

Table 10.1 (continued)

No neuroscientific expert testimony versus neuroscientific expert testimony with neuroimages									
Study	Number of participants	Sample type	Legal judgment	Case type	Condition: no expert testimony	Condition: expert testimony with neuroimages	Additional independent variables	Effect on verdict	Effect on sentencing
Schweitzer et al. (2011)—meta-analysis	1374	Meta-analysis	Meta-analysis	Criminal; meta-analysis	Meta-analysis	Meta-analysis	N/A	Yes: reduced guilty verdicts	No: no change in sentence
Neuroscientific expert testimony with neuroimages versus neuroscientific evidence without neuroimages									
Study	Number of participants	Sample type	Legal judgment	Case type	Condition: expert without neuroimages	Condition: expert with neuroimages	Additional independent variables	Effect on verdict	Effect on sentencing
Greene and Cahill (2012)	208	University students; "death qualified" mock jurors	Death/life sentence	Criminal; first degree murder	Psychosis diagnosis and reported cognitive deficits suggesting frontal lobe damage	Psychosis diagnosis + MRI and PET scans	Defendant dangerousness: low versus high risk	N/A	No: no change in sentence
LaDuke et al. (2018)	896	MTurk; U.S. adult mock jurors	Sentence	Criminal; burglary + aggravated assault	Facts of case: mock police report of the crime, defendant statement, defendant plea	Expert + MRI or fMRI	Structural [MRI] versus functional [fMRI] neuroimaging	N/A	No: no change in sentence

(continued)

Table 10.1 (continued)

Neuroscientific expert testimony with neuroimages versus neuroscientific evidence without neuroimages									
Study	Number of participants	Sample type	Legal judgment	Case type	Condition: expert without neuroimages	Condition: expert with neuroimages	Additional independent variables	Effect on verdict	Effect on sentencing
Marshall et al. (2017)—Experiment 1	758	MTurk, U.S. adult mock jurors	Sentence	Criminal; murder	Neuroscientist testified defendant exhibited under activation in amygdala and orbitofrontal cortex, explained characteristics of psychopathy	Expert + fMRI	N/A	N/A	No: no change in sentence
Mowle et al. (2016)	419	Adult community members summoned for jury duty	Verdict (guilty/ not guilty); sentence	Criminal; robbery + assault	Psychologist discusses defendant's mental disorder and traumatic brain injury to prefrontal cortex, predisposing him to impulsivity	Psychologist testimony + brain scan	Diagnosis: schizophrenia versus psychopathy	No: no change in verdict	No: no change in sentence

(continued)

Table 10.1 (continued)

Study	Number of participants	Sample type	Legal judgment	Case type	Condition: expert without neuroimages	Condition: expert with neuroimages	Additional independent variables	Effect on verdict	Effect on sentencing
Neuroscientific expert testimony with neuroimages versus neuroscientific evidence without neuroimages									
Phalen, Salerno, and Schweitzer (2021)—Experiment 2	353	Prolific participants, mock jurors	Total damage award, pain and suffering award	Civil; personal injury	Pain measurement assessment technique: neuroimaging without visual	Pain measurement assessment technique: neuroimaging + visual	Pain type: emotional versus physical	N/A	N/A to sentencing; No: no difference in awards
Saks et al. (2014)—Experiment 1	825	U.S. adult sample, "death qualified" mock jurors	Death/life sentence	Criminal; first-degree murder	Two neuroscience experts testified to the mental disorder diagnosis based on the defendant's fMRI scans	Neuroscientific expert + fMRI	Diagnosis: psychopathy versus schizophrenia versus healthy	N/A	No: no change in sentence

(continued)

Table 10.1 (continued)

Neuroscientific expert testimony with neuroimages versus neuroscientific evidence without neuroimages									
Study	Number of participants	Sample type	Legal judgment	Case type	Condition: expert without neuroimages	Condition: expert with neuroimages	Additional independent variables	Effect on verdict	Effect on sentencing
Saks et al. (2014)—Experiment 2[a]	882	U.S. adult sample, "death qualified" mock jurors	Death/life sentence	Criminal; frist degree murder	Neurological without an image	Neurological with neuroimage	Diagnosis: schizophrenia versus psychopathy; future dangerousness: high versus low; which party offers evidence: prosecution versus defense	N/A	Yes: reduced death sentences for neuroimage condition when presented by the defense
Schweitzer and Saks (2011)	1170	SSI, U.S. adult mock jurors	Verdict (guilty/ NGRI or GMBI)	Criminal; assault	Neurologist testified to MRI showing damage to frontal lobe, could cause defendant to act irrationally and uncontrollably	Expert + MRI	N/A	No: no change in verdict	N/A

(continued)

Table 10.1 (continued)

Neuroscientific expert testimony with neuroimages versus neuroscientific evidence without neuroimages									
Study	Number of participants	Sample type	Legal judgment	Case type	Condition: expert without neuroimages	Condition: expert with neuroimages	Additional independent variables	Effect on verdict	Effect on sentencing
Schweitzer et al. (2011)—Experiment 1	237	SSI, U.S. adult mock jurors	Verdict (first-degree/ second-degree/ manslaughter/ not guilty); sentence	Criminal; armed robbery + homicide	Neuroscientist testified the defendant suffered from structural frontal lobe damage preventing him from forming intent	Expert + fMRI	N/A	No: no change in verdict	No: no change in sentence
Schweitzer et al. (2011)—Experiment 2	294	SSI, U.S. adult mock jurors	Verdict (guilty/ not guilty); sentence	Criminal; robbery + assault	Neuroscientist testified the defendant suffered from structural frontal lobe damage preventing him from forming intent	Expert + fMRI	N/A	No: no change in verdict	No: no change in sentence

(continued)

Table 10.1 (continued)

Neuroscientific expert testimony with neuroimages versus neuroscientific evidence without neuroimages									
Study	Number of participants	Sample type	Legal judgment	Case type	Condition: expert without neuroimages	Condition: expert with neuroimages	Additional independent variables	Effect on verdict	Effect on sentencing
Schweitzer et al. (2011)—Experiment 3	512	SSI, U.S. adult mock jurors	Verdict (simple assault/ aggravated assault/not guilty); sentence	Criminal; assault	Neuroscientist testified the defendant suffered from structural frontal lobe damage preventing him from forming intent	Expert + fMRI and MRI	N/A	No: no change in verdict	No: no change in sentence
Schweitzer et al. (2011)—Experiment 4	433	SSI, U.S. adult mock jurors	Verdict (simple assault/ aggravated assault/not guilty); sentence	Criminal; assault	Neuroscientist testified the defendant suffered from structural frontal lobe damage preventing him from forming intent	Expert + MRI	N/A	No: no change in verdict	No: no change in sentence
Schweitzer et al. (2011) – Meta-analysis	1374	Meta-analysis	Meta-analysis	Criminal; meta-analysis	Meta-analysis	Meta-analysis	N/A	No: no change in verdict	No: no change in sentence

(continued)

Table 10.1 (continued)

Comparing neuroscientific expert evidence with varying neuroimages									
Study	Number of participants	Sample type	Legal judgment	Case type	Condition: expert testimony + Image 1	Condition: expert testimony + Image 2	Additional independent variables	Effect on verdict	Effect on sentencing
Appelbaum et al. (2015)—Experiment 1	960	YouGov participants, U.S. adult mock jurors	Sentence	Criminal; murder	Psychiatrist testified to rare gene that predisposed him to impulsive and violent behaviors + genetic test results image	Psychiatrist testified to defendant having abnormal activity in frontal lobes predisposed him to impulsive and violent behaviors + MRI scans	Heinousness of crime: low versus high; Criminal history: non-violent, violent, none	N/A	No: no change in sentence

(continued)

Table 10.1 (continued)

Comparing neuroscientific expert evidence with varying neuroimages									
Study	Number of participants	Sample type	Legal judgment	Case type	Condition: expert testimony + Image 1	Condition: expert testimony + Image 2	Additional independent variables	Effect on verdict	Effect on sentencing
Appelbaum et al. (2015)—Experiment 2	960	YouGov participants, U.S. adult mock jurors	Verdict (guilty or NGRI)	Criminal; kiddnapping and murder	Psychiatrist testified to genetic investigation: analysis of defendant's family tree + image	Psychiatrist testified to genetic investigation: genetic abnormalities lab test + image	Heinousness of crime: low versus high; diagnosis: bipolar disorder versus impulsivity; defendant history of mental illness: none versus mild versus severe	No: no change in verdict	N/A

(continued)

Table 10.1 (continued)

Comparing neuroscientific expert evidence with varying neuroimages

Study	Number of participants	Sample type	Legal judgment	Case type	Condition: expert testimony + Image 1	Condition: expert testimony + Image 2	Additional independent variables	Effect on verdict	Effect on sentencing
Appelbaum et al. (2015)—Experiment 3	763	YouGov participants, U.S. adult "death qualified" mock jurors	Death/life sentence	Criminal; first-degree murder	Psychiatrist testified to rare gene that predisposed him to impulsive and violent behaviors + genetic test results image	Psychiatrist testified to defendant having abnormal activity in frontal lobes predisposed him to impulsive and violent behaviors + MRI scans	Heinousness of crime: low versus high; criminal history: none, non-violent, violent	N/A	No: no change in sentence
Baker et al. (2013)	73	Undergraduate students; mock jurors	Verdict (guilty/not guilty); punishment	Criminal; assault	Neurologist testified that defendant's MRI revealed frontal lobe damage that could result in impulsivity. Presented bar graph comparing defendant's brain activity to normal brain activity	Neurologist testified that defendant's MRI revealed frontal lobe damage that could result in impulsivity. Presented MRI scans comparing defendant's brain activity to normal brain activity	N/A	No: no change in verdict	No: no change in degree of punishment

Note N/A = not available data or the variable was not included in the study, U.S. = United States, fMRI = functional magnetic resonance imaging, MRI = magnetic resonance imaging, NGRI = not guilty by reason of insanity, GBMI = guilty but mentally ill

[a]When clinical, genetic, or neurological without image evidence was presented by the defense, more death sentences were rendered, while reduced death sentences emerged when the prosecution presented these same forms of evidence, illustrating a backfire effect. The authors included an analysis that covaried mock jurors' dangerousness, responsibility, and treatability ratings and found that the Offering Party × Evidence interaction became non-significant when treatability was introduced into the model suggesting treatability mediated the impact of expert evidence on sentencing decisions [41]

[b]The authors included a mediation analysis between conditions and sentencing with mock jurors' ratings of the defendant's control over his behavior entered as the mediator, which they found to be significant. When the defendant's behavior was explained via the environmental condition, they perceived him as having more control over his behavior which led to longer sentence lengths [43]

[39] asked a national sample of participants to serve as mock judges in a case where a male defendant was convicted of sexually assaulting a female victim. They manipulated the mental status of the defendant—described by either an "expert", a neuroscientist, or a psychologist—as either being healthy (i.e., no mental health issues), having an impulse control disorder of neurobiological origins (i.e., a large tumor in the impulse control region of the defendant's brain), or the same disorder of psychological origins. Additionally, they manipulated the defendant's diagnosis treatability status as either seen to be completely treatable (i.e., successful removal of the tumor or cognitive-behavioral therapy was a success) or untreatable (i.e., tumor was inoperable, therapy was a failure [39]). Based on this design, the researchers reported that mock judges who were assigned to the neuroscientific expert condition rendered significantly shorter prison sentences compared to the psychological and "healthy" conditions.

Notably, other studies have not produced significant results pertaining to prison sentence recommendations. Mock jurors in Marshall and colleagues' [45] study read expert testimony presented by either a psychiatrist or neuroscientist in a murder trial. The psychiatrist testified that the defendant had a lack of impulse control, feelings of guiltlessness, and explained the characteristics of psychopathy he demonstrated. The neuroscientist also explained the characteristics of psychopathy demonstrated by the defendant but added that he exhibited under-activation in his amygdala and orbitofrontal cortex. Findings from this study did not reveal any differences in mock jurors' sentencing recommendations between expert testimony conditions.

Moreover, LaDuke and colleagues [46] presented participants with more ecologically valid materials—video footage of expert testimony. In this expert testimony condition, mock jurors viewed video footage of a psychologist describing the defendant's brain scans (either MRI or fMRI) suggesting neurological abnormalities which could explain his criminal behavior—he was convicted of burglary and aggravated assault. The expert also opined that the defendant was at a high risk for future violence. Mock jurors assigned to the control condition only read facts of the case including the police report of the crime, the defendant's statement and plea. Yet, no significant differences were observed between the expert and control conditions on mock jurors' sentencing recommendations [46].

A few studies have also been conducted to investigate the impact of neuroscience evidence on sentencing recommendations determined by legal experts. Aspinwall et al. [47] recruited 181 United States state trial judges and asked them to read a hypothetical case where the convict was diagnosed with psychopathy. When biomechanical evidence was presented during sentencing in support of this diagnosis, judges were less likely to rate psychopathy as an aggravating factor and sentencing was significantly reduced. The authors also conducted a content analysis of the judges' reasoning for their decisions and found that overall, they listed more aggravating factors, however, when the biomechanical evidence was presented, the proportion of mitigating factors also increased.

Interestingly, Fuss and colleagues [48] attempted a full replication of Aspinwall's study using a sample of German trial judges, but did not observe significant effects of biomechanical explanations for the defendant's behavior on sentencing recommendations. They did, however, report an increase in judicial orders of involuntary commitment in a forensic-psychiatric hospital for the defendant when the prosecution had presented the biomechanical evidence. Nevertheless, the authors suggest that biomechanical evidence presented to support a diagnosis of psychopathy affects judges' sentencing decisions differently in German courts compared to other legal systems [48].

Still, others have designed experiments to analyze various forms of evidence under the umbrella of "neuroscience." In a series of studies, Cheung and Heine [43] investigated the effects of genetic versus environmental explanations on mock jurors' legal decisions related to a defendant who fatally stabbed another individual. In the environmental condition, the defendant was described as having been reared in an environment that predisposed him to violence. In the genetic conditions, participants were told the defendant either had a (fictitious) genetic predisposition for responding to provocations with violence (genetic condition), a gene that created a stronger impulse to act violently compared to others without the gene (gene-impulse condition), or a gene that decreased the defendant's ability to control his violent impulses relative to those without the gene (gene-control condition [43]). Their results (Experiment 3) suggested mitigating effects of the genetic evidence explanations for the defendant's behavior compared to environmental explanations. Moreover, participants assigned to the gene-impulse condition recommended shorter sentence lengths compared to the genetic condition, thus suggesting that people pair genetic explanations with a lack of control and are more sensitive to evidence presenting an increase in engaging in impulsive behavior.

Lastly, in their aggregated dataset containing a large sample of both undergraduate and Amazon Mechanical Turk (MTurk) participants, Cheng and Heine [43] observed a mitigating effect of genetic explanations for the defendant's behavior on sentencing recommendations; wherein, participants rendered shorter sentences in the genetic versus the environmental condition. Notably, there was no change in verdict severity between forms of presented expert testimony.

Conversely, Appelbaum and Scurich [49] found signs of an aggravating effect of neuroscientific evidence. In their study, participants read about an apparently impulsive homicide case involving two men fighting over the attention of the perpetrator's girlfriend. Four different explanations for the defendant's behavior were proffered: (1) the defendant was simply behaving impulsively, (2) the defendant's impulsive and violent behavior was attributed to a history of child abuse, (3) his behavior was attributed to a gene formation that alters brain function, and (4) a combination of childhood abuse and irregular gene formation contributed to the defendant's impulsive and violent behavior. Participants who read about the defendant's history of childhood abuse paired with an expert's testimony of him having a genetic disposition to impulsive and violent behavior as an adult resulted in increased prison sentence recommendations. Similar findings were also rendered for the childhood abuse condition. Appelbaum and Scurich [49] posited the lack of significant findings for the standalone genetic evidence condition suggests fears that laypeople fall prey to strong beliefs in "genetic determinism" [50] may be overstated. More recently, Costa and colleagues [51] compared childhood abuse and genetic risk explanations for homicidal behavior to a jealousy motive (see Table 10.1 for experiment details). These researchers found no difference in verdicts between childhood abuse, genetic risk, or the jealousy motive explanations [51]. Therefore, these findings also do not support the notion that laypeople are particularly swayed by behavioral genetic evidence.

As stated previously, there have been null findings produced from studies examining the effects of neuroscientific expert evidence without images on verdict decisions. For example, Schweitzer and colleagues [52] performed four studies with similar methodologies to assess the effects of neuroscientific expert evidence on both verdict and sentencing recommendations for defendants accused of committing serious crimes (e.g., robbery plus assault, murder). They also quantitatively summarized the results from these four studies in a meta-analysis. In their neuroscientific expert evidence condition, a neuroscientist testified that the defendant suffered from structural frontal lobe damage, was prone to losing control when becoming angry, and could not form the intent necessary to be found guilty. In the control condition, the defense attorney argued the defendant's neurological defect prevented him from forming the requisite intent, yet no expert testimony was presented. Across all four studies, the presence of neuroscientific expert testimony failed to produce changes in verdict selection or sentencing recommendation [52]. Furthermore, the aggregated data they reported in their meta-analysis similarly found no effects of the neuroscientific evidence on verdict or sentencing [52]. Mowle and colleagues [53] also tested whether neuroscience evidence describing the defendant as having suffered brain damage to the frontal cortex, potentially explaining his impulsive actions, would mitigate mock jurors' verdict and sentencing recommendations. Their findings were also non-significant for both legal decisions.

An overall conclusion from the reviewed studies is that the presentation of neuroscientific evidence sans neuroimages may work to thwart a death sentence [44] or reduce prison sentences [39] but may not introduce enough reasonable doubt to influence mock jurors to render an acquittal [53]. Importantly, these effects may be moderated by defendant mental health status [41] (Experiment 1) and dangerousness [44]. Similar mitigating effects were detected for sentencing with a sample of United States trial judges [47], however these findings have not been replicated cross-culturally [48]. There is some evidence to suggest that behavioral genetic evidence sans images are influential in the courtroom, again resulting in more lenient sentencing recommendations [43]. However, there is some support for the notion that the presentation of genetic evidence coupled with a history of childhood abuse can elicit an aggravating effect concerning sentencing [49].

5.2 Neuroscientific Expert Testimony with Neuroimages

Several studies have examined the effects of neuroscientific expert testimony with neuroimages, however, again, only a few studies have reported significant findings. It is important to note that these studies investigated the combined effect of neuroscientific expert witness testimony accompanied by neuroimages rather than assessing the impact of either form of information alone. We review the studies examining these effects below and organize the results again by legal judgment.

In their series of four experiments, Schweitzer and colleagues [52] (see above and Table 10.1), explored the impact of neuroimages on mock jurors' verdict decisions and sentencing recommendations in a variety of criminal cases. In the neuroscientific expert plus neuroimage condition, participants reviewed expert testimony from a neuroscientist stating the defendant suffered from brain damage rendering him incapable of forming the intent required to be found guilty, paired with brain image scans. Participants in the control condition only read the defense's argument. Their fourth experiment produced significant results, wherein mock jurors in the expert evidence and neuroimage condition returned more lenient verdicts (e.g., simple assault versus aggravated assault) compared to the control condition, yet no effect on sentencing recommendations was observed [52]. This mitigating effect on verdict remained stable in their meta-analysis where they reported an overall 12% reduction in convictions for the neuroscientific expert plus neuroimages conditions compared to the control conditions. Further, the authors report null findings regarding sentencing.

Two studies have examined the effects of neuroimages and expert testimony on NGRI verdicts. First, Schweitzer and Saks [54] presented participants with an expert plus neuroimages condition involving a neurologist who testified that the defendant's brain scan (an MRI image) showed damage to his frontal lobe which could cause him to act irrationally and uncontrollably. The control condition included similar claims presented by the defense attorney but was devoid of expert testimony. After reviewing the study materials, mock jurors were instructed to render a verdict (i.e., guilty/NGRI or guilty but mentally ill [GBMI]) decision depending on the condition they were assigned. Results indicated mock jurors assigned to the expert plus neuroimage condition were more likely to render NGRI/GBMI verdicts compared to the control condition. Further, this mitigating effect was mediated by the amount of volitional control mock jurors believed the defendant had over his actions.

A few years prior, Gurley and Marcus [40] also examined the effects of expert evidence coupled with neuroimages on NGRI verdict decisions. In their study, undergraduate mock jurors read a case summary describing the murder of a postal official including three independent variables: mental disorder (schizophrenia-induced psychosis or psychopathy), onset of disorder (traumatic brain injury or not specified), and neuroimaging evidence (present or absent). Neuroscientific expert testimony concerning the defendant's diagnosis was delivered by a psychologist and psychiatrist in all conditions. In the neuroimage condition, four MRI scans were presented that showed extensive damage to the prefrontal cortex which the experts testified was likely to impair the defendant's abilities to control his actions. Results indicated a significant effect of the presence of neuroimages, with mock jurors rendering more NGRI verdicts if an MRI showing a brain lesion was present [40]. Notably, the defendant's mental disorder diagnosis did not moderate the mitigating effects of the expert plus neuroimaging evidence as it did for general neuroscientific expert evidence in other studies [41].

Investigations into the effects of neuroimages on sentencing recommendations have also been conducted. In addition to their diagnosis only and neuropsychological conditions, Greene and Cahill [44] (see above and Table 10.1) tested the effects of a neuroimaging condition where the defense expert described the defendant's psychosis diagnosis, summary and interpretation of neuropsychological tests, coupled with MRI and PET scan images showing damage to the left frontal area of the brain and reduced metabolic activity, respectively. They observed reduced death sentences for high-risk defendants meaning that defendant dangerousness moderated the mitigating effects of the neuroscientific expert evidence with neuroimages [44]. In fact, participants assigned to the high-risk condition were nearly 22 times more likely to sentence the defendant to death if they read the diagnosis-only information versus if they received neuroimaging evidence [44]. Mock jurors sentenced low-risk defendants to death at similar rates between the experimental and control conditions [44].

Appelbaum and colleagues [55] conducted two studies to assess the impact of neuroimages on sentencing in a non-capital and capital case. In the first case, participants read about a fatal stabbing while the second described a bank robbery gone wrong leading to the death of a police officer. In both cases the defendant's attorney claimed his client's behavior was an impulsive act. In the neuroscientific expert evidence plus images condition, participants reviewed MRI scan images and a psychiatrist's interpretation of the scan which identified an abnormality that led the defendant to act violently and on impulses without thinking through any requisite consequences. Results indicated no effect of the expert testimony plus neuroimages condition on prison sentence recommendations in the first case (Experiment 1) but did observe a significant reduction in recommended death sentences for the capital case compared with the control condition [55] (Experiment 3).

However, in another study, Saks et al. [41] (Experiment 1) failed to observe a mitigating effect of expert testimony with neuroimages on death sentence rates (see above and Table 10.1). Interestingly, there was an effect of neuroscientific expert testimony without neuroimages when the defendant was described as having been diagnosed with schizophrenia. This finding, however, has not been replicated by other researchers. Additional studies, Mowle et al. [53] and LaDuke et al. [46] have also failed to identify significant effects of neuroscientific expert evidence paired with neuroimages on legal decisions (see the previous section and Table 10.1).

Overall, the results from these studies suggest that the effects of neuroscientific expert testimony coupled with neuroimages may be mitigating in certain contexts. For example, while defendant mental health diagnosis did not seem to moderate the mitigating effects of this evidence type on NGRI verdicts rendered [40] the defendant's risk of future dangerousness did moderate mock jurors' tendency to sentence the defendant to death [44]. Similarly, Appelbaum et al. [55] observed reduced death sentences in their study, however, Saks et al. [41] found no change in death sentence rates between the neuroscientific expert plus neuroimage evidence and their control condition. Moreover, many studies did not observe a significant effect of neuroscientific expert plus neuroimage evidence on sentencing [46, 52, 53, 55]. Finally, the Schweitzer et al. [52] studies and their meta-analysis found that compared to the control conditions, the neuroscientific expert plus neuroimage evidence conditions lead to a reduction in verdict severity and overall guilty verdicts, respectively. However, it is important to note that other studies found null effects of the combined neuroscientific expert plus neuroimage evidence on verdict decisions [52, 53]. All in all, the effects of neuroscientific expert testimony coupled with neuroimages potentially seems to impact mock jurors' legal decision-making, yet to say these effects are unduly prejudicial would be an overstatement.

5.3 Comparing Neuroscientific Expert Testimony with and without Neuroimages

Critically, almost none of the studies regarding the third category where neuroscientific expert testimony *with* neuroimages is compared to neuroscientific expert testimony *without* neuroimages have reported significant findings. For example, neither Schweitzer et al. [52] nor Mowle et al. [53] found significant differences in guilty versus not guilty verdicts. Moreover, multiple studies found null effects of neuroscientific expert testimony accompanied by neuroimages compared to conditions including only the neuroscientific evidence on sentencing recommendations [45, 46, 52, 53]. While mitigating effects of neuroscientific expert evidence (with and without neuroimages) were observed in studies referencing capital punishment and NGRI/GBMI cases, these effects did not translate to conditions solely including neuroimages [41, 44, 54].

Only in Saks et al. [41] (Experiment 2), did the impact of expert testimony with neuroimages versus without exert a significant impact on mock jurors' sentencing decisions. Specifically, reduced death sentences were rendered for the neuroimage condition (versus neuroscientific evidence condition) *only* when presented by the defense compared to the prosecution (see above and Table 10.1). Notably, when the authors included an analysis that covaried mock jurors' dangerousness, responsibility, and treatability ratings, this interaction disappeared when *treatability* was introduced into the model, thus suggesting *treatability* mediated the mitigating impact of expert evidence on sentencing decisions. Again, this raises concerns about the degree to which the neuroimage bias alone can impact legal decision-making.

Overall, this suggests that the seductive allure of neuroimages does not add incremental value over and above the initial presentation of neuroscientific expert testimony. This finding is not surprising as the efforts to replicate the findings reported in McCabe and Castel [32] on the seductive allure of neuroimages has not be replicated in non-legal contexts [30, 33, 34].

5.4 Comparing Various Forms of Neuroimaging

Lastly, Baker and colleagues [42] conducted a study to investigate whether people visually attend to neuroimages differently given various stimuli. In their study, participants read about a defendant who assaulted his neighbor after the neighbor's lawnmower threw a stone and cracked the defendant's window. In the neuroscientific expert evidence condition, a neurologist presented results from the defendant's MRI scan which revealed significant damage to the frontal lobe of the defendant's brain and could impair a person's ability to regulate their impulses, suggesting such brain damage may have caused the defendant to lose control over his actions. Participants were also shown either a black and white MRI image of a normal brain compared to another image depicting the defendant's abnormal brain or a black and white bar graph comparing the "localized brain activity" of a normal brain to the defendant's brain which signaled lower activity levels [42]. Results indicated that while participants visually attended to the MRI image differently than the bar graph, no significant differences between neuroimage conditions on verdict or degree of punishment ratings were observed. However, participants who reviewed the MRI scan judged the defendant to be more likely to recidivate than those who reviewed the bar graph [42].

A second study compared various images under the umbrella of neuroscience. Appelbaum and colleagues [55] presented either an MRI scan showing impairment in the frontal lobes, which a psychiatrist explained could potentially be associated with increased impulsivity, or images of genetic test results (i.e., a "Manhattan plot" and a figure showing a deletion in one of the defendant's chromosomes) explained to be associated with impulsivity or a psychiatric disorder, respectively; or a genogram, displaying a family tree of relationships explaining the likelihood that a clinical condition will occur when a particular genotype is present [55]. The authors assessed the effects of each type of neuroimage across three separate murder cases involving multiple legal decisions (i.e., prison sentence, verdict, death sentence, see Table 10.1). Results indicated that the effects of each form of neuroimage did not differentially impact mock jurors' legal decisions when compared to one another.

Taken together, these findings suggest that one form of neuroimaging does not outweigh the other regarding how laypeople are rendering their legal decisions. Notably, the Appelbaum et al. [55] study demonstrated the negligible difference between various forms of neuroimages across multiple legal judgments for murder trials. Yet, Baker et al. [42] did notice differences in how laypeople viewed the defendant based on the form of neuroimage they reviewed, suggesting the type of neuroimage may impact attitudes toward the defendant but not necessarily legal decisions.

6 Neuroscience Evidence and Jurors' Decisions for Civil Cases

As mentioned earlier in this chapter, neuroscience evidence has been used in a variety of civil cases, including personal injury claims, disability assessments, and medical malpractice lawsuits. For example, Desmoulin-Canselier [56] reported on the use of neuroscience evidence to support claims of mental incompetence related to entering into contracts and creating valid wills. de Kogel and colleagues [57] also reported on neuroscience evidence's utility with regard to establishing legal capacity, describing recent Dutch cases that have called into question capacity to marry and form a legal partnership. Neuroscience can be relevant to establishing injury existence and severity [57, 58], and Rosenthal [59] argued that this is especially true for hidden or invisible injuries of which jurors have historically been skeptical.

Most of the scholarship on the use of neuroscience evidence in civil cases has been legal in nature, generally focusing on admissibility under *Daubert* and relevant evidence law. For example, Liu [60] examined prior cases that have considered the admission of neuroscientific evidence and classified the courts' approaches to the admissibility decision. Murphy [61] took a slightly more forward-thinking approach, considering how courts might evolve to handle the admissibility decision as neuroscience evidence becomes more common.

While questions about admissibility are certainly important, this decision is a judicial one, rather than by jury. In the area of jury decision-making in civil law cases, a number of empirical questions remain to be answered or even tested at all. While it should be acknowledged that only a subset of civil claims are decided by juries, there are many types of cases that are commonly decided by this group. For example, personal injury and medical malpractice claims are often decided by juries. Some empirical work has been conducted investigating how juries approach determining awards for injuries that are difficult to assign concrete values to, like loss of enjoyment of life [62] and pain and suffering [63, 64], but most of this work does not directly examine the use of neuroscience evidence.

Of particular note are two studies conducted by Phalen and colleagues [65] in which they directly compared participants' evaluations of three different types of evidence, including neuroimaging, to support a claim of chronic pain. In the second experiment (see Table 10.1), participants evaluated vignettes describing legal claims that included as evidence either a clinical battery, neuroimaging evidence without a visual, or neuroimaging evidence with a visual. They found an effect for neuroscientific evidence compared to a clinical battery, regardless of whether a visual was included. In short, participants saw the pain as more severe and granted larger damage awards when they were given neuroscientific evidence.

This research is an important first step to answering questions about how the use of neuroscientific evidence might impact jurors' judgments about the legal issues at hand, whether they be about injury severity or mental capacity. However, much more work must be done in this area to truly understand whether this evidence influences civil jurors differently than other types of evidence seeking to support the same type of claims.

7 General Discussion

The increase in neuroscience evidence being discussed in law review articles over the past 28 years suggests that this form of evidence is often used in the modern courtroom (Fig. 10.1). Moreover, recent research supports this notion particularly in criminal court proceedings [66] wherein both neuroscientific expert testimony evidence and neuroimages are routinely presented [22]. Similarly, it is becoming more commonplace to see neuroscientific expert evidence introduced in civil court proceedings concerning claims of mental incompetence [56], establishing legal capacity [57], and supporting injury claims [57, 58]. With the ever-growing presence of this form of evidence in the courtroom comes questions about its influence on legal decision-making. Likewise, some studies have further probed the seductive allure of neuroscientific explanations in non-legal contexts [28, 29] and have tested the incremental influence of neuroimages [32]. For example, McCabe and Castel [32] indicated a significant neuroimage bias, but more recent research suggests this bias is not as persuasive as initially thought [30, 33, 34].

Although the methods for experimentally analyzing the influence of neuroscientific evidence in the courtroom vary in the works described above (and Table 10.1), there were a few trends that are worth noting. Regarding the studies that used criminal cases, there were only three that observed changes in verdict decisions. Both Schweitzer and Saks [54] and Gurley and Marcus [40] tested the effects of neuroscientific evidence with neuroimages on NGRI verdicts and reported a reduction in guilty verdicts when this evidence was presented. Additional support for the mitigating effect of neuroimages was observed in Schweitzer et al. [52] where the researchers found a reduction in verdict severity (Experiment 4) and overall guilty verdicts (Meta-analysis). However, many studies we reviewed did not find a neuroimage bias on mock jurors' verdict decisions [42, 52, 53] nor an impact of general neuroscientific evidence sans images [52, 53]. Moreover, no significant changes in verdict decisions were observed in studies presenting behavioral genetic evidence [43, 49, 51].

In all three studies that involved capital punishment cases, the presentation of neuroscientific expert evidence (coupled with or without images) worked to successfully reduce death sentences [41, 44, 55]. However, it is important to note that some of these effects were dependent on defendant characteristics such as mental health diagnosis [41] and dangerousness [44]. Furthermore, with a few exceptions [39, 43], neuroscientific expert evidence did not consistently reduce prison sentences [43, 45, 46, 52, 53, 55]. Interestingly, conflicting results were found for the studies incorporating behavioral genetic evidence; wherein Appelbaum and Scurich [49] found an aggravating effect on prison sentences, while Cheung and Heine [43] found a mitigating one. Notably, jurors rarely make sentencing decisions, except for the death penalty, so studies that find effects on sentencing but not verdict are not really finding consequential effects. Lastly, mitigating effects on sentencing were also observed in a study of recruited trial judges [47], however, these effects may be specific to Western court systems [48].

Importantly, while the presence of neuroscientific evidence influenced legal decisions in some studies with or without the inclusion of neuroimages (Table 10.1), only one study detected an incremental effect of neuroimages over that of general neuroscientific expert evidence without images, resulting in a reduction in death sentence recommendations [41] (Experiment 2). Yet, this effect depended on the specific party (i.e., the defense) presenting such evidence. Furthermore, no significant effects comparing various forms of neuroscientific evidence (i.e., genetic test results image versus MRI scans) emerged for verdict decisions or sentencing recommendations [42, 55].

To our surprise, there is a paucity of published empirical work on how neuroscience evidence is viewed by jurors in civil court contexts. We found only one study to investigate the influence of neuroscientific evidence in a civil court context [65]. Compared to a clinical battery assessment, the presence of neuroscientific evidence to support a claim of chronic pain increased damage awards whether this evidence was coupled with or without a visual. Making further inferences on how neuroscience evidence effects civil jurors' decisions would be premature at this point in time. Nevertheless, additional work must be conducted to further learn the extent to which neuroscientific evidence influences civil jurors' decisions.

8 Future Directions

Importantly, research must be conducted to assess additional variables that may moderate the influence of neuroscientific expert evidence we have just reviewed. Aono and colleagues [38] suggest the type of criminally relevant legal decision could have moderated the mitigating effects of neuroscientific evidence, although this was not directly assessed in any study they reported on, nor was this variable explored in studies included in our review. The authors explain that when mock jurors are asked to decide between verdict choices including NGRI and GBMI [40], they are primed to think about the defendant's mental (and neural) health. Conversely, a more thorough evaluation of the expert testimony may not be primed when mock jurors are asked to render sentencing recommendations. On the other hand, Saks et al. [41] proposed that the effects of neuroimages may be negligible in the case of *mens rea* or insanity defenses because jurors are more concerned with public safety and a swift punishment, thus, neuroimages are ineffective in swaying jurors' verdict decisions. Therefore, the mitigating effects of the neuroimage bias may be more prominent during the sentencing phase of a trial after it is guaranteed the defendant will be removed from society and serve some form of punishment. Taken together, these propositions demonstrate the effects of neuroscientific expert evidence (with or without neuroimages) may exist but operate according to varying contexts. Additional research should explore how criminal legal decision moderates the impact of neuroscientific evidence.

Researchers should also incorporate various case types into their research designs to determine which presentation of neuroscientific evidence elicits the greatest influence. In her analysis of judicial opinions, Farahany [20] found over half of criminal cases that used neurobiological evidence were identified as non-homicide felony cases including drug and assault charges. These case types also warrant empirical attention to examine the impact of this evidence form on jurors' legal decisions. Further, the introduction of neuroscientific evidence at various trial phases should be investigated. For example, much attention has been paid to the legal decisions coming out of the guilt and post-conviction phases of trials, however, neuroscientific evidence has frequently been used to challenge defendants' competency during criminal proceedings, particularly during pre-trial phases [20].

Other defendant-specific variables such as race, age, gender, physical attractiveness, and socioeconomic status may also play a role in the rates for neuroscientific evidence to successfully influence legal decisions in both criminal and civil courts. Especially since these variables have been documented to influence legal decisions in a wide range of case contexts [67–70]. Similarly, including juror-specific variables into the equation would further our knowledge about how the effects of neuroscience in the courtroom operate. For example, jurors who hold strong views about mental health, personal responsibility, free will, or have extensive knowledge concerning neuroscience may render judgments differently compared to those without such views or knowledge [28, 55].

Lastly, another interesting avenue to explore would be how the framing and delivery of neuroscience evidence (e.g., jury instructions versus expert testimony) affect legal decision-making. One study found defense expert witness testimony had greater success in educating mock jurors on invalid forensic testimony presented in a criminal trial which resulted in an increase of acquittals compared to the same information delivered through jury instructions [71]. Framing and delivery method of specialized knowledge has yet to be explored in the context of neuroscience evidence, thus warranting further examination.

9 Conclusion

The integration of neuroscience evidence in legal decision-making represents a significant advancement in the understanding of human behavior and its implications for justice. However, the mitigating influence neuroscientific evidence has on death sentences and some criminal verdict decisions (e.g., NGRI) seems to be highly context-dependent. Further, there is not enough empirical evidence to propose trends for the influence of neuroscientific evidence in civil cases, however, the current work suggests this form of evidence will assist jurors in understanding pain and suffering as well as awarding damages. Additional research is necessary to further understand the effects of neuroscience in both criminal and civils court contexts as these tests will likely continue to be admissible. Finally, while Dr. Diamond commented that "…if you show them [jurors] the X-ray, they're convinced," (as cited in DeBenedicts [72], p. 30), we would like to counter with 'while an image may be worth a thousand words, it is not the case with neuroimaging presentation… most—but not all laypeople—are seductively allured by the general expert evidence rather than images.'

References

1. Hughes V (2010) Science in court: head case. Nature 464(7287):340–343. https://doi.org/10.1038/464340a
2. Darby WC, MacIntyre M, Cockerill RG, Stephens DB, Weinstock R, Darby RR (2023) In the courts: ethical and legal implications of emerging neuroscience technologies used for forensic purposes. In: Roberts LW (eds) Ethics and clinical neuroinnovation. Springer, Cham. https://doi.org/10.1007/978-3-031-14339-7_10
3. Jones OD, Marois R, Farah MJ, Greely HT (2013) Law and neuroscience. J Neurosci 33(45):17624–17630. https://doi.org/10.1523/JNEUROSCI.3254-13.2013
4. Morse S (2018) 9. Neuroscience in forensic contexts: ethical concerns. In: Griffith E (ed) Ethics challenges in forensic psychiatry and psychology practice. Columbia University Press, New York Chichester, West Sussex, pp 132–158. https://doi.org/10.7312/grif18330-010
5. *State of Tennessee v. Davis Bradley Waldroup, Jr.* (Tenn. Crim. App. 2013)
6. *Koch v. Western Emulsions, Inc.,* Az. Sup.Ct., Pima County, 2008
7. Shen FX (2016) The overlooked history of neurolaw. Fordham Law Rev 85(2):667–695
8. *U.S. v. Hinckley,* 672F.2d 115 (D.C. Cir. 1982)
9. *People v. Weinstein,* 156 Misc. 2d 34, 591 N.Y.S.2d 715 (Sup. Ct. 1992)
10. *Daubert v. Merrell Dow Pharmaceuticals, Inc.,* 509 U.S. 579, 113 s. Ct. 2786, 125 L. Ed. 2d 469 (1993)
11. Miller G (2010) Brain exam may have swayed jury in sentencing convicted murderer. Science 14. https://doi.org/10.1126/article.29520
12. *Mobley v. State,* 455 S.E.2d 61 (Ga. 1995)
13. Byrd AL, Manuck SB (2014) MAOA, childhood maltreatment, and antisocial behavior: meta-analysis of a gene-environment interaction. Biol Psychiatry 75(1):9–17. https://doi.org/10.1016/j.biopsych.2013.05.004
14. Caspi A, McClay J, Moffitt TE, Mill J, Martin J, Craig IW, Poulton R et al (2002) Role of genotype in the cycle of violence in maltreated children. Science 297(5582):851–854. https://doi.org/10.1126/science.1072290
15. Kim-Cohen J, Caspi A, Taylor A, Williams B, Newcombe R, Craig IW, Moffitt TE (2006) MAOA, maltreatment, and gene–environment interaction predicting children's mental health: new evidence and a meta-analysis. Mol Psychiatry 11(10):903–913. https://doi.org/10.1038/sj.mp.4001851
16. Taylor A, Kim-Cohen J (2007) Meta-analysis of gene–environment interactions in developmental psychopathology. Dev Psychopathol 19(4):1029–1037. https://doi.org/10.1017/S095457940700051X
17. Bernet W, Vnencak-Jones CL, Farahany N, Montgomery SA (2007) Bad nature, bad nurture, and testimony regarding MAOA and SLC6A4 genotyping at murder trials. J Forensic Sci 52(6):1362–1371. https://doi.org/10.1111/j.1556-4029.2007.00562.x
18. Feresin E (2009) Lighter sentence for murder with 'bad genes.' Nature (published online Oct. 30). https://doi.org/10.1038/news.2009.1050
19. Denno DW (2011) Courts' increasing consideration of behavioral genetics evidence in criminal cases: results of a longitudinal study. Mich St L Rev 967
20. Farahany NA (2016) Neuroscience and behavioral genetics in US criminal law: an empirical analysis. J Law Biosci 2(3):485–509. https://doi.org/10.1093/jlb/lsv059
21. Farahany N (ed) (2011) The impact of behavioral sciences on criminal law. Oxford University Press

22. Denno DW (2015) The myth of the double-edged sword: an empirical study of neuroscience evidence in criminal cases. Boston Coll Law Rev 56(2):493–551
23. *Strickland v. Washington,* 466 U.S. 668, 104 S. Ct. 2052, 80 L. Ed. 2d 674 (1984)
24. Gaudet LM, Marchant GE (2016) Under the radar: neuroimaging evidence in the criminal courtroom. Drake L Rev 64:577
25. Calvillo M, Irimia A (2020) Neuroimaging and psychometric assessment of mild cognitive impairment after traumatic brain injury. Front Psychol 11:1423. https://doi.org/10.3389/fpsyg.2020.01423
26. Du Y (2020) The application of neuroscience evidence on court sentencing decisions: suggesting a guideline for neuro-evidence. Seattle J Soc Justice 18(2):19
27. Boyd v. Bert Bell/Pete Rozelle NFL Players, 410 F.3d 1173 (9th Cir. 2005)
28. Weisberg DS, Keil FC, Goodstein J, Rawson E, Gray JR (2008) The seductive allure of neuroscience explanations. J Cogn Neurosci 20(3):470–477. https://doi.org/10.1162/jocn.2008.20040
29. Weisberg DS, Taylor JC, Hopkins EJ (2015) Deconstructing the seductive allure of neuroscience explanations. Judgm Decis Mak 10(5):429–441. https://doi.org/10.1017/S193029750000557X
30. Michael RB, Newman EJ, Vuorre M, Cumming G, Garry M (2013) On the (non) persuasive power of a brain image. Psychon Bull Rev 20(4):720–725. https://doi.org/10.3758/s13423-013-0391-6
31. Fernandez-Duque D, Evans J, Christian C, Hodges SD (2015) Superfluous neuroscience information makes explanations of psychological phenomena more appealing. J Cogn Neurosci 27:926–944. https://doi.org/10.1162/jocn_a_00750
32. McCabe DP, Castel AD (2008) Seeing is believing: the effect of brain images on judgments of scientific reasoning. Cognition 107(1):343–352. https://doi.org/10.1016/j.cognition.2007.07.017
33. Gruber D, Dickerson JA (2012) Persuasive images in popular science: testing judgments of scientific reasoning and credibility. Public Underst Sci 21(8):938–948. https://doi.org/10.1177/0963662512454072
34. Hook CJ, Farah MJ (2013) Look again: effects of brain images and mind–brain dualism on lay evaluations of research. J Cogn Neurosci 25(9):1397–1405. https://doi.org/10.1162/jocn_a_00407
35. Schweitzer NJ, Baker DA, Risko EF (2013) Fooled by the brain: re-examining the influence of neuroimages. Cognition 129(3):501–511. https://doi.org/10.1016/j.cognition.2013.08.009
36. Hopkins EJ, Weisberg DS, Taylor JC (2016) The seductive allure is a reductive allure: people prefer scientific explanations that contain logically irrelevant reductive information. Cognition 155:67–76. https://doi.org/10.1016/j.cognition.2016.06.011
37. Weisberg DS, Hopkins EJ, Taylor JC (2018) People's explanatory preferences for scientific phenomena. Cogn Res Principles Implications 3(1):44. https://doi.org/10.1186/s41235-018-0135-2
38. Aono D, Yaffe G, Kober H (2019) Neuroscientific evidence in the courtroom: a review. Cogn Res Principles Implications 4(1):40. https://doi.org/10.1186/s41235-019-0179-y
39. Allen CH, Vold K, Felsen G, Blumenthal-Barby JS, Aharoni E (2019) Reconciling the opposing effects of neurobiological evidence on criminal sentencing judgments. PloS One 14(1). https://doi.org/10.1371/journal.pone.0210584
40. Gurley JR, Marcus DK (2008) The effects of neuroimaging and brain injury on insanity defenses. Behav Sci Law 26(1):85–97. https://doi.org/10.1002/bsl.797
41. Saks MJ, Schweitzer NJ, Aharoni E, Kiehl KA (2014) The impact of neuroimages in the sentencing phase of capital trials. J Empir Leg Stud 11(1):105–131. https://doi.org/10.1111/jels.12036
42. Baker DA, Schweitzer NJ, Risko EF, Ware JM (2013) Visual attention and the neuroimage bias. PLoS ONE 8(9):e74449. https://doi.org/10.1371/journal.pone.0074449

43. Cheung BY, Heine SJ (2015) The double-edged sword of genetic accounts of criminality: causal attributions from genetic ascriptions affect legal decision making. Pers Soc Psychol Bull 41(12):1723–1738. https://doi.org/10.1177/0146167215610520
44. Greene E, Cahill BS (2012) Effects of neuroimaging evidence on mock juror decision making. Behav Sci Law 30(3):280–296. https://doi.org/10.1002/bsl.1993
45. Marshall J, Lilienfeld SO, Mayberg H, Clark SE (2017) The role of neurological and psychological explanations in legal judgments of psychopathic wrongdoers. J Forensic Psychiatry Psychol 28(3):412–436. https://doi.org/10.1080/14789949.2017.1291706
46. LaDuke C, Locklair B, Heilbrun K (2018) Neuroscientific, neuropsychological, and psychological evidence comparably impact legal decision making: implications for experts and legal practitioners. J Forensic Psychol Res Pract 18(2):114–142. https://doi.org/10.1080/24732850.2018.1439142
47. Aspinwall LG, Brown TR, Tabery J (2012) The double-edged sword: does biomechanism increase or decrease judges' sentencing of psychopaths? Science 337(6096):846–849. https://doi.org/10.1126/science.1219569
48. Fuss J, Dressing H, Briken P (2015) Neurogenetic evidence in the courtroom: a randomised controlled trial with German judges. J Med Genet 52(11):730–737. https://doi.org/10.1136/jmedgenet-2015-103284
49. Appelbaum PS, Scurich N (2014) Impact of behavioral genetic evidence on the adjudication of criminal behavior. J Am Acad Psychiatry Law 42(1):91
50. Dar-Nimrod I, Heine SJ (2011) Genetic essentialism: on the deceptive determinism of DNA. Psychol Bull 137(5):800
51. Costa ET, Pate M, Gibson CL (2017) Can molecular genetic evidence influence jurors' perceptions of a defendant? Results from a randomized experiment. Crim Justice Behav 44(8):1101–1119. https://doi.org/10.1177/0093854817715882
52. Schweitzer NJ, Saks MJ, Murphy ER, Roskies AL, Sinnott-Armstrong W, Gaudet LM (2011) Neuroimages as evidence in a mens rea defense: no impact. Psychol Public Policy Law 17(3):357
53. Mowle EN, Edens JF, Clark JW, Sörman K (2016) Effects of mental health and neuroscience evidence on juror perceptions of a criminal defendant: the moderating role of political orientation. Behav Sci Law 34(6):726–741. https://doi.org/10.1002/bsl.2251
54. Schweitzer NJ, Saks MJ (2011) Neuroimage evidence and the insanity defense. Behav Sci Law 29(4):592–607. https://doi.org/10.1002/bsl.995
55. Appelbaum PS, Scurich N, Raad R (2015) Effects of behavioral genetic evidence on perceptions of criminal responsibility and appropriate punishment. Psychol Public Policy Law 21(2):134
56. Desmoulin-Canselier S (2020) Another perspective on "neurolaw": the use of brain imaging in civil litigation regarding mental competence. Neurosci Law Complicated Crossings New Perspect 529–547
57. de Kogel CH, Schrama WM, Smit M (2014) Civil law and neuroscience. Psychiatry Psychol Law 21(2):272–285. https://doi.org/10.1080/13218719.2013.808978
58. Catley P, Claydon L (2015) The use of neuroscientific evidence in the courtroom by those accused of criminal offenses in England and Wales. J Law Biosci 2(3):510–549. https://doi.org/10.1093/jlb/lsv025
59. Rosenthal H (2018) Scanning for justice: using neuroscience to create a more inclusive legal system. Colum Hum Rts L Rev 50:290
60. Liu CT (2014) Scanning the evidence: the evidentiary admissibility of expert witness testimony on MRI brain scans in civil cases in the post-Daubert era. NYU Ann Surv Am L 70:479
61. Murphy E (2016) Neuroscience and the civil/criminal Daubert divide. Fordham L Rev 85:619
62. Poser S, Bornstein BH, McGorty EK (2003) Measuring damages for lost enjoyment of life: the view from the bench and the jury box. Law Hum Behav 27:53–68. https://doi.org/10.1023/A:1021626928063

63. Hans VP, Helm RK, Reyna VF (2018) From meaning to money: translating injury into dollars. Law Hum Behav 42(2):95
64. McAuliff BD, Bornstein BH (2010) All anchors are not created equal: the effects of per diem versus lump sum requests on pain and suffering awards. Law Hum Behav 34(2):164
65. Phalen HJ, Salerno JM, Schweitzer NJ (2021) Can neuroimaging prove pain and suffering?: The influence of pain assessment techniques on legal judgments of physical versus emotional pain. Law Hum Behav 45(5):393
66. Meixner JB (2016) The use of neuroscience evidence in criminal proceedings. J Law Biosci 3(2):330–335. https://doi.org/10.1093/jlb/lsw028
67. Mazzella R, Feingold A (1994) The effects of physical attractiveness, race, socioeconomic status, and gender of defendants and victims on judgments of mock jurors: a meta-analysis. J Appl Soc Psychol 24(15):1315–1338. https://doi.org/10.1111/j.1559-1816.1994.tb01552.x
68. Mitchell TL, Haw RM, Pfeifer JE, Meissner CA (2005) Racial bias in mock juror decision-making: a meta-analytic review of defendant treatment. Law Hum Behav 29:621–637. https://doi.org/10.1007/s10979-005-8122-9
69. Pozzulo JD, Dempsey J, Maeder E, Allen L (2010) The effects of victim gender, defendant gender, and defendant age on juror decision making. Crim Justice Behav 37(1):47–63. https://doi.org/10.1177/0093854809344173
70. Winters GM, Collins CM, Kaylor LE, Jeglic EL (2020) The impact of defendant gender and attractiveness on juror decision-making in a sexual offense case. Deviant Behav 43(4):507–524. https://doi.org/10.1080/01639625.2020.1844364
71. Eastwood J, Caldwell J (2015) Educating jurors about forensic evidence: using an expert witness and judicial instructions to mitigate the impact of invalid forensic science testimony. J Forensic Sci 60(6):1523–1528. https://doi.org/10.1111/1556-4029.12832
72. DeBenedictis (1990) Criminal minds: PET scans used to prove accused killers' brain abnormalities. ABA J 76(1):30

Alissa Anderson Call Ph.D., is an Associate Professor of Psychology at Black Hills State University. Her research focuses on jury decision-making for cases involving sexual abuse and assault. She is also interested in factors that affect reporting domestic violence and campus sexual assault.

Liz Wilson M.A., is a researcher whose focus is on substance stigma in juror perceptions, jury deliberations and decision-making, and teacher/student sexual misconduct. Liz also has research interests in mental health issues faced by marginalized groups, specifically support provided within the criminal legal system.

Twila Wingrove J.D., Ph.D., is a Professor of Psychology at Appalachian State University. Her research focuses on legal decision-making and lay perceptions in cases involving sexual violence towards children and adults.

Groups and Decision Making: How We Choose Groups and How Groups Choose for Us

11

Laura Prislei, Erica Molinario, and Stefano Livi

"Men work together," I told him from the heart,
"Whether they work together or apart."
The Tuft of Flowers
Robert Frost (1915)

Abstract

This chapter examines how individuals select the social groups they choose to join and how these groups, in turn, influence individual behavior and decision-making. Using key concepts from social psychology, we explore three primary factors that guide group selection: (1) perceived similarity and affinity with other group members; (2) the group's reputation and social standing; (3) the anticipated sense of being accepted within the group. Additionally, we discuss how social groups influence individual decision-making through the enforcement of social norms, i.e., the collectively shared expectations of appropriate behavior within a group. Through this lens, the chapter aims to explore how these psychosocial mechanisms operate across various aspects of life,

L. Prislei (✉) · S. Livi
Sapienza University of Rome, Rome, Italy
e-mail: laura.prislei@uniroma1.it

E. Molinario
Jagiellonian University, Krakow, Poland

Florida Gulf Coast University, Fort Myers, USA

© The Author(s), under exclusive license to Springer Nature Switzerland AG 2025

T. McElroy (ed.), *Decision-Making in Life and Work*, Integrated Science 36,
https://doi.org/10.1007/978-3-031-88415-3_11

from romantic and friendship relationships to work and political participation, demonstrating their generalizability across diverse contexts and highlighting the pervasive influence of group dynamics in shaping decision making processes.

Graphical Abstract

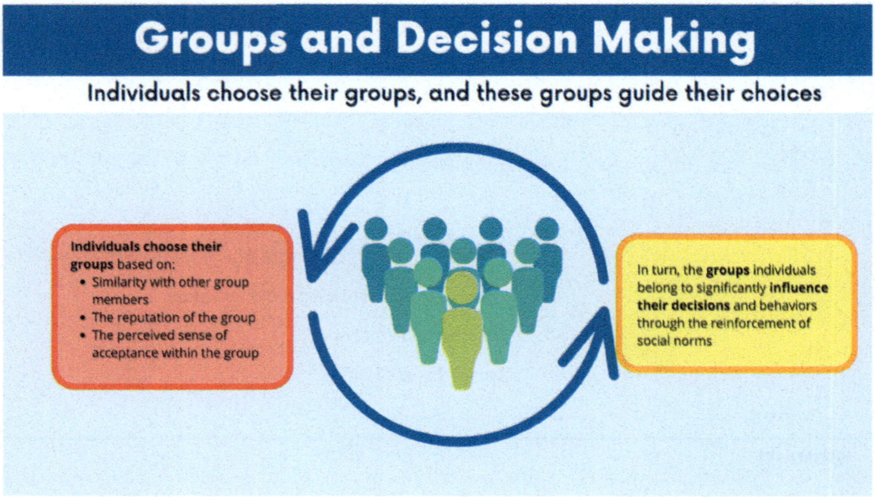

Keywords

Acceptance • Decision making • Groups • Members' similarity • Reputation • Social influence

1 Introduction

Since prehistoric times, humans have displayed a natural inclination to form groups. Early humans relied on collaboration for survival, undertaking tasks such as hunting large prey and defending against predators—challenges that were insurmountable for individuals but achievable through coordinated effort. Groups, defined as "two or more interdependent individuals who influence one another through social interactions" [1], have thus played a fundamental role in human evolution. As societies evolved, cooperation enabled increasingly complex forms of organization. For example, ancient Greece saw the rise of πόλεις (city-states), where structured communities adhered to shared laws and responsibilities. Over the centuries, humans expanded their collective efforts to monumental projects, such as the establishment of nation-states during the Renaissance and revolutions that reshaped history.

The human drive to form relations with other people is deeply rooted in evolutionary necessity. Group membership historically offered critical survival advantages, such as access to shared resources, food, and assistance in child-rearing. Groups provided protection during times of illness, danger, or disaster, enhancing survival odds. For children, proximity to adults ensured food, safety, and guidance, increasing the likelihood of reaching reproductive age. For adults, forming long-term attachments enhanced reproductive success and improved off-spring survival. Further, competition for scarce resources reinforced the need for alliances, as group connections often secured better access to vital necessities. At its core, this drive to form and sustain meaningful relationships is one of the most powerful and universal human motivations. The *belongingness hypothesis* [2] posits that individuals are driven to establish and maintain a minimum number of enduring and significant relationships in order to fulfill essential emotional and psychological needs. Scholars have demonstrated that friendships provide both mental and physical benefits to those involved: for instance, a stronger social network decreases depression risk [3] and age-related cognitive decline [4], but also protects against mortality risk from diseases such as Alzheimer, diabetes, or cardiovascular issues [5].

In addition to evolutionary advantages, *Social Identity Theory* [6] provides a psychological explanation for the importance of groups in human life. According to this theory of intergroup relations, individuals derive a significant part of their self-concept from the groups to which they belong. Indeed, people naturally categorize themselves and others into "in-groups" (to which they belong) and "out-groups" (to which they do not), which influences behavior and decision-making. This categorization offers a source of identity and self-esteem, leading individuals to a preference for one's in-group and a tendency to view out-groups more critically. As a consequence, groups show a tendency to compete over evaluatively positive distinctiveness, because such distinctiveness is reflected in one's social identity. Some decades later, as a development of the motivational component of the Social Identity Theory, Hogg [7] proposed the *Uncertainty-Identity Theory* to describe how feelings of self-uncertainty drive people to strongly identify themselves with social groups in order to protect from the uncertainty itself. In other words, according to this theory, group identification buffers uncertainty, offering a sense of structure through providing behavioral guidelines.

Moreover, groups provide a shared sense of reality. Humans rarely think in isolation from others; their understanding of reality is rather constructed interactively within the groups they value. Through these shared processes, members collectively construct a framework of agreed-upon categories and beliefs that form their perception of reality. This role of groups in shaping members' sense of reality has been a cornerstone of classical theories in social psychology [8–10] and remains central in contemporary studies of social behavior [11]. Given that groups influence individuals' thoughts, decisions, and actions, it is critical to understand how and why individuals choose to belong to one group over another.

2 How Do Individuals Choose Their Groups?

Each of us belongs to multiple groups throughout our lives. Some of these groups are assigned to us by circumstance, without any explicit choice on our part (for instance, our nationality, family of origin, or the neighborhood we are born into). Others, however, are the result of deliberate decisions based on personal preferences, such as selecting a circle of friends, choosing a workplace, or joining a community organization. Such decisions profoundly shape our behavior by influencing how we interpret and engage with the world around us.

This raises a fundamental question: what drives individuals' decision to select certain groups as their own? The literature identifies several key factors influencing an individual's decision to join or remain in a particular group. We argue that these factors can be broadly categorized into three main elements (see Table 11.1 for a synthesis): (1) the sense of similarity and affinity with other group members; (2) the perceived reputation of the group itself; (3) the feeling that one would be accepted within the group. In what follows, we will discuss each of these factors through the lens of existing literature in social psychology.

Table 11.1 Key factors influencing an individual's decision

Factors influencing group membership	Description
Similarity and affinity with group members	People are naturally drawn to those with similar traits, beliefs, and behaviors, a tendency known a "homophily." This phenomenon is evident in various contexts, such as romantic relationships, friendships, work collaborations, social activism and even within radicalized groups
Perceived reputation of the group	People tend to avoid associating with those who have poor reputations and seek connections with those who enhance their own social standing. Further, groups also consider reputation when selecting new members, preferring those seen as competent or reliable
Feelings of being accepted within the group	Humans are inherently motivated to seek acceptance from those who show affection, as rejection or ostracism historically posed survival risks. Social support and recognition are critical for psychological well-being, influencing decisions to maintain relationships

3 Similarity and Affinity with Group Members

An English famous proverb says "birds of a feather flock together", meaning that beings are naturally attracted to those who share similar characteristics. Just as birds of the same species tend to fly together in groups, people often form connections and communities with those who share common traits, beliefs, or behaviors. Whether for animals or humans, flocking together with similar others offers both practical and social benefits.

Lazarsfeld and Merton [12] coined the term "homophily" to refer to this natural tendency for individuals to gravitate toward those who share similar traits. More in general, the *homophily principle* summarizes the observation that relationships are more likely to form between individuals who share similarities than between those who differ. This principle is supported by a range of evidence across various aspects of daily life. For instance, online dating platforms and social media frequently match users based on shared interests (e.g., users that like the same favorite music or movies), as such similarity increases the likelihood of mutual initial attraction. In scientific literature, homophily is documented across a wide range of relational contexts, including marriage [13], friendship [14], religion education [15], and professional partnerships [16]. These studies consistently demonstrate that people with shared beliefs and backgrounds are more likely to form meaningful connections, value each other's contributions, and communicate effectively [17]. Research has further validated the principle of homophily through a variety of methodologies, including experimental studies and longitudinal analyses. For instance, Campigotto et al. [18] examined the factors influencing friendship networks among 10,000 secondary school students across four countries (England, Germany, the Netherlands, and Sweden) using a two-wave survey. Their findings revealed that students' gender, country of origin and religious affiliation significantly shaped the formation of their social networks, with participants tending to nominate peers who shared these characteristics as their best friends. Further, this effect proved to be stable over time, highlighting the lasting impact of homophily markers on social connections.

Evidence from neurosciences suggests that homophily can also be understood as the genetic similarity among individuals. In this vein, research showed that molecular mechanisms may play a role in mutual social preferences. In a longitudinal study, Fowler et al. [19] analyzed sibling and twin pair data to examine genetic homophily among friends, demonstrating that the genotype of one individual is predictive of the genotype of his/her friends. Accordingly, scholars indicate that friends are genetically more similar to one another than they are to non-friends, exhibiting levels of genetic "relatedness" comparable to that of 4th cousins [20]. These findings point to potential genetic underpinnings of social bonding, suggesting that shared genetic traits may facilitate friendship by promoting compatibility or mutual affinity. However, alternative explanations also account for these patterns of genetic similarity among friends. For instance, social structures such as racial stratification or economic inequality may limit the diversity of social circles, inadvertently creating genetic homophily [21]. Additionally, geographical proximity is a significant factor in friendship formation, as people who live near each other are more likely to interact and share genetic similarities due to localized population structures [22].

Beyond its influence on decision-making processes in the formation of friendships, homophily also appears to play a significant role in the selection of work partners. For instance, Hinds et al. [23] investigated the factors influencing individuals' choices of work collaborators, employing a combined model to examine individual attributes, relational attributes, and prior structural ties. The research analyzed four years of data from participants in 33 small project groups. The findings revealed homophily biases in group member selection, with individuals tending to prefer colleagues of the same ethnicity.

Network formation is often studied in terms of demographic factors such as gender, age, ethnicity, religion, and occupation. Harrison et al. [24] classify these as surface-level characteristics and differentiate them from deep-level traits, such as shared goals and values. Interestingly, their research demonstrates that while surface-level similarities may play a role initially, deep-level compatibility becomes more important over time, having a greater impact on team cohesion, satisfaction, and performance. In this vein, Rapallini and Rustichini [25] demonstrated the role of personality and cognitive skills in predicting best friend nominations in a sample of 7-year-old children. Their findings revealed that children who both scored above or both scored below the median for extraversion, conscientiousness, and openness to experience were more likely to nominate each other as best friends compared to pairs with differing personality profiles. This suggests that similarity in personality characteristics plays a significant role in shaping close social bonds, even from an early age. Additionally, similarity in cognitive abilities (measured by children's most recent grades reported in schools) also predicted best friend nominations. Along these lines, Osgood et al. [14] conducted a longitudinal analysis over seven waves, examining the link between similarity and friendship among a sample of over 15,500 adolescents. Their findings revealed that similarity in school grades exhibited a stronger association with friendship formation than most other attributes, including gender and ethnicity.

Research on relationships based on shared personality traits has also found evidence of socialization on depression [26, 27]. For instance, Shaefer et al. [28] identified a pattern of homophily in friendship among depressed adolescents. Analyzing data from a sample of 1820 students in nine different schools, the Authors found that those suffering from depression, as they were socially withdrawn, were more likely to form connections with peers who were also depressed and similarly socially excluded.

But similarity in deep-level traits does not only influence friendship bonds. Evidence from the organizational context demonstrated that the affinity between an individual's personal values and those of an organization significantly influences their choice of employer. This phenomenon, often referred to as *person-organization fit* [29], highlights how psychological compatibility in terms of goals and value alignment affects both the decision to join a company and subsequent job satisfaction. For example, individuals are more likely to select organizations whose values [30, 31] and goals [32] align with their own.

A similar dynamic is observed in activism, where individuals often join groups or movements that reflect their deeply held values and commitment to specific social causes, such as environmental sustainability, social justice, or human rights. This process aligns with the Greenber and Mollick [33] concept of *activist choice homophily*. The Authors argue that homophilic support is not merely based on individual similarity, but is rather driven by the perception of shared structural challenges linked to a common group-level social identity. In such situations, individuals often feel a sense of solidarity with other group members who face similar disadvantages, motivating them to affiliate with or support the group. In a similar vein, Thomas et al. [34] proposed the *normative alignment model,* which suggests that commitment to collective action depends on creating a social identity that includes a coherent pattern of norms for emotion, efficacy, and behavior.

This discussion can also be extended to radical political groups, such as terroristic movements. In the context of violent extremism, Webber and Kruglanski [35] proposed a socio-psychological model of radicalization which posits that individuals who adopt an ideological *narrative* justifying violence as a means to restore a *need of being socially recognized* are often motivated to seek out a *network* of others who share similar beliefs (see Kruglanski et al. [36], for a comprehensive discussion on the *3N Model of Radicalization*). According to the model, individuals who endorse violence as a means to regain social acknowledgment are often driven to seek out others who share their beliefs. Aligning with a group that upholds similar ideologies offers a sense of social validation, fulfilling both epistemic and relational needs. These needs are met by the predictable and controllable environment provided within a community of like-minded individuals. Consequently, those who justify violence are likely to gravitate toward the company of others with similar ideological beliefs [37].

Notably, not only are people naturally drawn to those who are similar to themselves, but they also tend to become increasingly similar to the individuals with whom they share close relationships. For instance, in their longitudinal research in school contexts, Shin and Ryan [38] used a stochastic actor-based model of social network analysis to demonstrate that friends influence each other over time, leading to greater similarity in various aspects of academic adjustment, such as effortful and disruptive behavior, and achievement (i.e., GPA calculated from report card grades). More recently, Shin [39] found evidence that early adolescents become more similar to their connected social relations over time also in terms of academic engagement. Interestingly, just as we tend to become more similar to those with whom we share close relationships, we also tend to perceive them as being more similar to ourselves. In this vein, Kenny and Acitelli [40] described the *bias effect* which refers to the tendency of individuals to attribute their own characteristics to their partners—essentially projecting themselves onto their partners. Building on this, Theodorou et al. [41] found that this bias effect may impact relationship satisfaction. Specifically, analyzing data from 161 heterosexual couples, the Authors found that a stronger bias effect was associated with greater relationship satisfaction. In other words, the positive impact of an individual's positivity on their own relationship satisfaction was mediated by their heightened perception of their partner's positivity.

To summarize, the concept of homophily underscores the impact of similarity in promoting social connections. From friendships and professional collaborations to activism and even radicalization, the tendency to gravitate toward those who share common traits, beliefs, or values reflects both practical and psychological motivations. However, the tendency to seek out similar individuals is not the sole factor influencing how people choose which groups to join or avoid. In what follows, we will explore two additional key criteria that shape these decisions: the perceived reputation of the group and the individual's sense of being accepted within it.

4 Perceived Reputation of the Group

As part of his character training, young George Washington had transcribed into his school copybook the maxim "Associate yourself with men of good quality if you esteem your own reputation; for it is better to be alone than in bad company", which was one of the 110 Rules of Civility and Decent Behavior in Company and Conversation originally composed by French Jesuits during the late sixteenth century. Although this saying may seem antiquated, contemporary social psychological research remarks its enduring relevance: a "good" reputation is pivotal in shaping social interactions, and the people we associate with play a crucial role in shaping whether others perceive us in a positive or conversely negative way. The critical role of reputation in shaping decisions about whom to interact with is strongly supported by the social function of gossip, which serves as a means to share information [42]. Ethnographic research, for instance, suggests that group members readily disseminate reputational information about those who do not conform to normative levels of cooperation [43]. In this vein, scholars have coined the term *stigma by association* to describe the process by which individuals close to stigmatized persons are themselves discredited [44, 45]. Thus, on one hand, people are innately motivated to distance themselves from those with poor reputations to avoid potential social repercussions. On the other hand, they actively seek to associate with individuals of good reputation, leveraging these connections to enhance their own social standing and credibility.

Extending this idea to romantic relationships, the love model derived from the *Quest for Significance Theory* by Kruglanski and colleagues [46] proposes that individuals—particularly those who feel socially unrecognized or disrespected—seek partners who possess social value, such as being powerful, wealthy, attractive, or famous. To capture this phenomenon, the Authors use the term "trophy qualities" to describe attributes in a partner that, when present, enhance the perceived social status of the individual in the relationship. These qualities serve as a form of symbolic capital, allowing individuals to indirectly acquire the recognition and validation they may lack through their partner's esteemed traits. The concept of "trophy qualities" garnering attraction finds support in empirical literature. For instance, Regan et al. [47] observed that in mate selection, individuals emphasized socially desirable traits in potential partners, such as sexual attractiveness and social status. Similarly, research on speed dating has shown that participants are more likely to pursue relationships with individuals who are physically attractive or have strong earning potential [48]. More recently, the theoretical love model proposed by Kruglanski and colleagues was further validated by both experimental and longitudinal studies conducted by Contu et al. [49], which demonstrated that perceiving romantic partners as possessing socially valued traits—along with being admiring and caring (as will be discussed in the next paragraph)—enhances an individual's sense of significance (i.e., the feeling of being important in the eyes of others) gained through the romantic relationship. This, in turn, increases both the likelihood of entering a romantic relationship and sustaining it over time. Additionally, the Authors found that individuals experiencing a loss of significance, and thus a heightened desire to regain it, were particularly attuned to their partner's ability to confer significance through their socially valued traits. These findings align with Aron and Aron's [50] *inclusion of others in the self principle*, which is central to their self-expansion model. According to this framework, romantic relationships are motivated by the drive to enhance one's self-efficacy, defined as the ability to achieve goals. To fulfill this need, individuals seek to expand their self-concept by incorporating a partner's resources, traits, and perspectives into their own identity. Thus, the Authors propose that people are drawn to and maintain relationships based on their partner's desirability, which reflects the qualities the partner possesses that enhance this self-expansion process.

Along these lines, research in organizational contexts has shown that a key factor influencing an organization's ability to attract new talent is its reputation, which reflects the relative standing of a firm's name compared to its competitors [51–53]. In the previously mentioned study by Hinds and colleagues [23], which investigated the factors influencing individuals' choices of work collaborators, the Authors found that, in addition to homophily with other group members, another key criterion in selecting collaborators was the perception of their reputation for competence. Accordingly, Cable and Turban [54] found that job seekers' perceptions of an organization's reputation significantly impacted their interest in pursuing employment there. Moreover, the results showed that this effect arises because reputation serves as a signal of desirable job attributes and it influences the pride individuals anticipate deriving from being associated with the organization. Notably, the study also revealed that, on average, job seekers were willing to accept approximately 7% lower salaries for identical roles at firms with positive reputations compared to those with negative reputations.

While individuals consider a group's reputation when deciding whether to adhere or not to the group, groups likewise evaluate the reputation of a potential new candidate member. For instance, job candidates with strong reputations for competence, reliability, or expertise are often favored, as they are perceived to bring value to the organization and contribute positively to its success [55, 56]. Stiff and Van Vugt [57] showed evidence of this through an experimental study in which they manipulated the candidate's reported self-sacrifices to enter an imaginary group and the source of this information. Participants performed a collaborative task involving solving simple anagrams under conditions mimicking a goods dilemma game (i.e., an anagram puzzle [58]). The groups were led to believe that their success depended on collective effort. After several rounds of the task, participants were presented with the opportunity to admit a new member to their group, with the decision based on specific information provided about the candidate. In more detail, candidates were described as they could earn considerably more money in their current task, and thus gave up a lot to join the new group (high sacrifice condition) or as they could earn considerably less in their current task (low sacrifice condition). Additionally, the information about the candidate's sacrifice came either from a current group member (reputation condition) or directly from the candidate (non-reputation condition). Results showed that participants rated candidates in the high-sacrifice condition as more committed than those in the low-sacrifice condition, but only when the information came from a group member (reputation condition). Additionally, candidates in the high-sacrifice condition received more votes for admission than low-sacrifice candidates, but again this effect was observed only in the reputation condition. Consistent with these findings about the role of reputation in shaping decisional processes related to the acceptance of newcomers, Feinberg et al. [59] conducted an experiment in which participants played a public-goods game in small groups. Each participant played 18 rounds, with group members changing each time, and was given the opportunity to exchange information about their teammates' abilities with others. The results revealed that participants readily shared reputational information about others. Crucially, those who received this gossip used it to form judgments about reputation and to select interaction partners. Specifically, participants preferred to engage with individuals known to be cooperative and excluded those identified as defectors.

Reputation, therefore, plays a fundamental role not only when we decide whether to join a group (be it a dyad or a larger collective), but also when we consult with others about accepting someone into our team. The third key factor influencing group-related decision-making processes is the feeling of being wanted or valued within the group. This aspect will be explored in the next paragraph.

5 Feelings of Being Accepted Within the Group

Imagine moving to a new city to start a new job and meeting a group of peers at work who share your interests. You learn that you share with them the same interests (for example, they are fans of your favorite musician and are planning to attend an upcoming concert you have been longing to see). This group also enjoys an excellent reputation, both socially and professionally, and you know that working alongside them could teach you a great deal. Based on what we have discussed so far (and extremely simplifying the deductions for the sake of brevity), joining this group would seem like an obvious and appealing choice. Shared interests provide an easy foundation for connection, while the group's positive reputation enhances its attractiveness. But what happens if you perceive that the group is not interested in welcoming you, or worse, actively excludes you? Probably, such feelings of ostracism would likely discourage you from pursuing a connection with these individuals. The reaction to avoid individuals that are perceived as unwelcoming is deeply rooted in human psychology and has strong evolutionary roots. From early childhood, humans demonstrate a fundamental need to form relationships with those who show affection and attention toward them [60], as seeking acceptance (or avoiding rejection) represents a survival mechanism. Historically, belonging to a group meant access to critical resources, protection, and social learning, while exclusion increased the risks of vulnerability and isolation. Further, research shows that individuals tend to gravitate toward others who make them feel accepted, reinforcing their sense of belonging and self-worth. This phenomenon is well captured by *Sociometer Theory* by Leary and Baumeister [61], which posits that self-esteem—the degree to which self-concept's qualities are perceived as positive—functions as an internal gauge of social inclusion. According to this theory, when individuals perceive themselves as appreciated by others, their self-esteem rises, signaling a secure position within their social group. Conversely, a lack of perceived appreciation triggers feelings of rejection, motivating efforts to repair or replace the lost connection.

In line with this perspective, the seminal work of Berkman and Syme [62] revealed that social support from close relationship partners is a critical predictor of lower mortality rates. This discovery has since spurred extensive research into the mechanisms linking social support with psychological well-being and physical health. Supportive relationships mitigate loneliness and depressive symptoms [63, 64] while enhancing happiness and overall well-being [65, 66]. For instance, Shin and Gyeong [67], in a study of 1033 adults from South Korea, found that support from spouses and friends was linked to greater happiness and well-being, as well as reduced loneliness, while strain in these relationships was associated with increased depressive symptoms. Although the relationship between social support and psychological well-being is reciprocal, growing longitudinal evidence suggests that social support actively promotes improved well-being and health outcomes over time [68, 69]. Taken together, this body of research highlights why individuals are inherently motivated to seek relationships (with both romantic partners and friends) where they feel supported and accepted.

However, humans do not seek just acceptance within a group, but they also need to feel recognized and admired for their value. Social psychologists refer to this need for acknowledgment as the *quest for significance* (see Kruglanski et al. [36] for an in-depth analysis of the phenomenon). This need motivates individuals to engage more deeply with others that validate their worth and make them feel appreciated. Such validation promotes a stronger commitment to the group, a process referred to as the *collectivistic shift*, where personal goals become closely aligned with the group's objectives in exchange for social recognition and esteem [70, 71].

The role of perceived appreciation in influencing the willingness to initiate or maintain relationships has been extensively studied across various contexts. In the realm of romantic relationships, Riela et al. [72] found that individuals were more likely to pursue romantic relationships when a desirable other expressed appreciation for them. Similarly, research by Marigold et al. [73] demonstrated that experimentally manipulating a partner's expressions of admiration led participants to report higher self-esteem compared to those in a control condition. Furthermore, Contu et al. [49] showed that individuals facing a loss of significance (e.g., experiencing a major personal failure, or feeling unrecognized in their achievements) are more inclined to enter and sustain romantic relationships when their potential or current partners possess significance-enhancing qualities. These qualities include not only the partner's good reputation (as discussed in the previous paragraph), but also their expressed appreciation and admiration for the individual.

Further, the perception of being valued and appreciated by others appears to be closely linked to the motivations that drive individuals to join extremist groups. According to the *Quest for Significance Theory* by Kruglanski et al. [70], when individuals experience social rejection, or feel they have been somehow disrespected, they often face a profound need to be acknowledged by others. This desire for recognition can lead individuals to seek out radical groups that make them feel recognized.

In essence, the pursuit of appreciation and acknowledgment is a powerful driver of human behavior, shaping decisions about relationships and group memberships. However, the need to feel accepted and valued often motivates individuals to conform to group norms and expectations. This tendency to align oneself with specific values, behaviors, or beliefs will be explored in the following section.

6 Group Influence in Decision Making Processes

As discussed thus far, individuals have a natural tendency to align themselves with social groups. According to the *Social Identity Theory* by Tajfel and Turner [6], people instinctively classify themselves into social categories, and these categories shape their self-concept and subsequent behaviors. In other words, during decision-making processes, the behaviors, expectations, and attitudes of those around us play an essential role in guiding our choices. This phenomenon is referred to as *social influence*, defined by Cialdini et al. [74] as the psychological change that occurs as a result of the behavior or presence of others. Importantly, this process can occur both consciously and unconsciously: that is, individuals might explicitly adapt their behavior to fit in with a group or implicitly mimic others without deliberate intent [75]. Thus, decision-makers are motivated not only by their material self-interest, but also by a preference for conforming to *social norms*, i.e. individual's beliefs about the commonly accepted and expected behaviors within a group [76]. Research has consistently demonstrated that individuals often conform their behavior to align with the actions of others. This is exemplified in studies like Asch's [77] seminal experiment on conformity, which highlighted the role played by group consensus in shaping individual judgments, even in situations where the correct answer is objectively obvious.

The impact of social norms has been extensively studied in the context of pro-social behaviors such as recycling, energy conservation, and charitable donations. For instance, Cialdini, Reno, and Kallgren [78] demonstrated the power of perceptions of what others typically do (i.e., *descriptive norms*) in influencing littering behavior. In their study, participants were less likely to litter in a clean environment, where the absence of litter signaled that cleanliness was the norm. Further, in their experiment involving 290 Californian households where a smart energy meter had been installed, Schultz et al. [79] demonstrated that providing descriptive normative feedback (i.e., their energy use compared to neighbors) reduced consumption among high-energy users but caused a boomerang effect, increasing usage among low-energy users. Interestingly, adding messages of approval (i.e., the drawing of a smiley face) or disapproval (the drawing of a sad face) mitigated this effect, encouraging low consumers to maintain their behavior. Thus, both descriptive norms (i.e., perceptions of what most people typically do) and injunctive norms (i.e., perceptions of what behaviors are socially approved or disapproved) play a role in shaping individuals' decisions and actions. Similarly, De Dominicis et al. [80] showed that real-time feedback on energy use, paired with strong community identification, significantly reduced electricity consumption. Along these lines, Bergquist and colleagues [81] analyzed results from 91 different studies ($N = 227{,}730$), finding a positive main effect of social norms on pro-environmental behaviors compared to no-treatment control conditions. Further, Schultz et al. [82] found personalized water reports, including normative comparisons and visual cues, decreased water usage, particularly among high-consumption households. These studies underscore the interplay of descriptive and injunctive norms, amplified by community connection, in shaping sustainable behaviors.

Scholars have also investigated the impact of social norms on health outcomes. For example, Christakis and Fowler [83] demonstrated with a 32-years longitudinal study involving 12,067 participants that individuals were significantly more likely to become obese if they had close social ties with obese individuals, suggesting that perceived norms around body weight and health behaviors propagate through social connections. On the other hand, social norms have been leveraged to promote healthier behaviors, such as stopping binge drinking [84], smoking cessation [85] and increased physical activity [86]. For instance, Haines and Spear [84] conducted a 5-year study at a residential campus with 23,000 students, finding an association between the media campaign designed to change perceptions of drinking norms and the reduction of students' binge drinking. More recently, Graupensperger et al. [87] found that students' perceptions of changes in their peers' alcohol consumption were strongly linked to self-reported changes in their own drinking behavior. Further, social norms also play a critical role in shaping responses to public health crises. During the COVID-19 pandemic, for instance, multiple studies highlighted how adherence to preventive measures such as mask-wearing and social distancing was influenced by the influence of social ties and group norms. Bicchieri et al. [88], in a study spanning nine countries with varying levels of pandemic impact (namely, China, Colombia, Germany, Italy, Mexico, South Korea, Spain, the United Kingdom, and the United States) found that both beliefs about what others are doing (i.e., descriptive norms) and beliefs about what others approve of (i.e., injunctive norms) significantly influenced adherence to health guidelines, and this influence persisted even when accounting for other individual or group characteristics.

Research has also highlighted the effects of normative influence in promoting compliance with prosocial (vs. antisocial) behaviors. For instance, in an online experiment by Isler and Gächter [89], participants engaged in honesty and cooperation tasks after observing the behavior of their peers. The findings revealed both positive and negative peer influences on honesty and cooperation, with negative influences generally exerting a stronger effect. Similarly, Kolle and Quercia [90], in two experimental studies, investigated behavior in a public goods game as well as participants' normative and empirical expectations regarding cooperation. When others' contributions were visible, the Authors identified a strong norm of conditional cooperation, where individuals deemed it socially appropriate to match the contributions of others and expect others to adhere to this behavioral rule. In contrast, in scenarios involving strategic uncertainty about others' actions, empirical and normative expectations diverged significantly. Participants believed that fully contributing to the public good was the most socially appropriate choice but anticipated that others would contribute only half of their resources. This misalignment rendered normative expectations less predictive of actual behavior, underscoring the necessity for a close alignment between expectations about what others are believed to do and what others are believed to approve to effectively influence behavior through social norms.

Research has also explored the role of social influence within the context of friendships. Empirical evidence suggests that strong social connections, such as close friendships formed within cohesive social groups, play a significant role in the spread of behaviors through social contagion [91]. For instance, Shin [39], in a study involving a sample of 542 school students, demonstrated that both close friends and admired peers exert social influence on early adolescents' academic engagement. These findings align with Bandura's *Social Learning Theory* [92], which argues that people learn behaviors, attitudes, and values through observing and imitating others, particularly role models or influential individuals. Additionally, adolescents' individual perceptions of admiration shape their understanding of what is desirable and valuable in academic engagement. This process motivates them to adjust their own academic behaviors, increasingly mirroring those of the peers they hold in high regard.

However, social influence does not always lead to positive outcomes. Research on peer dynamics among children and adolescents has shown that peer influence can also promote harmful behaviors. A review by Dishion and Tipsord [93] explored the concept of peer contagion, defined as "a mutual influence process that occurs between an individual and a peer" [93, p. 190]. Their findings highlighted how problematic behaviors such as aggression [94], delinquency [95], and substance use [96] can propagate within peer groups through influence processes. The review further revealed that friendships characterized by recurrent engagement in deviant topics (e.g., discussions or shared activities that reinforce rule-breaking or risky behaviors) were associated with escalating levels of problematic behavior during adolescence, such antisocial tendencies [97]. This aligns with the principle of homophily—the tendency for individuals to associate with others who share similar characteristics or behaviors—emphasized earlier, and it demonstrates how peer group composition can amplify both positive and negative developmental trajectories.

However, teenagers are influenced by more than just their peers. In this vein, Cavalli-Sforza and colleagues [98] identified three primary modes of cultural trait transmission: vertical (i.e., from parent to child), horizontal (i.e., among peers of the same generation), and oblique (i.e., from non-parental individuals in the older generation, such as teachers, to younger individuals). Each mode of transmission plays a distinct role in cultural evolution. Horizontal transmission often drives rapid behavioral changes within a generation, while vertical transmission promotes stability across generations. Notably, the Authors observed strong vertical cultural transmission for traits such as political tendencies and religion. Oblique transmission, on the other hand, tends to reinforce societal norms and values by promoting homogeneity within groups while preserving diversity between them.

In recent years, scholars have sought to understand the factors that influence the intergenerational transmission of norms. For example, Livi et al. [99] examined the role of the epistemic motivation of the need for cognitive closure (i.e., the need for quick, firm answers to reduce uncertainty and ambiguity [100]) in maintaining cultural norms. The research underscores that individuals with high need for closure exhibit a heightened urgency to conform to existing norms and resist changes, driven by the psychological mechanisms of "urgency" and "permanence." These tendencies compel newcomers to adopt the prevailing group standards rapidly and maintain them over time, thereby promoting cultural stability. Moreover, high need for closure individuals are more likely to align their behaviors with those of old-timers, ensuring the persistence of shared practices and values across generations. This conformity not only strengthens group cohesion but also minimizes behavioral variability within the group, promoting the opinion uniformity characteristic of group centrism.

To conclude, the studies discussed above illustrate how, when making decisions, individuals are guided by what they believe others would do in their place or by what they perceive others consider the right thing to do. This holds true for individual behaviors, such as energy conservation [79], as well as for collective actions, such as adherence to public health guidelines during the COVID-19 pandemic [88]. Adherence to social norms can be leveraged by researchers to promote virtuous behaviors, such as reducing littering through descriptive normative cues [78], or to discourage detrimental behaviors, such as excessive water usage reduced through personalized normative reports [82]. However, it is important to recognize that normative social influence also plays a role in perpetuating non-constructive or deviant behaviors, such as the escalation of substance use and delinquency within peer groups through processes of peer contagion [95, 96]. Ultimately, when considering what drives decision-makers, it is impossible to overlook the role played by others and the significance of social groups in shaping choices and behaviors—often even without individuals' conscious awareness. Understanding this is essential for designing effective strategies to positively influence behavior, whether to promote social change or to address the perpetuation of harmful practices.

7 Conclusion

In light of the discussion so far, it appears that a self-reinforcing cycle emerges: on one hand, individuals choose the groups they join based on some specific key-factors (i.e., perceived similarity with other group members; group reputation; and the sense of being accepted within the group itself). On the other hand, these very groups, through their social norms and collective dynamics, exert an influence on the individual's decision-making processes. This interplay highlights the reciprocal nature of the relationship between individuals and groups. Groups not only serve as a reflection of individual preferences, but also act as powerful agents of socialization, shaping behavior, attitudes, and choices. Understanding this dynamic sheds light on why group membership is such a central factor in decision-making and how the norms and expectations of these collectives can reinforce or challenge individual autonomy. This cyclical relationship is not purely abstract but plays out in everyday scenarios, from workplace teams shaping professional choices to social movements driving cultural and political change. Ultimately, decision-making within groups transcends individual cognition, becoming a product of ongoing social negotiation and influence. In this sense, the act of choosing a group (whether it happens consciously or not) and being shaped by it is emblematic of the deeply social nature of human beings, whose decisions are rarely made in isolation but rather emerge from the elaborated tailoring of collective life. To put it simply, no personal decision can ever be entirely individual, and to choose a group is to choose a source of influence.

Acknowledgements Erica Molinario's work on this chapter was supported by the project SONATA 16, registration number 2020/39/D/HS6/02647, funded by the Polish National Science Center.

References

1. APA Dictionary of Psychology. https://dictionary.apa.org/. Accessed 30 Dec 2024
2. Baumeister RF, Leary MR (1995) The need to belong: desire for interpersonal attachments as a fundamental human motivation. Psychol Bull 117:497–529. https://doi.org/10.1037/0033-2909.117.3.497
3. Jensen MP, Smith AE, Bombardier CH, Yorkston KM, Miró J, Molton IR (2014) Social support, depression, and physical disability: age and diagnostic group effects. Disabil Health J 7:164–172. https://doi.org/10.1016/j.dhjo.2013.11.001

4. Hikichi H, Tsuboya T, Aida J, Matsuyama Y, Kondo K, Subramanian SV, Kawachi I (2017) Social capital and cognitive decline in the aftermath of a natural disaster: a natural experiment from the 2011 Great East Japan Earthquake and Tsunami. Lancet Planet Health 1:e105–e113. https://doi.org/10.1016/S2542-5196(17)30041-4
5. Tan J, Wang Y (2019) Social integration, social support, and all-cause, cardiovascular disease and cause-specific mortality: a prospective cohort study. Int J Environ Res Public Health 16:1498. https://doi.org/10.3390/ijerph16091498
6. Tajfel H, Turner JC. An integrative theory of intergroup conflict. In: The social psychology of intergroup relations. Brooks/Cole., Monterey, CA, pp 33–47
7. Hogg MA (2007) Uncertainty–identity theory. In: Advances in experimental social psychology. Academic Press, pp 69–126
8. Festinger L (1950) Informal social communication. Psychol Rev 57:271–282. https://doi.org/10.1037/h0056932
9. Lewin K (1948) Field theory. In: A review of personality theories
10. Sherif M (1936) The psychology of social norms
11. Hardin CD, Higgins ET (1996) Shared reality: how social verification makes the subjective objective. In: Handbook of motivation and cognition, vol 3: the interpersonal context. The Guilford Press, New York, NY, US, pp 28–84
12. Lazarsfeld P, Merton KR (1954) Friendship as a social process: a substantive and methodological analysis
13. Luo S, Chen H, Yue G, Zhang G, Zhaoyang R, Xu D (2008) Predicting marital satisfaction from self, partner, and couple characteristics: is it me, you, or us? J Pers 76:1231–1266. https://doi.org/10.1111/j.1467-6494.2008.00520.x
14. Osgood DW, Ragan DT, Dole JL, Kreager DA (2022) Similarity of friends versus nonfriends in adolescence: developmental patterns and ecological influences. Dev Psychol 58:1386–1401. https://doi.org/10.1037/dev0001359
15. Louch H (2000) Personal network integration: transitivity and homophily in strong-tie relations. Soc Netw 22:45–64. https://doi.org/10.1016/S0378-8733(00)00015-0
16. Ahuja G, Polidoro F Jr, Mitchell W (2009) Structural homophily or social asymmetry? The formation of alliances by poorly embedded firms. Strateg Manage J 30:941–958. https://doi.org/10.1002/smj.774
17. McPherson M, Smith-Lovin L, Cook JM (2001) Birds of a feather: homophily in social networks. Annu Rev Sociol 27:415–444. https://doi.org/10.1146/annurev.soc.27.1.415
18. Campigotto N, Rapallini C, Rustichini A (2022) School friendship networks, homophily and multiculturalism: evidence from European countries. J Popul Econ 35:1687–1722. https://doi.org/10.1007/s00148-020-00819-w
19. Fowler JH, Settle JE, Christakis NA (2011) Correlated genotypes in friendship networks. Proc Natl Acad Sci 108:1993–1997. https://doi.org/10.1073/pnas.1011687108
20. Christakis NA, Fowler JH (2014) Friendship and natural selection. Proc Natl Acad Sci 111:10796–10801. https://doi.org/10.1073/pnas.1400825111
21. Boardman JD, Domingue BW, Fletcher JM (2012) How social and genetic factors predict friendship networks. Proc Natl Acad Sci 109:17377–17381. https://doi.org/10.1073/pnas.1208975109
22. Preciado P, Snijders TAB, Burk WJ, Stattin H, Kerr M (2012) Does proximity matter? Distance dependence of adolescent friendships. Soc Netw 34:18–31. https://doi.org/10.1016/j.socnet.2011.01.002
23. Hinds PJ, Carley KM, Krackhardt D, Wholey D (2000) Choosing work group members: balancing similarity, competence, and familiarity. Organ Behav Hum Decis Process 81:226–251. https://doi.org/10.1006/obhd.1999.2875
24. Harrison DA, Price KH, Bell MP (1998) Beyond relational demography: time and the effects of surface- and deep-level diversity on work group cohesion. Acad Manage J 41:96–107. https://doi.org/10.5465/256901
25. Rapallini C, Rustichini A (2016) Elective affinities matter as much as ethnicity in multi-ethnic schools. J Econ Behav Organ 131:243–262. https://doi.org/10.1016/j.jebo.2016.09.001

26. Hogue A, Steinberg L (1995) Homophily of internalized distress in adolescent peer groups. Dev Psychol 31:897–906. https://doi.org/10.1037/0012-1649.31.6.897
27. Prinstein MJ (2007) Moderators of peer contagion: a longitudinal examination of depression socialization between adolescents and their best friends. J Clin Child Adolesc Psychol 36:159–170. https://doi.org/10.1080/15374410701274934
28. Schaefer DR, Kornienko O, Fox AM (2011) Misery does not love company: network selection mechanisms and depression homophily. https://journals.sagepub.com/doi/full/10.1177/0003122411420813?casa_token=TqjYLR4QNSgAAAAA%3ASzXWJWn2H1UuItafc61Uv-4AMPg_mEqJgjWY10ndIgtj1qD-360cO1a3RWDAphLwDU3DQajpOvIOUak. Accessed 31 Dec 2024
29. Kristof AL (1996) Person-organization fit: an integrative review of its conceptualizations, measurement, and implications. Pers Psychol 49:1–49. https://doi.org/10.1111/j.1744-6570.1996.tb01790.x
30. Adkins CL, Ravlin EC, Meglino BM (1996) Value congruence between co-workers and its relationship to work outcomes. Group Organ Manage 21:439–460. https://doi.org/10.1177/1059601196214005
31. Andrews MC, Baker T, Hunt TG (2011) Values and person-organization fit. Leadersh Organ Dev J 32:5–19. https://doi.org/10.1108/01437731111099256
32. Witt LA (1998) Enhancing organizational goal congruence: a solution to organizational politics. J Appl Psychol 83:666–674. https://doi.org/10.1037/0021-9010.83.4.666
33. Greenberg J, Mollick E (2017) Activist choice homophily and the crowdfunding of female founders—Jason Greenberg, Ethan Mollick, 2017. https://journals.sagepub.com/doi/full/10.1177/0001839216678847?casa_token=RAbxhoX0PbIAAAAA%3ABt8b65Lrwq6PYcdrNvnAua6Hy9N1Sme1fktbjathIR8P27FRaR8L5q_3vGuJiBJX6TjX0HOdoG4hwI8. Accessed 31 Dec 2024
34. Thomas EF, McGarty C, Mavor KI (2009) Aligning identities, emotions, and beliefs to create commitment to sustainable social and political action. Pers Soc Psychol Rev 13:194–218. https://doi.org/10.1177/1088868309341563
35. Webber D, Kruglanski AW (2016) Psychological factors in radicalization. In: The handbook of the criminology of terrorism. Wiley, pp 33–46
36. Kruglanski AW, Molinario E, Ellenberg M, Di Cicco G (2022) Terrorism and conspiracy theories: a view from the 3N model of radicalization. Curr Opin Psychol 47:101396. https://doi.org/10.1016/j.copsyc.2022.101396
37. Bélanger JJ, Moyano M, Muhammad H, Richardson L, Lafrenière M-AK, McCaffery P, Framand K, Nociti N (2019) Radicalization leading to violence: a test of the 3N model. Front Psychiatry 10. https://doi.org/10.3389/fpsyt.2019.00042
38. Shin H, Ryan AM (2014) Early adolescent friendships and academic adjustment: examining selection and influence processes with longitudinal social network analysis. Dev Psychol 50:2462–2472. https://doi.org/10.1037/a0037922
39. Shin H. Social contagion of academic behavior: comparing social networks of close friends and admired peers. PLOS ONE
40. Kenny DA, Acitelli LK (2001) Accuracy and bias in the perception of the partner in a close relationship. J Pers Soc Psychol 80:439–448. https://doi.org/10.1037/0022-3514.80.3.439
41. Theodorou A, Livi S, Alessandri G (2020) The role of individuals' and their partners' positivity in the satisfaction with the relationship: an actor-partner interdependence mediation model. TPM Test Psychom Methodol Appl Psychol 27:331–359. https://doi.org/10.4473/TPM27.3.3
42. Craik KH. Reputation: a network interpretation. Oxford University Press
43. Wilson DS, Wilczynski C, Wells A, Weiser L (2000) Gossip and other aspects of language as group-level adaptations. https://doi.org/10.7551/mitpress/2872.003.0026
44. Goldstein SB, Johnson VA (1997) Stigma by association: perceptions of the dating partners of college students with physical disabilities. Basic Appl Soc Psychol 19:495–504. https://doi.org/10.1207/s15324834basp1904_6

45. Pryor JB, Reeder GD, Monroe AE (2012) The infection of bad company: stigma by association. J Pers Soc Psychol 102:224–241. https://doi.org/10.1037/a0026270
46. Kruglanski AW, Ellenberg M, Yu H, Lemay EP, Moskalenko S, Szumowska E, Molinario E, Pierro A, Contu F (unpublished). A formula for love: partner trophy and fandom beget actor significance
47. Regan PC, Levin L, Sprecher S, Christopher FS, Gate R (2000) Partner preferences: what characteristics do men and women desire in their short-term sexual and long-term romantic partners? J Psychol Hum Sex 12:1–21. https://doi.org/10.1300/J056v12n03_01
48. Eastwick PW, Finkel EJ (2008) Sex differences in mate preferences revisited: do people know what they initially desire in a romantic partner? J Pers Soc Psychol 94:245–264. https://doi.org/10.1037/0022-3514.94.2.245
49. Contu F, Kruglanski AW, Ellenberg M, Yu H, Pierro A (2024) Romantic relationships as a source of significance. Eur J Soc Psychol ejsp.3112. https://doi.org/10.1002/ejsp.3112
50. Aron A, Aron EN (1986) Love as the expansion of self: understanding attraction andsatisfaction
51. Belt JA, Paolillo JGP (1982) The influence of corporate image and specificity of candidate qualifications on response to recruitment advertisement. J Manage 8:105–112. https://doi.org/10.1177/014920638200800107
52. Story J, Castanheira F, Hartig S (2016) Corporate social responsibility and organizational attractiveness: implications for talent management. Soc Responsib J 12:484–505. https://doi.org/10.1108/SRJ-07-2015-0095
53. Gandasari D, Tjahjana D, Dwidienawati D, Ichsan M (2024) How to attract talents? The role of CSR, employer brand, benefits and career development. Cogent Bus Manage 11:2323774. https://doi.org/10.1080/23311975.2024.2323774
54. Cable DM, Turban DB (2003) The value of organizational reputation in the recruitment context: a brand-equity perspective. J Appl Soc Psychol 33:2244–2266. https://doi.org/10.1111/j.1559-1816.2003.tb01883.x
55. Ferris GR, Blass FR, Douglas C, Kolodinsky RW, Treadway DC. Personal reputaion in organizations
56. Kilduff M, Krackhardt D (1994) Bringing the individual back in: a structural analysis of the internal market for reputation in organizations. Acad Manage J 37:87–108. https://doi.org/10.5465/256771
57. Stiff C, Van Vugt M (2008) The power of reputations: the role of third party information in the admission of new group members. Group Dyn Theory Res Pract 12:155–166. https://doi.org/10.1037/1089-2699.12.2.155
58. Yamagishi T (1988) Exit from the group as an individualistic solution to the free rider problem in the United States and Japan. J Exp Soc Psychol 24:530–542. https://doi.org/10.1016/0022-1031(88)90051-0
59. Feinberg M, Willer R, Schultz M (2014) Gossip and ostracism promote cooperation in groups. Psychol Sci 25:656–664. https://doi.org/10.1177/0956797613510184
60. Bowlby J (1979) The Bowlby-Ainsworth attachment theory. Behav Brain Sci 2:637–638. https://doi.org/10.1017/S0140525X00064955
61. Leary MR, Baumeister RF (2000) The nature and function of self-esteem: sociometertheory. Adv Exp Soc Psychol 33
62. Berkman LF, Syme SL. Social networks, host resistance, and mortality: a nine-year follow-up study of Alameda county residents. Am J Epidemiol
63. Chen Y, Feeley TH (2014) Social support, social strain, loneliness, and well-being among older adults: an analysis of the health and retirement study. J Soc Pers Relatsh 31:141–161
64. Kuwert P, Knaevelsrud C, Pietrzak RH (2014) Loneliness among older veterans in the United States: results from the National Health and Resilience in Veterans Study. Am J Geriatr Psychiatry 22:564–569. https://doi.org/10.1016/j.jagp.2013.02.013
65. Chopik WJ (2017) Associations among relational values, support, health, and well-being across the adult lifespan. Pers Relatsh 24:408–422. https://doi.org/10.1111/pere.12187

66. Secor SP, Limke-McLean A, Wright RW (2017) Whose support matters? Support of friends (but not family) may predict affect and wellbeing of adults faced with negative life events. J Relatsh Res 8:e10. https://doi.org/10.1017/jrr.2017.10
67. Shin H, Gyeong S (2023) Social support and strain from different relationship sources: their additive and buffering effects on psychological well-being in adulthood. J Soc Pers Relatsh 40
68. Pierce KDR, Quiroz CS (2019) Who matters most? Social support, social strain, and emotions. https://journals.sagepub.com/doi/full/10.1177/0265407518817400?casa_token=nuP26u_b5WEAAAAA%3AHQlRogay7YRXJXf1NVs8L7rFByN6utbxuicbqzOFyCnyR6yb4tdotid7OcIfsAa07M2DgfVDxGkhkNw
69. Seeman TE, Lusignolo TM, Albert M, Berkman L (2001) Social relationships, social support, and patterns of cognitive aging in healthy, high-functioning older adults: MacArthur studies of successful aging. Health Psychol 20:243–255. https://doi.org/10.1037/0278-6133.20.4.243
70. Kruglanski AW, Chen X, Dechesne M, Fishman S, Orehek E (2009) Fully committed: suicide bombers' motivation and the quest for personal significance. Polit Psychol 30:331–357. https://doi.org/10.1111/j.1467-9221.2009.00698.x
71. Kruglanski AW, Bélanger JJ, Gelfand M, Gunaratna R, Hettiarachchi M, Reinares F, Orehek E, Sasota J, Sharvit K (2013) Terrorism—a (self) love story: redirecting the significance quest can end violence. Am Psychol 68:559–575. https://doi.org/10.1037/a0032615
72. Riela S, Rodriguez G, Aron A, Xu X, Acevedo BP (2010) Experiences of falling in love: investigating culture, ethnicity, gender, and speed. J Soc Pers Relatsh 27:473–493. https://doi.org/10.1177/0265407510363508
73. Marigold DC, Holmes JG, Ross M (2007) More than words: reframing compliments from romantic partners fosters security in low self-esteem individuals. J Pers Soc Psychol 92:232–248. https://doi.org/10.1037/0022-3514.92.2.232
74. Cialdini RB, Goldstein NJ (2004) Social influence: compliance and conformity. Annu Rev Psychol 55:591–621. https://doi.org/10.1146/annurev.psych.55.090902.142015
75. Cialdini RB (2003) Crafting normative messages to protect the environment. Curr Dir Psychol Sci 12:105–109. https://doi.org/10.1111/1467-8721.01242
76. Cialdini RB, Kallgren CA, Reno RR (1991) A focus theory of normative conduct: a theoretical refinement and reevaluation of the role of norms in human behavior. In: Zanna MP (ed) Advances in experimental social psychology. Academic Press, pp 201–234
77. Asch SE (1956) Studies of independence and conformity: I. A minority of one against a unanimous majority. Psychol Monogr Gen Appl 70:1–70
78. Cialdini RB, Reno RR, Kallgren CA (1990) A focus theory of normative conduct: recycling the concept of norms to reduce littering in public places. J Pers Soc Psychol 58:1015–1026. https://doi.org/10.1037/0022-3514.58.6.1015
79. Schultz PW, Nolan JM, Cialdini RB, Goldstein NJ, Griskevicius V. The constructive, destructive, and reconstructive power of social norms. Psychol Sci 18
80. De Dominicis S, Sokoloski R, Jaeger CM, Schultz PW (2019) Making the smart meter social promotes long-term energy conservation. Palgrave Commun 5:1–8. https://doi.org/10.1057/s41599-019-0254-5
81. Berquist M, Nilsson A, Schultz PW (2019) A meta-analysis of field-experiments using social norms to promote pro-environmental behaviors. Glob Environ Change
82. Schultz W, Javey S, Sorokina A (2019) Social comparison as a tool to promote residential water conservation. Front Water 1. https://doi.org/10.3389/frwa.2019.00002
83. Christakis NA, Fowler JH (2007) The spread of obesity in a large social network over 32 years. New Eng J Med. https://doi.org/10.1056/NEJMsa066082
84. Haines M, Spear SF (1996) Changing the perception of the norm: a strategy to decrease binge drinking among college students. J Am Coll Hea 45:3
85. Dono J, Miller C, Ettridge K, Carlene W (2020) The role of social norms in the relationship between anti-smoking advertising campaigns and smoking cessation: a scoping review. Health Educ Res 35:3. https://doi.org/10.1093/her/cyaa008

86. Howie EK, Daniels BT, Guagliano JM (2020) Promoting physical activity through youth sports programs: it's social. Am J Life Med 14:1
87. Graupensperger S, Jaffe AE, Fleming CNB, Kilmer JR, Lee CM, Larimer ME (2021) Changes in college student alcohol use during the COVID-19 pandemic: are perceived drinking norms still relevant? Emerg Adulthood 9:531–540. https://doi.org/10.1177/2167696820986742
88. Bicchieri C, Fatas E, Aldama A, Casas A, Deshpande I, Lauro M, Parilli C, Spohn M, Pereira P, Wen R (2021) In science we (should) trust: expectations and compliance across nine countries during the COVID-19 pandemic. PLOS One 16:e0252892
89. Isler O, Gächter S (2022) Conforming with peers in honesty and cooperation. J Econ Behav Organ 195:75–86. https://doi.org/10.1016/j.jebo.2021.12.026
90. Kölle F, Quercia S (2021) The influence of empirical and normative expectations on cooperation. J Econ Behav Organ 190:691–703. https://doi.org/10.1016/j.jebo.2021.08.018
91. Guilbeault D, Becker J, Centola D (2018) Complex contagions: a decade in review. In: Lehmann S, Ahn Y-Y (eds) Complex spreading phenomena in social systems: influence and contagion in real-world social networks. Springer International Publishing, Cham, pp 3–25
92. Bandura A (1977) Social learning theory
93. Dishion TJ, Tipsord JM (2011) Peer contagion in child and adolescent social and emotional development. Annu Rev Psychol 62:189–214. https://doi.org/10.1146/annurev.psych.093008.100412
94. Sijtsema JJ, Veenstra R, Lindenberg S, Salmivalli C (2009) Empirical test of bullies' status goals: assessing direct goals, aggression, and prestige. Aggress Behav 35:57–67. https://doi.org/10.1002/ab.20282
95. Dishion TJ, Spracklen KM, Andrews DW, Patterson GR (1996) Deviancy training in male adolescent friendships. Behav Ther 27:373–390. https://doi.org/10.1016/S0005-7894(96)80023-2
96. Dishion TJ, Capaldi D, Spracklen, Kathleen M, Li F. Peer ecology of male adolescent drug use. Dev Psychopathol 7:803–824
97. Dishion TJ, Nelson SE, Winter CE, Bullock BM (2004) Adolescent friendship as a dynamic system: entropy and deviance in the etiology and course of male antisocial behavior. J Abnorm Child Psychol 32:651–663. https://doi.org/10.1023/B:JACP.0000047213.31812.21
98. Cavalli-Sforza LL, Feldman MW, Chen KH, Dornbusch SM (1982) Theory and observation in cultural transmission. Science 218:19–27. https://doi.org/10.1126/science.7123211
99. Livi S, Kruglanski AW, Pierro A, Mannetti L, Kenny DA (2015) Epistemic motivation and perpetuation of group culture: effects of need for cognitive closure on trans-generational norm transmission. Organ Behav Hum Decis Process 129:105–112. https://doi.org/10.1016/j.obhdp.2014.09.010
100. Webster DM, Kruglanski AW (1994) Individual differences in need for cognitive closure. J Pers Soc Psychol 67:1049–1062. https://doi.org/10.1037/0022-3514.67.6.1049

Laura Prislei is a Ph.D. candidate in Social Psychology at the Department of Psychology of Developmental and Socialization Processes, Sapienza University of Rome, Italy. Her research centers on group dynamics, with a particular emphasis on political participation and susceptibility to radicalization, especially among youth. During her doctoral studies, she conducted research stays at Jagiellonian University in Krakow (Poland) and Claremont Graduate University (California, USA), where she collaborated with leading experts in Social Psychology.

Erica Molinario is an Assistant Professor at the Department of Psychology and the Water School at Florida Gulf Coast University (USA) and a visiting Assistant Professor at the Jagiellonian University (Poland). Her research focuses on understanding the psychological processes that drive human behavior in response to pressing social issues, such as pro-environmental actions, migration intentions, political extremism, radicalization, and intergroup conflicts. She examines these topics from a motivational perspective, exploring needs, goals, social identity, values, and social norms.

Stefano Livi Ph.D., is a full professor of Social Psychology at the Department of Social and Developmental Psychology, Sapienza University of Rome, Italy. His current research focuses on social and cognitive factors involved in group processes, with particular emphasis on group socialization, marginalization, and intergenerational transmission. These research areas are explored through classical models of social cognition, such as epistemic motivations (e.g., the need for cognitive closure).

Attribute Framing and Decision Making: An Expanded Typology

12

Mathew S. Isaac and Irwin P. Levin

An essential condition for a theory of choice that claims normative status is the principle of invariance: different representations of the same choice problem should yield the same preference. That is, the preference between options should be independent of their description...The observed failures of invariance indicate that people do not spontaneously aggregate concurrent prospects or transform all outcomes into a common frame.

Tversky and Kahneman [1] (pp. 168–169)

Abstract

We build this chapter on the influential research of Tversky and Kahneman, which showed that decision making is not only driven by objective information but also by psychological biases and heuristics. One pervasive source of bias is *framing*, which can occur when the same objective information is conveyed in either positive or negative terms. We extend earlier work that denoted three categories of framing effects: risky choice framing, goal framing, and attribute framing. Our primary emphasis in this chapter is on the latter category because of its commonality in everyday judgments and decisions. We review potential psychological mechanisms for attribute framing effects and explore their

M. S. Isaac (✉)
Department of Marketing, Seattle University Albers School of Business and Economics, Seattle, Washington, USA
e-mail: isaacm@seattleu.edu

I. P. Levin
Department of Psychology, University of Iowa, Iowa City, Iowa, USA

Department of Marketing, University of Iowa Tippie College of Business, Iowa City, Iowa, USA

© The Author(s), under exclusive license to Springer Nature Switzerland AG 2025

T. McElroy (ed.), *Decision-Making in Life and Work*, Integrated Science 36,
https://doi.org/10.1007/978-3-031-88415-3_12

affective and cognitive bases, incorporating insights provided by decision neuroscience. In addition, we extend the operational definition of attribute framing to include numerical frames that have the same valence as well as semantically-similar labels. We conclude by calling for more research using these expanded types of attribute framing effects to better understand the sources of decision biases.

Keywords

Framing • Attribute framing • Valence-consistent shift • Framing effects • Decision making • Judgment • Affect • Cognition

1 Introduction

The pioneering work of Amos Tversky and Daniel Kahneman has had a profound effect on the study of human judgment and decision making, the details of which are described in other chapters of this volume. Nevertheless, we wanted to start this chapter by acknowledging how Tversky and Kahneman's revelations concerning the importance and commonality of biases and heuristics in judgment and decision making have shaped the work of many researchers, including both authors of this chapter.

In the quote with which we open this chapter, Tversky and Kahneman [1] noted the tendency of human beings to make different decisions when the same problem is framed in different ways—a violation of the normative *principle of invariance*. The failures of invariance were first reported in the context of risky decision making, such as the famous "Asian Disease problem" [2] in which participants learned about an outbreak of an unusual disease and chose between two programs, one which was risk-averse and the other which was risky. When the programs were described in terms of potential lives *saved*, the risk-averse option was preferred; however, when they were framed in terms of potential lives *lost*, the risky option was favored. This preference reversal—which occurred even though the underlying information provided to participants was identical—was central to Tversky and Kahneman's subsequent work on prospect theory that ultimately culminated in Kahneman winning the 2002 Nobel Prize in Economics.

Although these violations of invariance in the context of risky choices are fascinating, the decision problems that Tversky and Kahneman studied were relatively complex and somewhat stylized. Participants were tasked with choosing from options that were described in ways that do not necessarily correspond to the descriptions people encounter on a regular basis. Over the past five decades, many researchers—including both authors of this chapter—have sought to understand if and when framing effects influence judgments and behaviors in simple but common everyday tasks, sometimes when frames differ by just a single word. This chapter provides a non-exhaustive synopsis of research findings related to such framing effects.

2 A Typology for Framing Effects

One of the first typologies of framing effects (which has been cited nearly 4000 times) was developed by the second author (in collaboration with Sandra Schneider and Gary Gaeth) over 25 years ago. The genesis of this typology was the authors' realization that different framing effects that had been observed were based on different operational definitions of framing itself. The authors set out to preselect categories of framing effects and find examples of each from which they could identify clear differences [3]. The three categories that emerged from this typology were risky choice framing, goal framing, and attribute framing.

As already discussed, risky choice framing relates to Tversky and Kahneman's original framing research. There have been many demonstrations of preference reversals and choice shifts (see [3], Table 2) involving variants of the Asian Disease problem, gambling tasks, and bargaining tasks. The common theme in risky choice framing effects is that participants make choices between risky and riskless options (or "prospects"), and these preferences are malleable based on how the options are described. To reiterate the basic finding, people tend to be more risk-averse in the domain of gains but more risk-seeking in the domain of losses.

Goal framing (see [3], Table 4) refers to whether pursuing a particular goal is framed in terms of its potential to provide a benefit or gain (positive frame) or to avoid a loss (negative frame). Different ways of framing an implicit goal may be more or less persuasive. For example, all patients likely have the implicit goal of remaining cancer-free. Given this universal goal, a doctor who wishes to encourage regular mammograms among her patients could either highlight the positive consequences of getting a mammogram (e.g., "if you get a mammogram, you take advantage of the best method for early detection of breast cancer") or focus on the negative consequences of not getting one (e.g., "if you don't get a mammogram, you fail to take advantage of the best method for early detection of breast cancer") [4]. Although negative goal frames tend to have greater persuasive power, this finding has been somewhat inconsistent, which indicates that many moderating variables (e.g., goal commitment, relevance, etc.) may play an important role in goal framing effects.

Finally, attribute framing (see [3], Table 3) involves a modest change—typically to a single attribute—that leads to systematic differences in evaluations or choice. The attribute frames discussed in the original typology examine oppositely-valenced numerical values (e.g., 90% natural ingredients vs. 10% artificial ingredients in food products) and assess how evaluative judgments and behavioral intentions are affected by exposure to one frame versus the other. Typically, positively-valenced attribute frames outperform negatively-valenced frames in terms of evaluations. Examples abound of attribute framing effects, spanning contexts such as performance assessments, potential purchases, business outcomes, and gambling decisions. As the focus of this chapter is on attribute framing, we will discuss this particular framing category in much more detail in the next section.

According to Levin, Schneider, and Gaeth [3], the three framing categories (i.e., risky choice, goal, attribute) can be distinguished based on what is framed (i.e., options with different risk levels, consequences of a behavior, characteristics) and what is affected (i.e., risk preference, persuasion likelihood, evaluation). Yet, despite these valid differences, the examples of framing effects identified in the original typology were similar in that they mainly included numerical information that was framed in opposite ways.

In the remainder of this chapter, we dive deeper into attribute framing and suggest an expanded typology that is not restricted to oppositely-valenced numerical frames. We will proceed by first highlighting how research on attribute frames—consistent with how it was conceptualized in the original typology—has progressed over the past 25+ years. Rather than attempting to perform a comprehensive literature review or meta-analysis, we will focus primarily on laying out a few psychological explanations that have been proposed for attribute framing effects, along with a brief discussion of insights that can be gleaned from neuroscience research. The question of *why* attribute framing effects occur will lead us to consider the relative influence of affect versus cognition and to review both decision-maker characteristics and contextual factors that have been shown to attenuate attribute framing effects. Finally, we will introduce our expanded typology for attribute framing effects and discuss three additional types of framing effects that ought to be classified as attribute frames.

3 Attribute Framing Revisited

Even when the actual liquid contained in a glass is the same, will people perceive the glass differently if it is described as "half full" versus "half empty"? This philosophical question may seem rather trivial, but consider these more consequential examples: "When encouraging a patient to have an elective surgery, should a doctor describe the procedure as having an 80% success rate or a 20% failure rate?"; "On a scholarship application, should a student communicate her latest exam performance as getting 90% correct or 10% incorrect?"; and "To maximally impress her coach, should a basketball player express her free throw shooting as 70% made or 30% missed?". To the extent that exposure to one frame versus the other produces disparate evaluations or judgments, a communicator's choice of frame may have significant downstream consequences. For instance, consider the serious health risk that a patient assumes if the frame (e.g., success rate vs. failure rate) that her doctor uses to describe a medical procedure leads her to forego the procedure altogether.

Of course, objectively speaking, the frame should not matter—each of the above pairings are *logically equivalent* in that they, "according to accepted normative theory, convey the same information" [4]. The concept of logical equivalence is easiest to explain in the domain of numerical information, particularly when describing probabilistic values for binary outcomes (e.g., wins/losses, successes/failures, etc.) with fractions, decimals, or percentages. Yet, even when numerical frames are logically equivalent (e.g., *X%* in one frame vs. *100-X%* in the opposite frame), substantial prior research has produced strong evidence for a systematic *valence-consistent shift*: attributes framed positively generate more favorable evaluations than attributes framed negatively [5–9].

Perhaps the best-known attribute framing example [10] is that the subjective ratings of ground beef are higher when the beef is described as 75% lean (positive valence) rather than 25% fat (negative valence). Even when it is actually consumed, consumers evaluate beef to be of higher quality, better tasting, less greasy, and more lean if it is labeled as 75% lean [11]. As another example, consumers indicate greater willingness to spend $30 on a restaurant if they learn there is a "50% chance they would be satisfied with the restaurant" as opposed to a "50% chance they would be dissatisfied with the restaurant" [12].

The valence-consistent shift has been observed in an impressive array of contexts, including medical treatment decisions [13, 14], consumer product evaluations [11, 15], life expectancy judgments [16, 17], and punishment decisions [18]. The Payne et al. [16] example literally addresses a life-or-death issue. Those people who were asked how much longer they expected to live gave significantly longer life expectancies than those who were asked when they expected to die. Even experts are susceptible to attribute framing effects: medical students were more likely to advocate a surgery described using survival rates (positive valence) as opposed to equivalent mortality rates (negative valence) [19]. Similarly, engineers allocated more funds to projects whose prior performance was framed in terms of successes (positive valence) rather than an equivalent percentage of failures (negative valence) [20].

3.1 Psychological Explanations

Why does the valence-consistent shift occur? The second author and his colleagues originally proposed a simple associative explanation. They suggested that positively-valenced frames evoked relatively more positive associations, whereas negatively-valenced frames evoked relatively more negative associations; as a result, evaluations tended to be more positive among individuals who encountered a positive (vs. negative) frame.

Since the original framing typology was published, a number of additional theories have been proposed to explain the effect. According to a query theory account [21], the valence of a frame changes the order in which individuals retrieve supporting evidence from memory about the claim. When given a positive frame, people first attempt to recruit evidence supporting the positive outcome and then attempt to recruit evidence supporting the negative outcome. When given a negative frame, the order is reversed. According to query theory, the initial "query" one makes from memory generates the most retrievals; thus, the relative weight of positive and negative information changes depending on frame valence and yields the valence-consistent shift.

A third explanation for the valence-consistent shift raised the provocative question of whether two opposite frames, even when they are numerical, can ever truly be logically equivalent [22]. According to this inferential "information leakage" explanation [23], the choice of a frame invariably carries reference point information that subtly reveals meaning about the communicator or the situation at hand. For example, in one lab experiment, participants were provided with two plastic cups, one full of water and another one that was empty. They were instructed to pour water from one cup to the other and to "set a half-full cup" at the edge of their desk. Consistent with an information leakage account, significantly greater than 50% of participants placed the cup that was originally empty at the edge of their desk. The authors claimed that "reference point information is leaked by speakers and absorbed by listeners in descriptions of frequency, proportion, and probability—the basic ingredients of attribute framing experiments" [23, 24] also suggested that reference dependence may play a role in attribute framing effects, arguing that positive versus negative frames evoke subjectively-different reference sets.

3.2 Insights from Neuroscience

The second author and a team of co-authors [25] introduced the term "decision neuroscience" to describe an emerging area of research designed to integrate research in neuroscience and behavioral decision making by showing that knowledge in neuroscience can enrich research in decision making. Although decision neuroscientists have tended to focus more on risky choice framing than attribute framing (e.g., [26, 27]), a few published articles have attempted to examine the neural underpinnings of attribute frame effects. For instance, Ma et al. [28] exposed participants to a series of logically-equivalent attribute frames (e.g., lives lost vs. lives saved) and measured participants' event-related brain potentials (ERPs) using electroencephalogram (EEG) recordings while they processed and evaluated messages with different frames. The valence-consistent shift was replicated in participants' behavioral responses. Additionally, the researchers evaluated whether feedback-related negativity (FRN), which is an ERP component known to be a neural index of outcome processing, was differentially activated by positive versus negative attribute frames. This ERP component was found to be larger following exposure to negative frames as compared to positive frames. Based on this evidence, the authors concluded that oppositely-valenced frames are processed differently in the brain—in a manner that is consistent with the associative explanation discussed earlier. More recent work also using ERPs [29] largely corroborated these earlier findings, while providing additional insights in an online shopping context. Specifically, participants not only indicated a higher purchase intention after exposure to a positive (vs. negative) frame but anticipated that positively-framed products would perform better in the future. Furthermore, ERP measurements indicated that positive frames resulted in shorter reaction times, whereas negative frames required more attentional resources at the early stage of automatic processing and produced greater cognitive conflict and decision difficulty. The authors concluded, based on ERP evidence, that the processing of positive and negative messages was fundamentally different.

As a final example, Murch and Krawczyk [30] examined attribute frames in a social cognition context using another neuroscience technique, functional magnetic resonance imaging (fMRI). Participants were successively shown personal attributes that were framed positively or negatively and asked to evaluate if these attributes applied to them (from multiple perspectives). For example, participants might indicate whether their best friend would agree that the positive statement "I am loyal at least 75% of the time" or the negative statement "I am not kind 25% of the time" was descriptive (or not descriptive) of them. The fMRI whole-brain analysis allowed researchers to identify functionally-defined brain regions of interest (ROIs) and assess activity in these regions when participants were processing different frames. Based on this brain mapping exercise, positive frames activated known reflexive regions in the brain associated with quick and automatic responses, including the angular gyrus and the orbitomedial prefrontal cortex. In contrast, negative frames activated known reflective regions in the brain that guide greater cognitive control, such as the bilateral insula. The authors found considerable variation in framing susceptibility, with some participants' decisions almost fully influenced by the frame (as reflected by brain region activity) while others were highly resistant to framing effects, except for statements with very high percentages of either positive or negative attributes.

Collectively, neuroscience research using ERPs and fMRI has demonstrated different patterns of brain activation stemming from oppositely-valenced attribute frames. Although it cannot tease apart (or rule out) nuanced psychological mechanisms that have been proposed for the valence-consistent shift, neuroscience may be helpful in understanding the role that affect and cognition play in attribute framing. We turn to this topic next.

3.3 Affect Versus Cognition: Individual and Situational Differences

Is the valence-consistent shift an affective bias or a cognitive bias? Consistent with an affective account, prior research has shown that emotions play an important role in many framing effects [31, 32]. In their previously-mentioned study that examined both behavioral responses and neurological activity, Murch and Krawczyk [30] measured individual differences among participants using a number of established scales and identified a number of personality factors that affect neurological response and framing susceptibility. Specifically, higher intelligence (as estimated by the Wechsler Test of Adult Reading) and greater cognitive reflection (as assessed by performance on the Cognitive Reflection Task) were correlated with lower susceptibility to attribute framing effects. The implication that intellectually rigorous individuals are less influenced by valenced frames suggests that affectively-based associations may drive the valence-consistent shift. McElroy and

Conrad [33] found evidence for unconscious valence processing in attribute framing effects. Further corroboration of this account comes from more recent research by the second author (in conjunction with co-authors Gaurav Jain, Gary Gaeth, and Dhananjay Nayakankuppam) [34] showing that the unusual use of extremely precise numbers (e.g., 81.03% success rate vs. 18.97% failure rate) led individuals to rely less on associations and engage in more effortful processing as compared to when they were exposed to more typical round (i.e., zero-ending) number frames (e.g., 80% success rate vs. 20% failure rate).

If the valence-consistent shift is primarily the result of emotional processing, then individuals who are more involved in the evaluation task should be less likely to exhibit the bias as they would engage in more systematic (and corrective) processing. An article by Donovan and Jalleh [35] examined the construct of involvement using the previously-mentioned "75% lean" versus "25% fat" ground beef stimuli. Because involvement is known to trigger cognitive elaboration and systematic processing that is deliberate and rule-based [36], the authors suspected that highly involved (i.e., health-conscious) consumers would be less likely to exhibit the valence consistent shift. This hypothesis was not supported; in fact, they observed a stronger valence consistent shift among highly involved consumers [35]. This finding raised the intriguing possibility that cognitive processing may not debias the valence consistent shift at all but may actually be responsible for attribute framing effects.

An indirect avenue to disentangle the role of affect and cognition involves examining whether the valence-consistent shift persists after *actual consumption* of the stimuli that is (either positively or negatively) framed. The underlying premise is that actual consumption tasks are naturally more likely to involve affective processing. Indeed, Kempf [37] demonstrated that during an actual consumption experience, consumers focus more on affective dimensions of their experience and rely more on this information (vs. cognitions) when making their evaluations. Because affective stimuli tend to capture and hold attention, this leaves fewer attentional resources to fully consider other, non-affective information [38–40]. Arguably, finding that the valence-consistent shift is attenuated in contexts involving actual consumption could be construed as evidence that attribute framing effects are a cognitive bias.

This area of inquiry (i.e., actual consumption) is interesting in its own right given that most attribute framing research has focused on how valenced frames influence judgments of hypothetical or prospective consumption (e.g., [41]). One exception is a study by the second author and Gary Gaeth [11] that examined the influence of lean/fat framing on evaluations made after participants actually consumed identical samples of ground beef. When evaluative judgments were made following consumption of the ground beef, the valence-consistent shift (e.g., [6]) was attenuated. The first author and Morgan Poor also examined how the provision of positive versus negative frames interact with actual consumption experiences to influence judgments [42]. However, this work not only examined judgments immediately following the consumption episode but also involved retrospective judgments taken a short time later. The authors found that the valence-consistent shift was weakened immediately after consumption, presumably because of the affective information from the consumption experience, but was observed again in retrospective judgments. These results, along with those of a follow-up study conducted by Poor and the first author in which participants were explicitly directed to focus on their cognitions or their emotions during an actual consumption experience, suggest that cognition plays an important role in the valence-consistent shift [43].

Reflecting on these cumulative findings, we agree that trying to determine if the valence-consistent shift is affective or cognitive is overly reductive and represents a false dichotomy since affect and cognition are intrinsically intertwined. Given the entire body of research examining boundary conditions and individual differences, both in the absence and presence of actual consumption, it seems likely that affect and cognition both play an explanatory role in the valence-consistent shift and may even act jointly to produce this effect.

4 An Expanded Typology for Attribute Framing Effects

Having highlighted developments in framing research based on the original conceptualization of attribute framing, we now offer a broader typology for attribute framing that allows previously unclassified framing effects to be housed under the "attribute framing" umbrella. Our expanded typology for attribute framing consists of four different types of attribute frames: (1) logically-equivalent numbers of opposite valence, (2) logically-equivalent numbers of the same valence, (3) semantically-similar labels of the same valence, and (4) semantically-similar labels of opposite valence. Figure 12.1 depicts how this expanded typology relates to the original typology of Levin, Schneider, and Gaeth [3].

The original typology only considered the first type of attribute frame (i.e., logically equivalent numbers of opposite valence), which we have just discussed in depth. Next, we describe the three other frame types and provide representative examples of research findings pertaining to each one.

4.1 Logically-Equivalent Numbers of the Same Valence

The second type of attribute frame entails logically-equivalent numbers of the same valence. For example, consider 0.5, ½, or 50%. All three are logically equivalent because they communicate the same numerical value in decimal, fraction, and percentage notation, respectively. Given that each number on its own conveys identical information, the valence of the three frames is the same—with the specific context determining if the valence is positive, negative, or neutral. For instance, if 0.5, ½, or 50% are used to denote a basketball player's 3-point shooting effectiveness, the valence would be positive (given that this represents extraordinarily high accuracy). However, if 0.5, ½, or 50% are used to denote performance on an exam, the valence would be negative (given that getting half the questions right on an exam

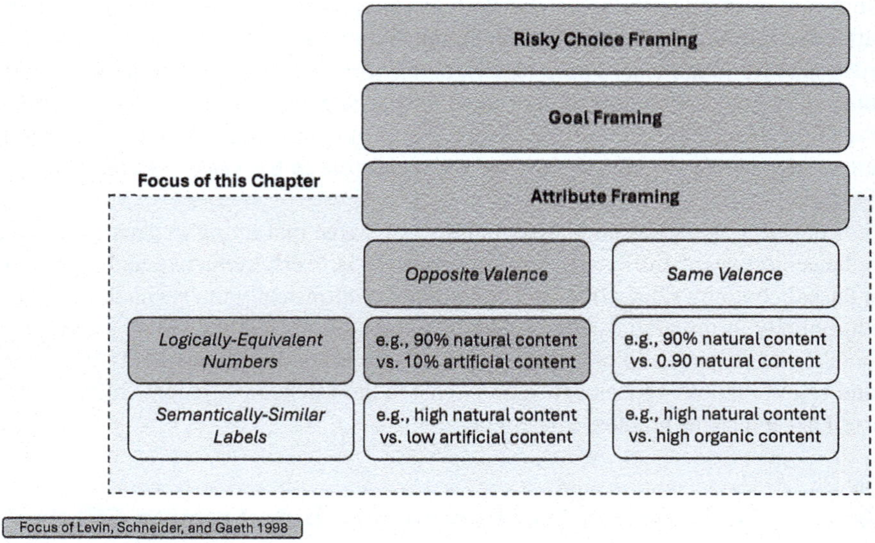

Fig. 12.1 Expanded typology of framing effects

usually translates to a failing exam score). In other contexts, 0.5, ½, or 50% may be considered average or neutral. In each case, the choice of frame (i.e., decimal, fraction, or percentage) does not affect the number's valence.

It would be overly simplistic to suggest that an attribute framed using one notation versus another will universally produce more favorable evaluations, as this will likely depend on the specific decision context. Indeed, research has shown that individuals prefer using fractions to represent certain relationship types (e.g., red marbles vs. all marbles) but favor decimals in other situations (e.g., partitioned displays of continuous masses) [44]. When the notation of numerical stimuli matches individuals' preferences in a particular context, this correspondence may yield metacognitive benefits (e.g., greater fluency, or processing ease) that engender more positive evaluations. In addition to the decision context, decision makers' numeracy (i.e., fluency with math concepts, or mathematical literacy) may dictate whether their evaluations are affected by the notational frame (e.g., decimal, fraction, or percentage). In fact, research has shown that compared to whole numbers, fractions, decimals, and especially percentages are non-intuitive [45, 46], and the facility of individuals to correctly convert one notational frame to another is likely lower if they are less numerate.

The first author of this chapter, in conjunction with co-authors Julio Sevilla and Rajesh Bagchi, demonstrated that framing a ranked-list claim using percentages (e.g., top 20%, top 50%) rather than whole numbers (e.g., top 20, top 50) influences evaluations and choices of the ranked item [47]. Specifically, they found that for small set sizes, expressing rank claims using logically-equivalent numbers (e.g., top 10 out of 50) led to more favorable evaluations than using percentages (e.g., top 20% out of 50). However, for large set sizes, the opposite patten was observed in that numbers (e.g., top 40 out of 200) led to less favorable evaluations than using logically-equivalent percentages (e.g. top 20% out of 200) led to more favorable evaluations. This framing effect was attributed to *format neglect*, wherein participants attended primarily to the face value (or nominal value) of the number and failed to adequately account for the claim's format (i.e., number vs. percentage). Because numerically-smaller face values represent more favorable ranks, evaluations were superior when nominal values were smaller. Largely in line with a face value explanation, other work has found that framing a price promotion in terms of dollars off versus percentage off affects consumers' perception of the deal and purchase intentions [48, 49] as well as post-promotion price expectations [50].

The finding that individuals emphasize face values has been corroborated by considerable research, mainly in the domain of consumer behavior, and used to explain several other framing effects that involve logically-equivalent numbers of the same valence. For instance, it is well-established that changing the scale or unit of measurement when displaying information affects subjective perceptions of product quantity. For instance, although 1000 m and 1 km represent the same distance, consumers may perceive these two distances differently due to the *numerosity heuristic* [51], which is essentially another instantiation of the prioritization of face values. According to this heuristic, increasing the number of units into which a stimulus has been divided makes the quantity or probability seem greater, particularly when cognitive resources are limited. Thus, 1000 m may seem larger than 1 km. On account of the numerosity heuristic, consumers' evaluations and behaviors may be impacted in predictable ways [52–54].

In the domain of pricing, there is substantial evidence that logically-equivalent frames of the same valence affect price perceptions [55–57], as well as subsequent choices and marketplace decisions. For example, research examining how consumers use different currencies has found that consumer spending is influenced by the face value of the currency [56, 57]. Raghubir and Srivastava [56] observed that consumers are likely to spend less in real terms when the face value of a foreign currency is a multiple of the home currency (e.g., 4 Malaysian ringgits = 1 U.S. dollar) but spend more when the face value is a fraction of the home currency (e.g., 0.4 Bahraini dinar = 1 U.S. dollar). This is because the same product price (e.g., 100 U.S. dollars) seems more expensive when expressed in larger units (e.g., 400 Malaysian ringgits) but less expensive when expressed in smaller units (e.g., 40 Bahraini dinar), in accordance with the numerosity heuristic. In the context of grocery store pricing (e.g., for fruits or vegetables), the first author, along with Sarah Whitley and Julio Sevilla, recently showed that unit-based prices (e.g., $0.50 per unit) are perceived more favorably by shoppers than weight-based prices (e.g., $1 per pound) for low-weight items (e.g., ½ pound), but that unit-based prices (e.g., $2 per unit) are perceived more favorably by shoppers than weight-based prices (e.g., $1 per pound) for high-weight items (e.g., 2 pounds), even when the item's weight is provided [58]. Once again, consumers seem to anchor on face value prices when subjectively evaluating a product's value and making purchase decisions.

A final example, known as a "pennies-a-day" strategy [59, 60], involves the temporal reframing of an expense into a smaller time period. To illustrate, a subscription that costs $120 per year can potentially be reframed as $10 per month. Splitting up a larger transaction into a series of smaller but more-frequently occurring transactions has been shown to be an effective marketing strategy. Although the author attributed the effectiveness of a pennies-a-day pricing approach to a more complex mechanism involving comparison retrieval and transaction evaluation, the lower face value price of each smaller (vs. larger) transaction clearly played a role.

Reverting to the three notations that were discussed at the start of this section (i.e., decimals, fractions, percentages), although prior research does not claim that one notation will be universally preferred, it does provide some guidance on how to most effectively communicate logically-equivalent representations of the same notation. For example, in the domain of fractions or ratios, researchers have found evidence of *denominator neglect* or *ratio bias*, whereby larger face values in the fraction numerator (i.e., ratio antecedent) seem more probable. For example, individuals expect the likelihood of randomly drawing a red ball from an urn to be greater when the urn contains 10 red balls out of 100 total, as compared to just 1 red ball out of 10 total. When comparing two numbers using percentage changes, higher face values again impact evaluations. For example, when comparing the numbers 1500 and 1000, the difference between the two numbers seems larger if 1500 is presented as being 50% more than the 1000, versus if 1000 is described as being 33.3% less than 1500 [45]. For this reason, consumers may prefer a product quantity increment of 50% over an economically equivalent price discount of 33.3% [61] even though the frames are logically equivalent and similar in valence (i.e., product quantity increases and price discounts are both positive).

In considering how salespeople perceive their own earnings, the first author (again with co-authors Julio Sevilla and Rajesh Bagchi) observed a *commission effect* where identical earnings seem larger when they are the product of a large commission rate and a small sales amount, as compared to a small commission rate and a large sales amount. For instance, salespeople mistakenly think they have earned more from a 10% commission on a $10,000 sale, as compared to a 1% commission on a $100,000 sale, even though their earnings are exactly $1000 in each case. The authors showed that individuals disproportionately attend to the face value of the percentage rather than the sales amount because it is easier to evaluate whether a particular commission rate is "good" or "bad," but that sales amounts are inherently less evaluable [62].

4.2 Semantically-Similar Labels of the Same Valence

The third type of attribute frame involves semantically-similar labels of the same valence. A recent journal article [63] raised the intriguing possibility that "a simple difference in frame" of a crowdfunding request might influence potential funders' willingness to contribute. Specifically, the authors provided evidence that using "want" instead of "need" was more persuasive in *reward-based* crowdfunding contexts (e.g., Kickstarter) for which funders seek a return on their contribution. On the other hand, using "need" (vs. "want") was more effective for *donation-based* crowdfunding campaigns (e.g., GoFundMe) where funders tend to be more altruistically motivated. This finding is an example of a semantically-similar frame, in that the words "want" and "need" are often used interchangeably even though they carry different connotations and subtly different meanings. Indeed, the scholars who conducted this study suggested that the effect arises because "want" implies that the fundraiser is more independent.

Many other instances of semantically-similar frames producing disparate judgments have been documented by social scientists. Patrick and Hagtvedt [64] found that using the words "I don't" (e.g., "I don't work on Sundays") rather than "I can't" is more persuasive because it signals greater self-determination and empowerment. Jia, Wyer, and Shen [65] demonstrated that when a service provider requests a customer to wait (e.g., for service or for product delivery), the request "will you wait?" is perceived more negatively than "can you wait?" among consumers with higher feelings of social power. This effect occurs because responding affirmatively to "will you wait?" feels restrictive, and high-power consumers resist such restrictions on their freedom. In the context of online reviews, Packard and Berger [66] showed that explicit endorsements such as "I recommend it" are more persuasive than implicit endorsements such as "I like it" or "I enjoyed it." This is because reviewers who provide explicit endorsements are perceived to have both a stronger preference and greater expertise. Mayer and Tormala [67] found that using "I think" versus "I feel" message framing in persuasive communications can affect information processing and decision making. Specifically, they proposed a matching hypothesis such that "I think" framing was more effective when the target attitude or message recipient was cognitively oriented, whereas "I feel" framing was more effective when there was an affective orientation. Finally, in research published by the first author, describing the same drug, medication, or therapy as a "cure" rather than a "treatment" altered individuals' price preferences [68]. Individuals expected lower prices when treatments were described as cures because they felt that cures should be made universally accessible given their high communal value.

Perhaps the most famous example of similar but non-equivalent frames affecting judgments is Loftus and Palmer's [69] demonstration that participants who viewed a film depicting an automobile accident estimated the speed of the collision to be greater when asked how fast the cars were travelling when they "smashed into" versus "hit" one other. More participants who encountered the "smashed into" wording even claimed to have seen shattered glass following the collision (even though there was none). Loftus and Palmer's [69] influential work spawned considerable research examining how leading questions in eyewitness testimony and survey questionnaires shape people's subsequent responses. In fact, the first author's doctoral dissertation showed that consumers sometimes respond differently when asked to rate their "satisfaction" with a product as compared to their "happiness" with the same product, in part because the happiness frame leads people to make presuppositional inferences about the communicator's decision to use this somewhat "loaded" or lexically-marked word [70, 71].

The designation of an attribute frame type featuring semantically-similar labels of the same valence can even accommodate numerical frames that are not logically equivalent. For instance, research on price precision has shown that observers judge people who make more precise estimates (e.g., 2611) to be more confident than those who make less precise estimates that use round (i.e., zero-ending) numbers (e.g., 2600), even when estimate magnitudes are about the same (i.e., they are semantically similar) [72]. Another finding is that large numbers seem smaller when they are more precise versus more round [73]. For example, $395,425 is perceived as smaller than $395,000, even though the magnitudes are similar (and in this case, the precise price is actually larger). This is because round numbers are normatively used to describe larger prices and quantities. Differences in number-based inferences and/or associations can affect evaluations and judgments, such as product durability perceptions [74] and marketplace behaviors. For instance, Schindler and Kibarian [75] found that consumer purchasing increased when prices featured 99-endings (e.g., $29.99) rather than 00-endings (e.g., $30.00), even though the price points were semantically similar.

Much research has shown that individuals make normatively incongruous judgments when considering values at scale endpoints; these "misbehaviors" could be considered attribute framing effects. For instance, the first author, along with Katie Spangenberg, documented a "perfection premium" in evaluative judgments such that individuals disproportionately rewarded perfection on an attribute compared to near-perfect values on the same attribute [76]. For instance, when comparing students who had scored similarly on the ACT, the perceived gap in intelligence between the two students was greater when the higher-scoring student earned a perfect score (i.e., a 36 out of 36). At the other end of the spectrum, individuals comparing a lower-quality chocolate to a higher-quality chocolate priced a penny higher were more likely to choose the lower-quality chocolate if it was free versus nominally priced but not free (e.g., 14 cents) [77]. The authors attributed this "zero-price" effect to affect-laden associations with the concept of "free."

4.3 Semantically-Similar Labels of Opposite Valence

The last type of attribute frame involves semantically-similar labels that are opposite in valence (e.g., high natural content vs. low artificial content). This type of framing deviates most from the typical conception of an attribute frame since the two frames use labels that are opposite in meaning and lack logical equivalence. As a result, studies that examine how individuals respond to attributes or characteristics having *semantically-similar labels of opposite valence* are rare. In one study ([78], Study 2), participants were exposed to either a vague positive frame about a firm (e.g., "most of our employees are satisfied") or a vague negative frame (e.g., "only few of our employees are dissatisfied"). In contrast to other conditions in the same study involving concrete frames with logically-equivalent numbers in which the authors observed the valence-consistent shift, there were no significant differences between participants' attitude towards the firm in the vague frame conditions. This may have been because these vague frames were open to substantial interpretation on the part of study participants.

However, in another study examining wording of hotel reviews, Kamoen, Mos, and Dekker [79] found that a hotel described as "good" (positive valence) was evaluated more favorably than a hotel described as "not bad" (negative valence), even though each review was accompanied by a 5-star rating. On the other hand, there was no difference in the evaluation of a hotel that was described as "bad" (negative valence) versus "not good" (positive valence) when each review was accompanied by a 1-star rating. This asymmetry was attributed to verbal politeness; message recipients interpreted the use of "not good" as expressing the same evaluation as "bad," just in a more polite manner.

Collectively, these results suggest that there may be considerable information leakage [23] when semantically-similar labels of opposite valence are used. This makes it very difficult to draw generalizable conclusions about how individuals will respond to frames of this type.

5 Conclusion

It has been over 25 years since the second author of this chapter and his co-authors introduced an influential typology of framing effects [3]. We wrote this chapter because we believe the time is right to revisit this typology, in light of the extensive and insightful research on framing that has been conducted over the past quarter century by behavioral researchers and neuroscientists. We began by highlighting advances in our understanding of logically-equivalent numbers that are opposite in valence—which is how attribute frames were originally conceptualized in the original typology. Subsequently, we introduced an expanded typology for attribute framing effects that includes three additional types of framing effects that should be classified as attribute frames: logically-equivalent numbers of the same valence, semantically-similar labels of the same valence, and semantically-similar labels of opposite valence. Considerable research has examined these additional attribute frames, particularly the first two types. In this chapter, we have attempted to summarize a few of the findings from this literature.

Going forward, we expect that it will be necessary to further update the entire typology surrounding framing effects for several reasons. First, the distinction between goal framing effects and certain attribute framing effects can sometimes be murky. According to Levin, Schneider, and Gaeth [3], a key difference between the two frame categories is that goal framing affects persuasion likelihood, while attribute framing affects evaluations. However, this may no longer be a clear point of differentiation since research on attribute frames has broadened its scope beyond evaluations to assess how different frames differentially persuade and motivate behavioral change. Furthermore, attributes are regularly described using imprecise gain or loss frames that have historically been a hallmark of goal framing. For instance, when crafting an advertising appeal for Welch's grape juice, Lee and Aaker [80] created promotion frames (e.g., "great tasting as well as energizing") and prevention frames (e.g., "help keep arteries clear") that could potentially be categorized as attribute frames (specifically, semantically-similar labels of the same valence) or goal frames.

Second, new variants of framing effects continue to emerge that do not fit cleanly into the current typology, even our expanded one. For instance, recent consumer research has focused on how framing the same product rating with numerical values (e.g., 3.5 out of 5) leads to different evaluations than using an analog rating (e.g., 3.5 shaded stars out of 5), or how different visual representations of the same information (e.g., stars vs. bars) can influence consumer judgments and behaviors (e.g., [81, 82]). Another recent paper [83] identified a number format effect whereby digits (e.g., "5 hours") positively affect consumer behavior as compared to number words (e.g., "five hours") by yielding metacognitive benefits. None of these examples can be fully accommodated by our expanded typology since at least one of the frames in each article does not involve numbers (i.e., Arabic numerals) and yet each pair of frames is logically equivalent. Thus, neither the classification of logically-equivalent numbers or semantically-similar labels applies perfectly.

In conclusion, Tversky and Kahneman [1] brilliantly observed the failure of invariance with regard to different descriptions of the same (or almost-the-same) option. This observation has spawned a rich literature on framing effects—risky choice framing effects, goal framing effects, and many different flavors of attribute framing effects. As proud framing researchers, we are gratified that so many skilled scientists have expertly investigated the topic and generated much more than a chapter's worth of theoretical and practical insights. At the same time, we are excited by the emergence of new frame types and the unanswered questions that remain for framing researchers to explore in the future.

References

1. Tversky A, Kahneman D (1988) Rational choice and the framing of decisions. In: Bell DE, Raiffa H, Tversky A (eds) Decision making: descriptive, normative, and prescriptive interactions. Cambridge University Press, Cambridge, England, pp 167–192
2. Tversky A, Kahneman D (1981) The framing of decisions and the psychology of choice. Science 211(4481):453–458
3. Levin IP, Schneider SL, Gaeth GJ (1998) All frames are not created equal: a typology and critical analysis of framing effects. Organ Behav Hum Decis Process 76(2):149–188
4. Sher S, McKenzie CR (2011) Levels of information: a framing hierarchy. In: Keren G (ed) Perspectives on framing. Psychology Press, New York, NY, pp 35–63
5. Banks SM, Salovey P, Greener S, Rothman AJ, Moyer A, Beauvais J, Epel E (1995) The effects of message framing on mammography utilization. Health Psychol 14(2):178–184
6. Levin IP (1987) Associative effects of information framing. Bull Psychon Soc 25(2):85–86

7. Levin IP, Johnson RD, Russo CP, Deldin PJ (1985) Framing effects in judgment tasks with varying amounts of information. Organ Behav Hum Decis Process 36(3):362–377
8. McNeil BJ, Pauker SG, Sox HC, Tversky A (1982) On the elicitation of preferences for alternative therapies. N Engl J Med 306:1259–1262
9. Quattrone GA, Tversky A (1988) Contrasting rational and psychological analyses of political choice. Am Polit Sci Rev 82(3):719–736
10. Levin IP, Johnson RD (1984) Estimating price-quality tradeoffs using comparative judgments. J Consum Res 11(1):593–600
11. Levin IP, Gaeth GJ (1988) How consumers are affected by the framing of attribute information before and after consuming the product. J Consum Res 15(3):374–378
12. Peterson RA, Wilson WR (1992) Measuring customer satisfaction: fact and artifact. J Acad Mark Sci 20(1):61–71
13. Krishnamurthy P, Carter P, Blair E (2001) Attribute framing and goal framing effects in health decisions. Organ Behav Hum Decis Process 85(2):382–399
14. Levin IP, Schnittjer SK, Thee SL (1988) Information framing effects in social and personal decisions. J Exp Soc Psychol 24(6):520–529
15. Beach LR, Puto CP, Heckler SE, Naylor G, Marble TA (1996) Differential versus unit weighting of violations, framing, and the role of probability in image theory's compatibility test. Organ Behav Hum Decis Process 65(2):77–82
16. Payne JW, Sagara N, Shu SB, Appelt KC, Johnson EJ (2013) Life expectancy as a constructed belief: evidence of a live-to or die-by framing effect. J Risk Uncertain 46:27–50
17. Davis MA, Bobko P (1986) Contextual effects on escalation processes in public sector decision making. Organ Behav Hum Decis Process 37(1):121–138
18. Dunegan KJ (1996) Fines, frames, and images: examining formulation effects on punishment decisions. Organ Behav Hum Decis Process 68(1):58–67
19. Marteau TM (1989) Framing of information: its influence upon decisions of doctors and patients. Br J Soc Psychol 28(1):89–94
20. Duchon D, Dunegan KJ, Barton SL (1989) Framing the problem and making decisions: the facts are not enough. IEEE Trans Eng Manage 36(1):25–27
21. Hardisty DJ, Johnson EJ, Weber EU (2010) A dirty word or a dirty world? Attribute framing, political affiliation, and query theory. Psychol Sci 21(1):86–92
22. Frisch D (1993) Reasons for framing effects. Organ Behav Hum Decis Process 54(3):399–429
23. Sher S, McKenzie CR (2006) Information leakage from logically equivalent frames. Cognition 101(3):467–494
24. Janiszewski C, Silk T, Cooke AD (2003) Different scales for different frames: the role of subjective scales and experience in explaining attribute-framing effects. J Consum Res 30(3):311–325
25. Shiv B, Bechara A, Levin I, Alba JW, Bettman JR, Dube L, McGraw AP et al (2005) Decision neuroscience. Mark Lett 16:375–386
26. Levin IP, Xue G, Weller JA, Reimann M, Lauriola M, Bechara A (2012) A neuropsychological approach to understanding risk-taking for potential gains and losses. Front Neurosci 6(15):1–11
27. Weller JA, Levin IP, Shiv B, Bechara A (2007) Neural correlates of adaptive decision making for risky gains and losses. Psychol Sci 18(11):958–964
28. Ma Q, Feng Y, Xu Q, Bian J, Tang H (2012) Brain potentials associated with the outcome processing in framing effects. Neurosci Lett 528(2):110–113
29. Jin J, Zhang W, Chen M (2017) How consumers are affected by product descriptions in online shopping: event-related potentials evidence of the attribute framing effect. Neurosci Res 125:21–28
30. Murch KB, Krawczyk DC (2014) A neuroimaging investigation of attribute framing and individual differences. Soc Cogn Affect Neurosci 9(10):1464–1471
31. De Martino B, Kumaran D, Seymour B, Dolan RJ (2006) Frames, biases, and rational decision-making in the human brain. Science 313(5787):684–687

32. Kahneman D, Frederick S (2007) Frames and brains: elicitation and control of response tendencies. Trends Cogn Sci 11(2):45–46
33. McElroy T, Conrad J (2009) Thinking about product attributes: investigating the role of unconscious valence processing in attribute framing. Asian J Soc Psychol 12(2):157–161
34. Jain G, Gaeth GJ, Nayakankuppam D, Levin IP (2020) Revisiting attribute framing: the impact of number roundedness on framing. Organ Behav Hum Decis Process 161:109–119
35. Donovan RJ, Jalleh G (1999) Positively versus negatively framed product attributes: the influence of involvement. Psychol Mark 16(7):613–630
36. Celsi RL, Olson JC (1988) The role of involvement in attention and comprehension processes. J Consum Res 15(2):210–224
37. Kempf DS (1999) Attitude formation from product trial: distinct roles of cognition and affect for hedonic and functional products. Psychol Mark 16(1):35–50
38. Anderson B (2005) Practices of judgement and domestic geographies of affect. Soc Cult Geogr 6(5):645–659
39. Derryberry D (1993) Attentional consequences of outcome-related motivational states: congruent, incongruent, and focusing effects. Motiv Emot 17:65–89
40. Vuilleumier P (2005) How brains beware: neural mechanisms of emotional attention. Trends Cogn Sci 9(12):585–594
41. LeBoeuf RA, Shafir E (2003) Deep thoughts and shallow frames: on the susceptibility to framing effects. J Behav Decis Mak 16(2):77–92
42. Isaac MS, Poor M (2016) The sleeper framing effect: the influence of frame valence on immediate and retrospective judgments. J Consum Psychol 26(1):53–65
43. Poor M, Isaac MS (2023) Affective debiasing: focusing on emotion during consumption attenuates attribute framing effects. J Behav Decis Mak 36(5):e2347
44. DeWolf M, Grounds MA, Bassok M, Holyoak KJ (2014) Magnitude comparison with different types of rational numbers. J Exp Psychol Hum Percept Perform 40(1):71–82
45. Kruger J, Vargas P (2008) Consumer confusion of percent differences. J Consum Psychol 18(1):49–61
46. Parker M, Leinhardt G (1995) Percent: a privileged proportion. Rev Educ Res 65(4):421–481
47. Sevilla J, Isaac MS, Bagchi R (2018) Format neglect: how the use of numerical versus percentage rank claims influences consumer judgments. J Mark 82(6):150–164
48. Chen SFS, Monroe KB, Lou YC (1998) The effects of framing price promotion messages on consumers' perceptions and purchase intentions. J Retail 74(3):353–372
49. Heath TB, Chatterjee S, France KR (1995) Mental accounting and changes in price: the frame dependence of reference dependence. J Consum Res 22(1):90–97
50. DelVecchio D, Krishnan HS, Smith DC (2007) Cents or percent? The effects of promotion framing on price expectations and choice. J Mark 71(3):158–170
51. Pelham BW, Sumarta TT, Myaskovsky L (1994) The easy path from many to much: the numerosity heuristic. Cogn Psychol 26(2):103–133
52. Bagchi R, Davis DF (2016) The role of numerosity in judgments and decision-making. Curr Opin Psychol 10:89–93
53. Burson KA, Larrick RP, Lynch JG Jr (2009) Six of one, half dozen of the other: expanding and contracting numerical dimensions produces preference reversals. Psychol Sci 20(9):1074–1078
54. Monga A, Bagchi R (2012) Years, months, and days versus 1, 12, and 365: the influence of units versus numbers. J Consum Res 39(1):185–198
55. Wertenbroch K, Soman D, Chattopadhyay A (2007) On the perceived value of money: the reference dependence of currency numerosity effects. J Consum Res 34(1):1–10
56. Raghubir P, Srivastava J (2002) Effect of face value on product valuation in foreign currencies. J Consum Res 29(3):335–347
57. Raghubir P, Srivastava J (2009) The denomination effect. J Consum Res 36(4):701–713
58. Whitley SC, Sevilla J, Isaac MS (2025) Units or pounds? How anchoring on salient price information influences perceived value of product offerings. J Mark Res (forthcoming). https://doi.org/10.1177/00222437251340367

59. Gourville JT (1998) Pennies-a-day: the effect of temporal reframing on transaction evaluation. J Consum Res 24(4):395–408
60. Gourville JT (1999) The effect of implicit versus explicit comparisons on temporal pricing claims. Mark Lett 10:113–124
61. Chen H, Marmorstein H, Tsiros M, Rao AR (2012) When more is less: the impact of base value neglect on consumer preferences for bonus packs over price discounts. J Mark 76(4):64–77
62. Isaac MS, Sevilla J, Bagchi R (2024) The commission effect: framing affects perceived magnitude of identical payouts. J Exp Psychol Appl 30(1):187–205
63. Su L, Sengupta J, Li Y, Chen F (2024) "Want" versus "need": how linguistic framing influences responses to crowdfunding appeals. J Consum Res 50(5):923–944
64. Patrick VM, Hagtvedt H (2012) "I don't" versus "I can't": when empowered refusal motivates goal-directed behavior. J Consum Res 39(2):371–381
65. Jia Y, Wyer RS Jr, Shen H (2021) "Will you?" versus "can you?": Verbal framing moderates the effect of feelings of power on consumers' reactions to waiting. J Exp Psychol Appl 27(2):213–227
66. Packard G, Berger J (2017) How language shapes word of mouth's impact. J Mark Res 54(4):572–588
67. Mayer ND, Tormala ZL (2010) "Think" versus "feel" framing effects in persuasion. Pers Soc Psychol Bull 36(4):443–454
68. Isaac MS (2023) The cure effect: individuals demand universal access for health treatments that claim to eliminate disease symptoms. J Exp Psychol Appl 29(3):544–556
69. Loftus EF, Palmer JC (1974) Reconstruction of automobile destruction: an example of the interaction between language and memory. J Verbal Learn Verbal Behav 13(5):585–589
70. Greenberg JH (1966) Language universals. Mouton, The Hague
71. Huttenlocher J, Higgins ET, Clark HH (1971) Adjectives, comparatives, and syllogisms. Psychol Rev 78(6):487–504
72. Jerez-Fernandez A, Angulo AN, Oppenheimer DM (2014) Show me the numbers: precision as a cue to others' confidence. Psychol Sci 25(2):633–635
73. Thomas M, Simon DH, Kadiyali V (2010) The price precision effect: evidence from laboratory and market data. Mark Sci 29(1):175–190
74. Pena-Marin J, Bhargave R (2016) Lasting performance: round numbers activate associations of stability and increase perceived length of product benefits. J Consum Psychol 26(3):410–416
75. Schindler RM, Kibarian TM (1996) Increased consumer sales response though use of 99-ending prices. J Retail 72(2):187–199
76. Isaac MS, Spangenberg K (2021) The perfection premium. Soc Psychol Pers Sci 12(6):930–937
77. Shampanier K, Mazar N, Ariely D (2007) Zero as a special price: the true value of free products. Mark Sci 26(6):742–757
78. Bartikowski B, Berens G (2021) Attribute framing in CSR communication: doing good and spreading the word–but how? J Bus Res 131:700–708
79. Kamoen N, Mos MB, Dekker WF (2015) A hotel that is not bad isn't good. The effects of valence framing and expectation in online reviews on text, reviewer and product appreciation. J Pragmat 75:28–43
80. Lee AY, Aaker JL (2004) Bringing the frame into focus: the influence of regulatory fit on processing fluency and persuasion. J Pers Soc Psychol 86(2):205–218. https://doi.org/10.1037/0022-3514.86.2.205
81. Abell A, Morgan C, Romero M (2024) The power of a star rating: differential effects of customer rating formats on magnitude perceptions and consumer reactions. J Mark Res 00222437241240694
82. Jia H, Wan EW, Zheng W (2023) Stars versus bars: how the aesthetics of product ratings "shape" product preference. J Consum Res 50(1):142–166
83. Romero M, Craig AW, Mormann M, Kumar A (2025) Are "10 grams of protein" better than "ten grams of protein"? How digits versus number words influence consumer judgments. J Consum Res 51(5):1006–1026

Mathew S. Isaac is Chair of the Marketing Department and Professor of Marketing at the Seattle University Albers School of Business and Economics. Dr. Isaac's research, which examines how contextual and motivational factors influence product evaluations and purchase intentions, has been published in leading Marketing and Psychology journals. He is a 2024 Marketing Science Institute (MSI) Scholar, Associate Editor for *Journal of Marketing Research* and *Journal of Consumer Psychology*, and past Associate Editor for *Journal of Consumer Research* who currently serves on six editorial review boards.

Irwin P. Levin is Professor Emeritus at the University of Iowa, where he taught in the Departments of Psychology and Marketing. Dr. Levin's research focuses on the area of judgment and decision making, with an emphasis on individual differences. Dr. Levin's published research is widely cited, and his work has been supported by grants from the National Science Foundation, the National Institute on Aging, the National Institutes of Mental Health, and MSI. He has served as President of the Society for Judgment and Decision Making and on editorial review boards for the leading journals in his area.

Index

© The Editor(s) (if applicable) and The Author(s), under exclusive license to Springer Nature Switzerland AG 2025
T. McElroy (ed.), *Decision-Making in Life and Work*, Integrated Science 36,
https://doi.org/10.1007/978-3-031-88415-3

The manufacturer's authorised representative in the EU is Springer Nature Customer Service Centre GmbH, Europaplatz 3, 69115 Heidelberg, Germany. If you have any concerns regarding our products, please contact ProductSafety@springernature.com

Printed and bound by CPI Group (UK) Ltd, Croydon, CR0 4YY
30/09/2026
02225501-0007